Electrochemical Energy Storage for the Grid

While renewable energy is vital to ensuring a sustainable power grid, it is highly variable and unpredictable, and electricity generation from renewables is usually small and distributed compared to traditional power generation. Energy storage is attracting attention as an effective approach to addressing the variable nature of renewable energy, as its implementation can improve grid stability, reliability, and resilience. This book focuses on the fundamentals, advances, and applications of electrochemical energy storage for the grid.

Key Features:

- Offers a comprehensive review of various types of electrochemical energy storage for the grid.
- Covers important scientific fundamentals and practical applied technologies.
- Discusses opportunities and challenges.

This book is aimed at researchers and practicing engineers in the energy, chemical, and related engineering disciplines.

Electrochemical Energy Storage for the Grid

Edited by
Min Yang, Wanzhong Shen,
and Junbo Hou

CRC Press
Taylor & Francis Group
Boca Raton London New York

CRC Press is an imprint of the
Taylor & Francis Group, an **informa** business

Designed cover image: © 2026 Shutterstock

First edition published 2027
by CRC Press
2385 NW Executive Center Drive, Suite 320, Boca Raton FL 33431

and by CRC Press
4 Park Square, Milton Park, Abingdon, Oxon, OX14 4RN

CRC Press is an imprint of Taylor & Francis Group, LLC

ISBN: 978-1-032-79525-6 (hbk)
ISBN: 978-1-032-79526-3 (pbk)
ISBN: 978-1-003-49251-1 (ebk)

DOI: 10.1201/9781003492511

Typeset in Times
by codeMantra

Contents

Preface

Due to environmental protection, climate change, fossil fuel depletion, and energy security concerns, renewable energy is penetrating the grid, and the occupation ratio is continuously increasing. The Paris Agreement on climate change even puts CO_2 emissions peak and carbon neutrality on the agenda over the main nations and economies, charting the course for the world to transition to green and low-carbon development. However, one might realize that renewable energy is highly variable and unpredictable. The electricity generation from renewables like wind and solar is usually small and distributed compared to the traditional power generation like thermal and hydropower. The generation is also location constrained and weather dependent. Thus, the power quality, transmission and distribution capability, the balance of power supply and demand, and the stability of the system will become the main issues for the new grid infrastructure, especially with an increasing level of renewable penetration. Recently, energy storage is attracting more and more attention to address the variable nature of renewable energy as the most effective approach. The implementation of energy storage in the new power grid can improve grid stability, reliability, and resilience.

Generally, the energy storage can be classified into three application scenarios. The first one is at the power generation side, which is mainly used to stabilize the fluctuations of new energy, track power generation plans, and alleviate wind and solar abandonment. The second one is at the transmission and distribution side, which is mainly used to ensure the safe and reliable operation of the power grid, such as peak regulation and frequency regulation. The third one is at the user side, which is mainly used for peak and valley arbitrage, demand management, dynamic expansion, and demand-side response. Although different application scenarios might have specific requirements for the energy storage, roughly three general aspects must be fulfilled regardless of the application scenarios: cost competitive, validated performance and safety, and customer and industry acceptance. For example, in the batteries or the flywheels, the storage component still constitutes only 30%–40% of the total system cost; thus, the focus needs to be on the entire system. Firstly, it is very important at the current stage while the cost of storage systems must decline to economically attractive levels. Secondly, the performance and safety of existing storage systems must be tested, evaluated, and validated. There is significant uncertainty over the match of the power and energy request and response for each scenario application, the usable life of batteries, and the length of time. Thirdly, the customers and industry must accept the employ of the storage technology into the grid. For example, the system operators and service providers must know how to deploy storage resources, and the stakeholders must know the profit points to employ storage technology effectively and profitably, thus encouraging investments.

The types of energy storage can be (1) mechanical energy technology or kinetic energy technology including pumped hydropower, compressed air, and flywheels; (2) thermal energy storage that is designed to store heat energy by cooling, heating, melting, condensing, or vaporizing a substance; (3) chemical energy technology including hydrogen, synthetic natural gas, and solar fuel storage systems, which is usually stored in the chemical bonds and the reactions to store heating or cooling capacity in chemical compounds; (4) electrochemical energy storage including various batteries, supercapacitors, and flow batteries, which is generally composed of three processes like mass transport and anode and cathode charge transfer; and (5) hybrid systems that can employ two or more above-mentioned technologies. So far, the most popular and interesting energy storage for the grid is the electrochemical one, and we will mainly introduce the electrochemical energy storage for the grid in this book.

Lithium battery is a type of battery that uses lithium metal or lithium alloy as positive/negative electrode materials and uses nonaqueous electrolyte solution. Lithium metal batteries were first proposed and studied by Gilbert N. Lewis in 1912. In the 1970s, M. S. Whittingham proposed and began to study lithium-ion batteries. Because the chemical properties of lithium metal are very active, the processing, storage, and use of lithium metal have very high environmental requirements. With the advancement of science and technology, lithium batteries have become mainstream. Lithium batteries can be roughly divided into two categories: lithium metal batteries and lithium-ion batteries. Lithium-ion batteries do not contain metallic lithium and are rechargeable. The fifth-generation product of rechargeable batteries, lithium metal batteries, was born in 1996. Its safety, specific capacity, self-discharge rate, and performance-price ratio are better than those of lithium-ion batteries. Due to its own high technical requirements, companies in only a few countries are producing this lithium metal battery. The Frenchman Plant invented the lead-acid battery in 1859. Lead-acid batteries have made great progress in terms of product types and varieties, product electrical performance, etc., whether in transportation, communications, electric power, military, or in the field of navigation and aviation. According to the differences in structure and use of lead-acid batteries, batteries can be roughly divided into four categories: (1) lead-acid batteries for starting; (2) lead-acid batteries for power; (3) fixed valve-controlled sealed lead-acid batteries; and (4) other categories, including small valve-regulated sealed lead-acid batteries and lead-acid batteries for miners' lamps. The nominal voltage of a single-cell lead-acid battery is 2.0 V; it can be discharged to 1.5 V and charged to 2.4 V. In applications, six single-cell lead-acid batteries are often connected in series to form a nominal 12 V lead-acid battery. There are also 24 V, 36 V, 48 V, etc. Nickel-metal hydride batteries are divided into high-voltage nickel-metal hydride batteries and low-voltage nickel-metal hydride batteries. Low-voltage nickel-metal hydride batteries have the following characteristics: (1) The battery voltage is 1.2–1.3 V, which is equivalent to nickel-cadmium batteries; (2) The energy density is high, more than 1.5 times that of nickel-cadmium batteries; (3) It can be charged and discharged quickly and provides good performance at low temperature; (4) It possesses sealed, strong resistance to overcharge and discharge; (5) No dendrites are generated, which can prevent short circuits in the battery; and (6) It is safe and reliable, and it will not cause any pollution to the environment or any memory effect. High-voltage nickel-metal hydride batteries have the following characteristics: (1) High reliability; it has good overdischarge and overcharge protection, can withstand higher charge–discharge rates, has no dendrite formation, has good specific characteristics, and its mass specific capacity is 60 A·h kg^{-1}, which is five times that of nickel-cadmium batteries; (2) Long cycle life, up to thousands of times; (3) Compared with nickel-cadmium batteries, they are fully sealed and require less maintenance; and (4) Excellent low-temperature performance; the capacity does not change significantly at −10°C. Flow battery is an electrochemical energy storage technology proposed by Thaller in 1974. Simply put, a flow battery consists of a stack unit, an electrolyte, an electrolyte storage and supply unit, and a management and control unit. A flow battery is a high-performance battery that uses positive and negative electrolytes to separate and circulate separately. It has the characteristics of high capacity, wide application area (environment), and long cycle life. Flow batteries realize the mutual conversion of electrical energy and chemical energy through reversible oxidation–reduction reactions (i.e., reversible changes in valence states) of active materials in the positive and negative electrolyte solutions. During charging, an oxidation reaction occurs at the positive electrode, causing the valence state of the active material to increase. A reduction reaction occurs at the negative electrode, causing the valence state of the active material to decrease. The discharge process is opposite to this. Different from general solid-state batteries, the positive and/or negative electrolyte solutions of flow batteries are stored in storage tanks outside the battery and are transported to the interior of the battery through pumps and pipelines for reaction.

In this book, we will firstly introduce a general overview of electrochemical energy storage for the grid and then comprehensively review various electrochemical energy storage for the grid including lead-acid batteries, lead-carbon batteries, Ni-MH batteries, Li-ion batteries, Li solid-state batteries, Na-ion batteries, all vanadium flow batteries, all Fe flow batteries, hydrogen energy by water electrolysis, fuel cells, supercapacitors, and metal-air batteries. We expect that this book covers the most important scientific fundamentals and practical applied technologies in grid electrochemical energy storage. We hope it would be very helpful for beginners and experts, scientists and engineers, and theorists and experimenters. Finally, we would like to express our gratitude to the efforts of Editor Allison Shatkin at Taylor & Francis for initiating this project.

About the Editors

Min Yang is now a Professor in Shanghai DianJi University, and Senior Director at the Central Research Institute, Shanghai Electric Group. She obtained her M.S. degree from Harbin Institute of Technology, China, and her Ph.D. degree from Dalian Institute of Chemical Physics, Chinese Academy of Sciences, in Chemical Engineering.

Wanzhong Shen is Chairman of Liyuan Technology, China. He has decades of experience in corporate management and in the industry of environmental technology, specifically in the development of new environmentally friendly materials.

Junbo Hou is Head of Advanced Energy Storage of Hefei Zero Carbon Technology Co., Ltd, in Hefei, China (SUNECO). He received his B.S. and M.S. degrees from Harbin Institute of Technology, China, and his Ph.D. degree from Dalian Institute of Chemical Physics, Chinese Academy of Sciences, all in Chemical Engineering, particularly Electrochemical Engineering.

Contributors

Kedi Cai
College of Chemistry
Bohai University
Jinzhou, Liaoning, China

Chuankai Fu
Harbin Institute of Technology
Harbin, Heilongjiang, China

Jianxin Gao
Dalian Institute of Chemical Physics
Dalian, Liaoning, China

Shanshan Gao
Dalian Institute of Chemical Physics
Dalian, Liaoning, China

Xing Hu
School of Aeronautics
Shanghai DianJi University
Pudong, Shanghai, China

Zhenye Kang
School of Marine Science and Engineering
Hainan University
Haikou, Hainan, China

Min Liu
Dalian Institute of Chemical Physics
Dalian, Liaoning, China

Qianfeng Liu
Dalian Institute of Chemical Physics
Dalian, Liaoning, China

Qinming Liu
School of Aeronautics
Shanghai DianJi University
Pudong, Shanghai, China

Zhiqiang Lou
Shuangdeng Group Co. Ltd
Taizhou, Jiangsu, China

Shaobo Ma
Shandong University of Science and Technology
Jinan, Shangdong, China

Yulin Ma
Harbin Institute of Technology
Harbin, Heilongjiang, China

Zhengpei Miao
School of Marine Science and Engineering
State Key Laboratory of Tropic Ocean Engineering Materials and Materials Evaluation
Hainan University
Haikou, Hainan, China

Erdong Wang
Dalian Institute of Chemical Physics
Dalian, Liaoning, China

Taiyuan Wang
Dalian Institute of Chemical Physics
Dalian, Liaoning, China

Xueliang Wang
Dalian Institute of Chemical Physics
Dalian, Liaoning, China

Zhan Yan
Dalian Institute of Chemical Physics
Dalian, Liaoning, China

Chuangang Yao
School of Materials and Environmental Engineering
Bohai University
Jinzhou, Liaoning, China

Haixia Zhang
School of Materials and Environmental Engineering
Bohai University
Jinzhou, Liaoning, China

Quan Zhang
Hubei Key Laboratory of Pollutant Analysis & Reuse Technology
College of Chemistry and Chemical Engineering
Hubei Normal University
Huangshi, Hunan, China

Pengjian Zuo
Harbin Institute of Technology
Harbin, Heilongjiang, China

1 State-of-the-Art Lead-Acid Batteries for the Application of Grid

Zhiqiang Lou

1.1 STRUCTURE CHARACTERISTICS AND PRINCIPLES OF LEAD-ACID BATTERY

Before introducing lead-carbon batteries, we should understand the structure and principles of lead-acid battery. Figure 1.1 shows the composition and structure of a valve-regulated lead-acid (VRLA) battery [1].

Table 1.1 shows the composition and function of the components of the VRLA battery.

The two-electrode reactions and total reactions of the lead-acid batteries are as follows.

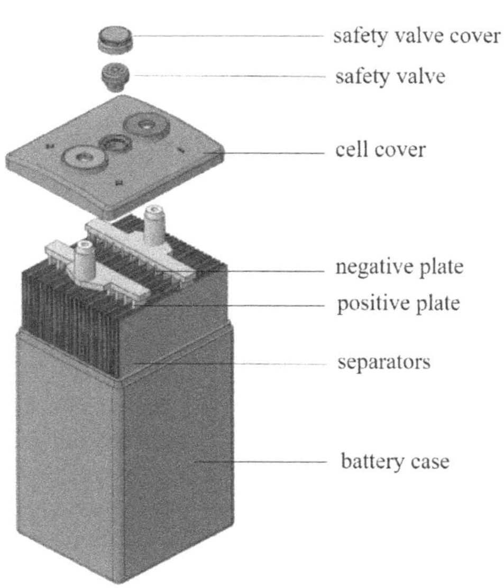

FIGURE 1.1 VRLA battery structure.

DOI: 10.1201/9781003492511-1

TABLE 1.1
The Composition and Function of the Components of VRLA Battery

Part	Composition and Function
Polar plate	It is composed of grids and active material. It is the carrier of electrochemical reactions and energy storage
Separators	The role is to prevent the positive and negative poles from short circuit. AGM diaphragm also has the effect of storing and maintaining electrolyte
Electrolyte	During the electrochemical reaction of the battery, it acts as an ionic conductor. The sulfate solution in the lead-acid battery not only acts as an ionic conductor but also participates in the electrochemical reaction
Battery case	The place where the battery response is provided, and there is a certain mechanical strength to withstand the internal pressure of the battery
Terminal	Interface connecting the external circuit
Safety valve	The sealing effect is to prevent internal acid mist leakage and prevent external air from entering; security effect, when the internal pressure of the battery is too large, the opening valve will release the pressure to prevent the battery from deforming and bursting

Discharge:

$$\text{Total reaction: } PbO_2 + Pb + 2H_2SO_4 = 2PbSO_4 + 2H_2O$$
$$\text{Negative pole: } Pb + SO_4^{2-} - 2e^- = PbSO_4$$
$$\text{Positive pole: } PbO_2 + 4H^+ + SO_4^{2-} + 2e^- = PbSO_4 + 2H_2O$$

Charge:

$$\text{Total reaction: } 2PbSO_4 + 2H_2O = PbO_2 + Pb + 2H_2SO_4$$
$$\text{Positive pole: } PbSO_4 + 2H_2O - 2e^- = PbO_2 + 4H^+ + SO_4^{2-}$$
$$\text{Negative pole: } PbSO_4 + 2e^- = Pb + SO_4^{2-}$$

When discharged, the chemical energy stored in the battery is converted into electrical energy output via an external circuit. The positive electrode active material PbO_2 is reduced to Pb^{2+} and then reacts with H_2SO_4 in the solution to form $PbSO_4$ and H_2O; the negative electrode active material sponge-shaped lead loses electrons, is oxidized to Pb^{2+}, and reacts with H_2SO_4 in the solution to form $PbSO_4$.

When charged, electrical energy is converted into chemical energy and stored in the battery. When charging, the positive electrode of the battery is connected to the positive electrode of the power supply, and the positive electrode undergoes an oxidation reaction. $PbSO_4$ on the positive electrode loses electrons and is oxidized into PbO_2, consumes H_2O, and H^+ is generated. $PbSO_4$ on the negative electrode obtains electrons and is then reduced to sponge-shaped lead.

In the later stage of charging, water is decomposed on the positive electrode plate, and oxygen is released. In the liquid-rich lead-acid battery, the oxygen produced is discharged from the outside of the battery, which inevitably leads to water loss.

In VRLA batteries, high-porosity fiberglass separators are used to achieve liquid lean design. There are some voids in the separators that are not filled with electrolyte, forming a gas channel, and the generated oxygen is transmitted to the negative plate to react with sponge-shaped lead through the gas channel in the separators. PbO is an alkaline oxide that immediately reacts with the sulfuric acid in the electrolyte to form lead sulfate. Lead sulfate is converted into sponge-shaped lead after charging, achieving oxygen circulation, reducing battery water loss, and achieving maintenance-free purpose.

Ordinary lead-acid batteries have the advantages of low cost, but the cycle life is short. The negative electrode is sulfated under high-rate partial-state-of-charge (HRPSoC) conditions, resulting in a higher energy storage cost of the unit. Due to the addition of activated carbon or other carbon materials, lead-carbon batteries prevented the phenomenon of negative sulfated, extended the battery life, reduced the cost of use, and developed great potential in the field of new energy storage.

1.2 LEAD-CARBON BATTERY CLASSIFICATION

The technology of lead-carbon batteries has been developed for nearly 20 years. According to the structure of the negative electrode, it can be divided into the following four types of lead-carbon batteries, and the structure is shown in Figure 1.2.

a. Capacitor lead-carbon battery

   All capacitive lead-carbon batteries have high-specific-surface-area carbon materials instead of negative electrode sponge-shaped lead, that is, full carbon negative electrode, with high power characteristics and cycle life [3]. This technology is represented by Axion Power in the United States. However, since all negative electrodes are made of carbon materials, the energy density of the battery will decrease.

b. Internal parallel lead-carbon battery

   The internal parallel lead-carbon battery is also called super battery [4,5]. Its negative electrode is composed of two parts in parallel: one is a capacitive carbon electrode, and the other is a sponge-shaped lead. Through structural design, the lead-acid battery and supercapacitor become a whole. It not only has the characteristics of high-energy lead-acid batteries but also inherits the advantages of high power of supercapacitors. In super batteries, capacitive carbon electrons can greatly slow down the impact of large current on lead electrodes and reduce the occurrence of lead negative sulfation. This technology is represented by East Penn. Because the super battery is

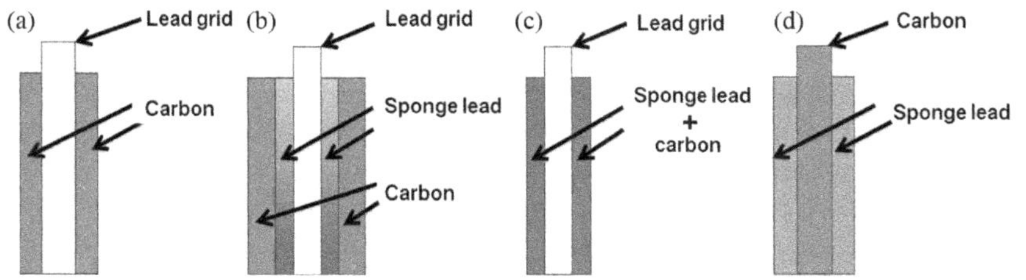

FIGURE 1.2 Structure of four kinds of lead-carbon batteries [2].

integrated with carbon electrodes and lead plates, it is still difficult to weld them.

c. Lead-carbon battery (internal hybrid)

Lead-carbon battery (internal hybrid) is to directly add carbon material with double-layer capacitance and high conductivity to the negative electrode during the paste process, so that the lead-active substance is in full contact with the carbon material. The weight ratio of the carbon material is approximately 1 wt.%. The internal mixed lead-carbon batteries and super batteries have similar principles, but since carbon materials can be directly added to the negative lead paste for mixing, the production process is mature, and the lead and carbon materials have synergistic effects, promoting smaller lead sulfate particles [6], which improve the cycle life of the battery and improve the charging and discharging capacity of the battery [7]. Large-scale production and assembly can be achieved without renovating the existing production lines.

d. Carbon current collector lead-carbon battery

The carbon replacement or partial replacement of the lead plate grid with a 3D structure is used as the negative electrode current collector, and the negative electrode current collector is filled with the negative active material, which reduces the weight of the electrode and increases the energy density of the battery [8,9]. This technology is less practically applicable.

Based on the manufacturing consideration of the battery, mass-produced lead-carbon batteries are generally internal mixed structures.

1.3 THE MECHANISM OF ACTION OF CARBON IN LEAD-CARBON BATTERIES

Under the charging and discharging conditions of small current, small lead sulfate particles are formed in the negative electrode during the discharge process. These particles are evenly distributed inside the electrode plate and are discontinuous. When charging later, since small lead sulfate particles can fully contact H^+ in the electrolyte and their own solubility is also high, both lead sulfate on the surface of the negative electrode plate and lead sulfate inside the electrode plate can be reduced to sponge-shaped lead [10].

Under the condition of high-rate discharge, the reaction rate of HSO_4^- is far greater than the diffusion rate of HSO_4^-, resulting in most of the reaction occurring on the surface of the electrode. $PbSO_4$ can easily gather on the negative pole surface and generate $PbSO_4$ small crystals. Small crystals are unstable and dissolve easily into large $PbSO_4$ particles. Because the solubility and electrochemical activity of large particles are low, and the charging becomes not easy, they cannot be completely transformed into sponge-shaped lead. After continuous circulation, $PbSO_4$ particles are getting larger and larger and eventually form a dense lead sulfate passivation layer on the surface of the negative plate. As shown in Figure 1.3, the porosity and effective surface area for electron transport are greatly reduced. This restricts the diffusion of HSO_4^- to the inside of the plate [11]. The reversible charge–discharge reaction becomes worse, resulting in battery failure [12].

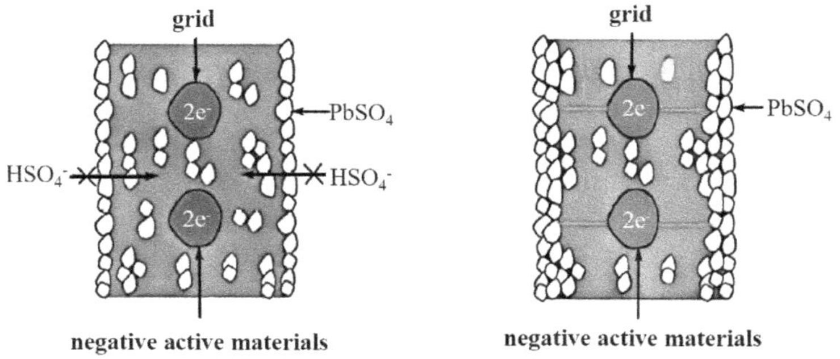

FIGURE 1.3 Schematic diagram of sulfate failure under negative electrode HRPSoC conditions [13]. (a) High rate discharge situation. (b) Low discharge rate situation.

The physical and chemical properties of negative electrode active materials include micromorphology, pore size distribution, capacitance, and conductivity. They can all be affected and improved by the specific surface area, shape, structure, conductivity, surface functional groups, and defect types of carbon materials, and so, the irreversible sulfation of the negative electrode can be inhibited. Since the action mechanism of different carbon materials in batteries is different [14], the action mechanism of carbon is generally recognized as the following four aspects:

1. Capacitance buffering effect [15]. The pores on the surface of carbon materials are relatively developed. During the charging and discharging process, it can form a relatively large electric double-layer capacitor at the electrolyte interface, which can buffer the instantaneous passing of large current and effectively suppress the surface of the negative electrode plate to form thick lead sulfate crystals.
2. Provide more electrode surfaces that can be used for charging and discharging reactions [16]. Electrochemical reactions occur not only on the lead surface but also on the carbon surface. Carbon materials can serve as new lead sulfate nucleus centers, providing more lead sulfate nucleus positions at high-current discharge, which is conducive to the uniform distribution of $PbSO_4$ particles in the plate.
3. Steric hindrance effect [17]. It prevents the formation of large lead sulfate particles and ensures that they can be reversibly converted into active lead. Carbon materials can add additional negative electrode structural frameworks to promote acid diffusion and provide sufficient H^+ and HSO_4^- ions for electrode reactions under high-rate partial-state-of-charge (HRPSoC) cycling conditions [18].
4. Improve the conductive network [19]. Carbon materials have a higher conductivity. The main component of the negative electrode plate under the discharge state of a lead-acid battery is $PbSO_4$ with poor conductivity. Carbon materials can build a three-dimensional conductive network in the negative electrode plate to promote the conversion of $PbSO_4$ to Pb during recharging.

1.4 MAIN FEATURES OF LEAD-CARBON BATTERIES

1. Safe and reliable. Lead-acid batteries have a long history and a mature process, and the battery electrolyte is a sulfuric acid solution. The battery temperature rises slowly, has no flammable substances, and has high safety performance. Lead-carbon battery technology is developed from traditional lead-acid battery technology. Its preparation process is basically the same as that of traditional lead-acid batteries and is considered to be one of the safest energy storage technologies.
2. Long cycle life. Since the negative electrode is added with special carbon materials, the negative electrode sulfate can be suppressed, which significantly extends the cycle life of lead-carbon batteries. It is especially suitable for partial-state-of-charge (PSoC) mode recycling. Figure 1.4 shows the comparison curve of 90%–30% PSoC cycle performance between lead-carbon battery and ordinary lead-acid battery. The cycle performance of the lead-carbon battery is more than 4 times that of the ordinary lead-acid battery.
3. Good scale. The maximum capacity of lead-carbon battery cells can reach 3,000 Ah, which is suitable for the grouping of large-scale energy storage systems.
4. High charging efficiency. The lead-carbon batteries have increased by about 1 times more than ordinary lead-acid batteries, and the charging time has been significantly shortened. Figure 1.5 shows a comparison of the charging

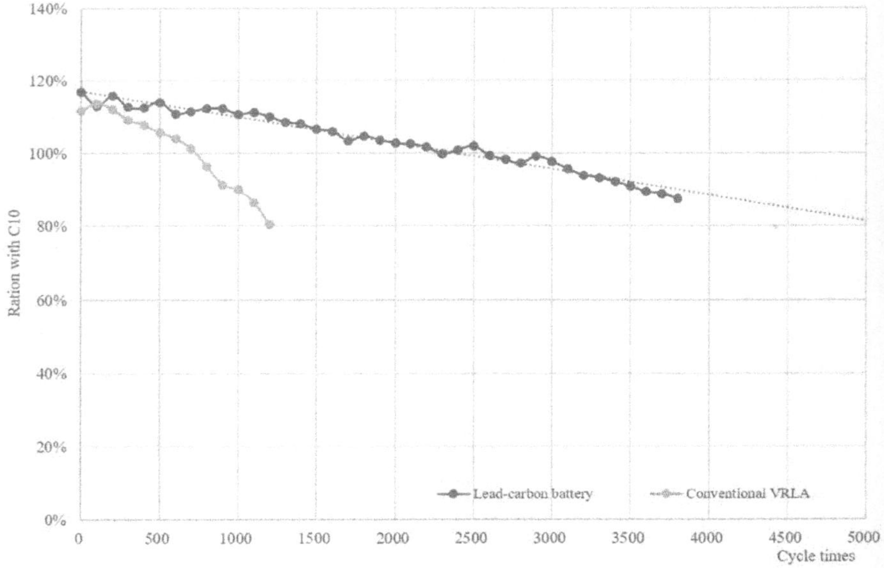

FIGURE 1.4 Lead-carbon batteries and ordinary lead-acid batteries 90%–30% PSoC cycle performance.

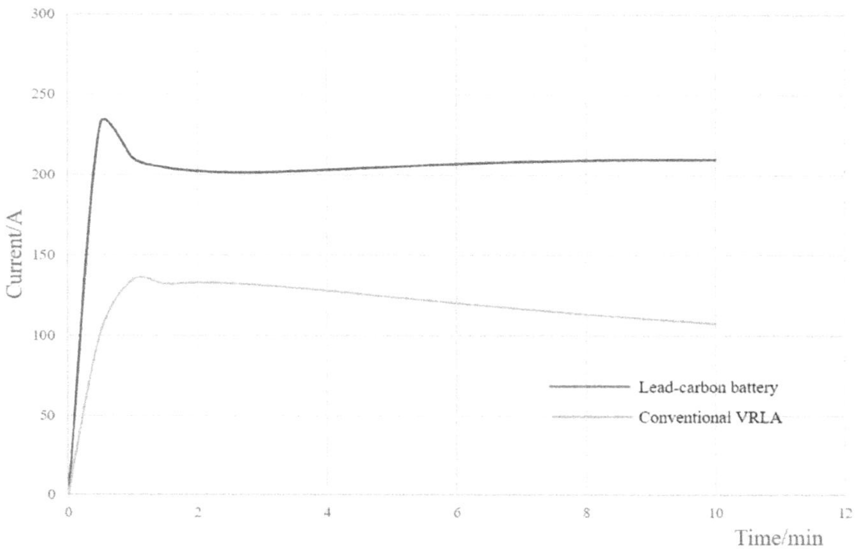

FIGURE 1.5 Comparison of the charging acceptance capacity of lead-carbon battery.

acceptability of a lead-carbon battery and an ordinary lead-acid battery. The charging current of a lead-carbon battery is nearly twice that of an ordinary battery.

5. It has a good economy and a high recovery rate. Based on the mature system of lead-acid batteries, lead-carbon batteries have mature recycling technology and recycling system, and the recycling efficiency of old batteries can reach 99%.

   However, lead-carbon batteries also have some shortcomings: compared with lithium-ion batteries, their energy density is relatively low, and they are mainly used in energy storage power plants that do not require high weight and volume and are not suitable for power batteries and mobile power supplies. Due to the relatively poor performance of high current, it is not suitable for use in high-power charging and discharging energy storage scenarios.

1.5 LEAD-CARBON BATTERY TECHNOLOGY DEVELOPMENT AND APPLICATION STATUS

Compared with traditional lead-acid batteries, lead-carbon batteries have the advantages of fast charging and discharging speeds, long cycle life (4–8 times that of ordinary lead-acid batteries), high energy efficiency, and wide applicable temperature range. As the most advanced technology in the field of lead-acid batteries, lead-carbon batteries are also the focus of the development of the international new energy industry, and they have broad application prospects. Many scientific research institutions and new energy companies have invested in the development and research of lead-carbon batteries.

Lead-carbon batteries have good circulating performance under the energy storage modes of solar energy, wind energy, and power grid regulation. At present, lead-carbon batteries have been used on large scale on the fixed energy storage devices for power grids and microgrids. In the United States, Australia, and Asia, lead-carbon batteries are used to smooth the frequency of renewable energy and increase the power generation utilization rate to improve the stability of the power grid. In 2011, PNM Grid accepted the East Penn grid solar storage integrated lead-carbon energy storage project (500 kWh photoelectric smoothing +1 MWh solar photovoltaic power storage). To optimize the hybrid power generation system on King Island in Australia, reduce dependence on diesel engines, and stabilize the power supply of wind power systems, Ecoult Australia developed and installed a lead-carbon battery energy storage system of 3 MW/1.6 MWh. Furukawa has also made some small-scale microgrid/grid dispersed energy storage in Japan and has a smart grid demonstration system for Shimizu Corporation in Maeda District, Kitakyushu.

China has also actively explored and studied lead-carbon batteries and achieved fruitful results. Compared with other energy storage batteries, lead-carbon batteries have unparalleled safety and reliability advantages and significant economic benefits. Lead-carbon batteries are used in different scenarios such as peak cutting and valley filling, power supply, and photovoltaic energy storage.

Lead-carbon batteries are used in energy storage projects in Hebei, Qinghai, Tibet, and other provinces. Among them, Zhangbei National Scenery Storage and Transportation Demonstration Power Station, Shoto, provided a total energy storage capacity of 6 MWh lead-carbon battery energy storage system, which is used for energy management of photovoltaic and wind power generation in the project, suppressing fluctuations, and comprehensive control and scheduling of grid frequency and voltage regulation. Qilian Qinghai Micro-Grid System consists of a hybrid energy storage system consisting of lithium batteries and lead-carbon batteries. The 4 MWh lead-carbon system is the main energy storage unit, responsible for guaranteeing nighttime power supply, whereas the lithium battery system is the daytime photovoltaic power generation support unit. The two complementary operations can completely solve the situation of no electricity or power shortage in most areas of Yanglong, Qilian County. In Dangxiong, Tibet, at an altitude of 4,800 meters, in order to solve the power limit of Tibet Yangyi photovoltaic power station and to smooth the photovoltaic fluctuation problem, Yangyi used a 19.2 MWh lead-carbon energy storage system. On Kaishan Island, Lianyungang, Jiangsu, the microgrid 600 kWh lead-carbon energy storage project provides a continuous power supply for the island's stable domestic electricity.

1.6 OPPORTUNITIES AND CHALLENGES OF LEAD-CARBON BATTERIES

With the transformation of the global energy structure and the rapid growth of energy storage demand, the lead-carbon battery market has ushered in unprecedented development opportunities. As the world's largest producer and consumer of lead-acid batteries, China's lead-carbon battery industry is also showing a booming trend.

The Chinese government attaches great importance to the development of the energy storage industry and has issued a series of policy documents to support the research and development and application of new energy storage technologies. For example, documents such as "Fourteenth Five-Year Plan: New Energy Storage Development Implementation Plan" and "Guiding Opinions on Promoting the Development of the Energy Electronics Industry" clearly listed lead-carbon batteries as one of the key directions for new energy storage core technology and equipment. These documents provide strong policy support for the development of lead-carbon battery technology.

Although the application of lead-carbon batteries in the field of energy storage shows very prominent advantages, to meet the needs of the rapid development of energy storage, the advancement of lead-carbon technology faces many challenges, mainly including the following three aspects.

1. The mechanism of action of carbon needs to be further clarified. Only by fully grasping the electrochemical behavior of carbon in the negative electrode and the contribution of battery performance can we be targeted to design and develop higher -performance lead-carbon batteries.
2. It is necessary to study carbon materials for lead-carbon batteries with low price and stable performance. The carbon materials currently used generally have problems such as expensive, low hydrogen evolution overpotential, and poor lead-carbon compatibility. Realizing the mass production of low-cost carbon materials can help further reduce the cost of lead-carbon battery, improve the overpotential of hydrogen evolution, and improve the compatibility with lead-active substances; this can effectively prevent the battery from losing water and extend its cycle life.
3. The battery is a complex electrochemical system. It must be optimized as a whole from the aspects of lead-carbon negative formula and battery structure to further improve the power performance of the lead-carbon battery and meet the diversified needs of energy storage.

1.7 CONCLUSION

Lead-carbon battery is a new energy storage technology that integrates the advantages of supercapacitors and lead-acid batteries, and it has broad application prospects in large-scale energy storage systems. This chapter discusses the principles and technical characteristics of lead-carbon batteries and their applications in the energy storage grid. The main conclusions are as follows:

1. Compared with ordinary lead-acid batteries, lead-carbon batteries have excellent charging and discharging performance, long cycle life in PSoC mode, and low cost and high performance. In addition, lead-carbon batteries have obvious advantages in safety, low-temperature performance, and production and recycling technologies.
2. Lead-carbon batteries still face some challenges in terms of power performance, energy density, etc.

3. The successful application of lead-carbon battery energy storage technology in photovoltaic energy storage, wind power energy storage, and microgrid demonstration projects shows that lead-carbon batteries have good application prospects in power energy storage.
4. With the rapid growth of the energy storage market and the rapid support of national policies, lead-carbon batteries are expected to become one of the main electrochemical energy storage technology routes with advantages.

REFERENCES

1. Fei C, Dianlong W. Failure mode analysis of lead-carbon batteries for energy storage. *Battery*, 2016, 53(05): 237–240.
2. Moseley PT, Rand DAJ, Peters K. Enhancing the performance of lead-acid batteries with carbon-in pursuit of an understanding. *Journal of Power Sources*, 2015, 295: 268–274.
3. Tao ZJ, Chen J. Lead-carbon battery energy storage technology. *Energy Storage Science and Technology*, 2015, 4(06): 546–555.
4. Cooper A, Furakawa J, Lam L, et al. The ultra-battery—a new battery design for a new beginning in hybrid electric vehicle energy storage. *Journal of Power Sources*, 2009, 188(2): 642–649.
5. Lam LT, Louey R. Development of ultra-battery for hybrid-electric vehicle applications. *Journal of Power Sources*, 2006, 158(2): 1140–1148.
6. Bolo L, Ferg E. ALABC project 1315R1, effect of carbon and carbon nanotube additives on the negative active mass of lead acid batteries at partial state of capacity cycling. Final Report. *Proceedings of Advanced Lead-Acid Battery Consortium*, Research Triangle Park, NC, USA, 2015.
7. Pavlov D, Nikolov P, Rogachev T. ALABC project 1012G, additives to negative active mass of VRLA batteries and mechanisms by which they suppress sulfation of the negative plates on high-rate partial-state-of-charge battery operation. Final Report. *Proceedings of Advanced Lead-Acid Battery Consortium*, Research Triangle Park, NC, USA, 2013.
8. Ma L, Chen B, Chen Y, et al. Pitch-based carbon foam electrodeposited with lead as positive current collectors for lead acid batteries. *Journal of Applied Electrochemistry*, 2009, 39(9): 1609–1615.
9. Jang Y, Dudney NJ, Tiegs TN, et al. Evaluation of the electrochemical stability of graphite foams as current collectors for lead acid batteries. *Journal of Power Sources*, 2006, 161(2): 1392–1399.
10. Rand DAJ, Holden LS, May GJ, et al. Valve-regulated lead-acid batteries. *Journal of Power Sources*, 1996, 59(1): 191–197.
11. Lam LT, Haigh NP, Phyland CG, et al. Failure mode of valve-regulated lead-acid batteries under high-rate partial-state-of-charge operation. *Journal of Power Sources*, 2004, 133(1): 126–134.
12. Nakamura K, Shiomi M, Takahashi K, et al. Failure modes of valve-regulated lead-acid batteries. *Journal of Power Sources*, 1996, 59(1–2): 153–157.
13. Lam LT, Haigh NP, Phyland CG, et al. Failure mode of valve-regulated lead-acid batteries under high-rate partial-state-of-charge operation. *Journal of Power Sources*, 2003, 133: 126–134.
14. Moseley PT, Rand DAJ, Monahov B. Designing lead–acid batteries to meet energy and power requirements of future automobiles. *Journal of Power Sources*, 2012, 219: 75–79.
15. Tashima D, Kurosawatsu K, Taniguchi M, et al. Basic characteristics of electric double layer capacitor mixing ketjen black as conductive filler. *Electrical Engineering in Japan*, 2008, 165(1): 1–8.

16. Boden DP, Loosemore DV, Spence MA, et al. Optimization studies of carbon additives to negative active material for the purpose of extending the life of VRLA batteries in high-rate partial-state-of-charge operation. *Journal of Power Sources*, 2010, 195(14): 4470–4493.
17. Pavlov D, Naidenov V, Milusheva Y, et al. Benzyl benzoate as an inhibitor of the sulfation of negative electrodes in lead-acid batteries. *Journal of Energy Storage*, 2018, 17: 336–344.
18. Pavlov D, Nikolov P, Rogachev T. Influence of expander components on the processes at the negative plates of lead-acid cells on high-rate partial-state-of-charge cycling. Part II. Effect of carbon additives on the processes of charge and discharge of negative plates. *Journal of Power Sources*, 2010, 195(14): 4444–4457.
19. Calabek M, Micka K, Ivak P, et al. Significance of carbon additive in negative lead-acid battery electrodes. *Journal of Power Sources*, 2006, 158(2): 864–867.

2 Thermal Runaway Mechanisms and Prevention Strategies in Lithium-Ion Batteries for Energy Storage

Xing Hu and Qinming Liu

2.1 INTRODUCTION

In the construction of a new power system dominated by renewable energy, large-scale energy storage has become a critical support to enhance grid stability. As the core component of centralized energy storage, the safe and stable operation of energy storage power stations is directly linked to the reliability of power supply and public safety. However, energy storage safety incidents have occurred frequently worldwide in recent years, among which thermal runaway of lithium-ion batteries has been identified as the primary cause of system failure and even catastrophic consequences.

As the core energy storage unit within these systems, lithium-ion batteries are widely adopted due to their advantages, such as high energy density and long cycle life. Nevertheless, their inherent thermal instability poses a potential risk to system safety. Under abusive conditions such as overcharging, short circuits, or high temperatures, a chain of exothermic reactions is triggered inside the battery. This process begins with the decomposition of the solid electrolyte interphase (SEI) film, progressively leading to separator collapse, electrolyte decomposition, and oxygen release from cathode materials, ultimately resulting in thermal runaway. This phenomenon is accompanied not only by a rapid temperature rise but also by the generation of large amounts of flammable gas, potentially causing fire or explosion.

Moreover, within the densely packed configuration of an energy storage system, thermal runaway in a single cell can readily propagate to adjacent batteries through heat conduction, radiation, and flame spread, forming a catastrophic thermal runaway propagation chain. This domino effect can cause a localized failure to escalate rapidly into a system-level accident, significantly amplifying safety risks. Therefore, a fundamental understanding of thermal runaway mechanisms in lithium-ion batteries is needed. This includes how it initiates and propagates. Effective strategies for early warning and prevention should also be developed. Solving these scientific and

DOI: 10.1201/9781003492511-2

technical challenges is essential, as it is required for the safe, large-scale application of energy storage technologies. This effort is urgently needed to ensure the safe operation of energy storage power stations, and it is a necessary step to support the energy transition strategy and achieve sustainable development.

2.2 BASIC STRUCTURE AND THERMAL SAFETY CHARACTERISTICS OF LITHIUM-ION BATTERIES FOR ENERGY STORAGE

2.2.1 Development and Energy Storage Applications of Lithium-Ion Batteries

2.2.1.1 Development of Lithium-Ion Batteries

The commercial success of lithium-ion batteries can be termed a quiet revolution in the energy storage field. Its conceptual origins date back to the 1970s oil crisis, when scientists were committed to developing a new generation of high-energy batteries. In 1975, Whittingham et al. successfully fabricated the first rechargeable lithium battery (Figure 2.1) using titanium disulfide (TiS_2) as the cathode and lithium metal as the anode, achieving an operating voltage of 2 V and a relatively low capacity decay rate. However, the use of lithium metal as the anode led to the formation of lithium dendrites during repeated charge–discharge cycles, posing significant safety risks. Consequently, such batteries were typically discouraged from being charged.

In 1980, Goodenough et al. first reported lithium cobalt oxide ($LiCoO_2$) (Figure 2.2), an oxide with a layered structure. As a novel cathode material for

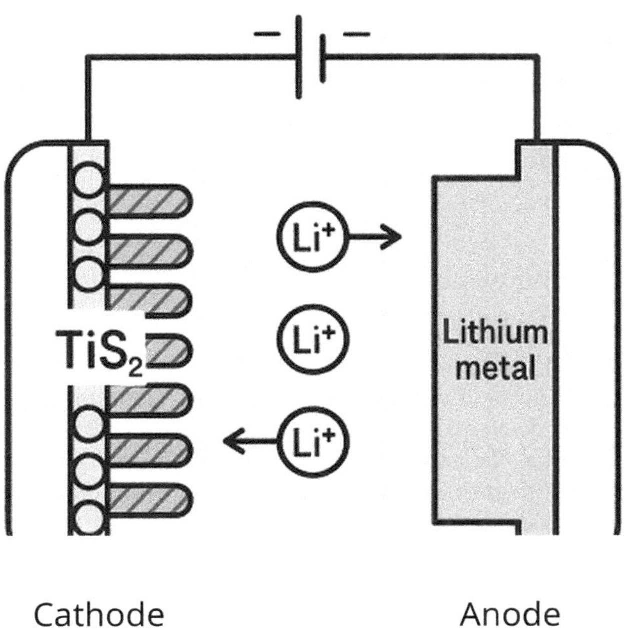

FIGURE 2.1 Whittingham's lithium metal battery.

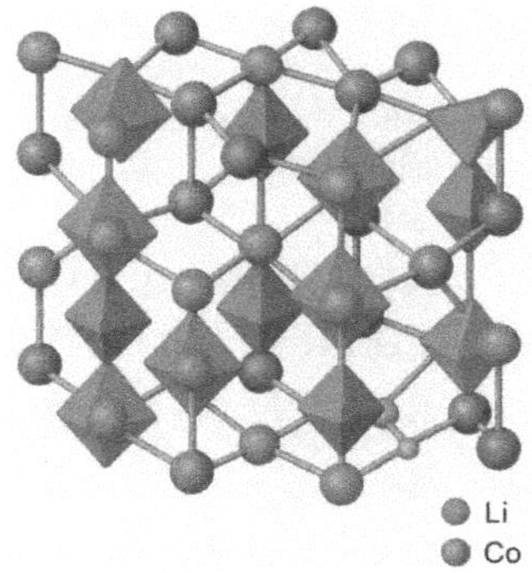

FIGURE 2.2 Goodenough's layered-structure $LiCoO_2$.

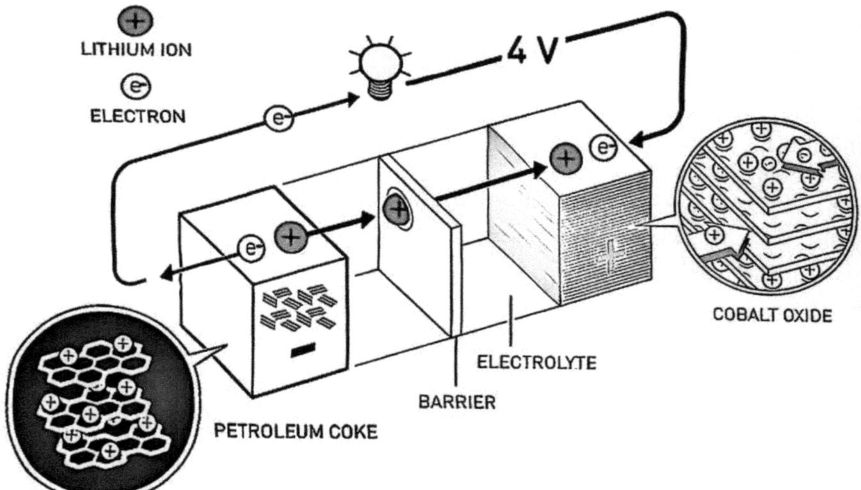

FIGURE 2.3 Yoshino's battery structure.

lithium-ion batteries, when paired with a lithium metal anode, it could generate a voltage as high as 4 V.

In 1985, Yoshino et al. discovered that heat-treated petroleum coke could reversibly intercalate and deintercalate lithium ions and exhibited a low potential (~0.5 V). Using this material as the anode, paired with a $LiCoO_2$ cathode, they assembled a novel rechargeable lithium battery (Figure 2.3). This configuration achieved a battery with higher energy density, which was named a lithium-ion battery for the first time.

In 1991, Sony Corporation pioneered the combination of Goodenough's $LiCoO_2$ cathode and Yoshino's carbon anode, launching the world's first commercial lithium-ion battery (Figure 2.4). This innovation fundamentally reshaped the landscape of consumer electronics, enabling the rapid growth of camcorders, mobile phones, and laptop computers.

FIGURE 2.4 Sony's commercial lithium-ion battery.

Entering the 21st century, with the diversified development of cathode material systems—from $LiCoO_2$ to the more cost-effective $LiFePO_4$ and higher energy-density ternary materials—the application boundaries of lithium-ion batteries have been continuously expanded. Their inherent advantages, such as high energy density, long cycle life, and relatively low self-discharge rate, align well with the requirements for large-scale energy storage. Particularly driven by the global energy transition and carbon neutrality goals, the grid integration of intermittent renewable energy sources, represented by wind and solar power, has surged dramatically, creating an extremely urgent need for large-scale, long-duration energy storage technologies within power grids.

Under this context, leveraging their technological maturity and rapid response capabilities, lithium-ion batteries have rapidly transitioned from the consumer electronics market to the broad energy storage stage. From grid-side applications like peak shaving, frequency regulation, and smoothing renewable energy output, to user-side applications such as arbitraging peak-off-peak electricity price differentials and providing backup power, lithium-ion battery energy storage systems have become an indispensable component in building the new power system.

2.2.1.2 Energy Storage Applications of Lithium-Ion Batteries

At the beginning of the 21st century, with the rapid development of renewable energy (wind and solar power) and the construction of smart grids, how to smooth the output of these intermittent energy sources and achieve peak shaving and valley filling became a new challenge. This provided a historic stage for lithium-ion batteries, whose performance was increasingly mature and costs were continuously declining. The ability of lithium-ion batteries to become the main force in energy storage systems is attributed to their unique advantages:

1. High energy density: Lithium-ion batteries can store a significant amount of electrical energy within a limited volume and weight. This makes the construction of compact and efficient megawatt-scale energy storage power stations possible, significantly reducing land and space costs.
2. High efficiency: The energy conversion efficiency during their charge and discharge processes can typically reach over 95%, far surpassing traditional energy storage methods. This means minimal energy loss during the charge–discharge cycle, enhancing the economic operational benefits of the entire system.
3. Long cycle life: High-quality lithium-ion batteries can achieve thousands to tens of thousands of deep charge-discharge cycles under suitable operating conditions, while their calendar life can reach 8–15 years, ensuring the economic feasibility of energy storage projects throughout their entire life cycle.
4. Rapid response: Lithium-ion batteries can respond to grid dispatch commands within milliseconds to seconds, quickly switching from charging to discharging states. This characteristic makes them a high-quality resource for grid frequency regulation, capable of offsetting minor frequency fluctuations in real time to maintain grid security.
5. Modular design: Based on the standard design concept of cell–module–system, lithium-ion energy storage systems possess great flexibility. They can scale seamlessly from kilowatt-hour-scale residential energy storage units to hundred-megawatt-hour-scale grid-side large-scale energy storage power stations, meeting customized needs across different scenarios.

Leveraging these core competencies, lithium-ion battery energy storage systems have become deeply embedded in every link of the energy value chain, playing multiple critical roles:

1. On the generation side, they are directly paired with photovoltaic power stations and wind farms, acting as stabilizers to smooth their short-term power fluctuations and as energy shifters to achieve temporal energy transfer—for example, storing surplus solar energy at noon and releasing it during the evening peak electricity consumption period. This significantly enhances the grid-friendliness and utilization hours of renewable energy.
2. On the grid side, they transform into rapid response resources, providing the grid with high-precision frequency regulation services that are difficult for traditional units to achieve, thereby maintaining grid frequency stability in real time. Simultaneously, they can serve as peak capacity resources, discharging during peak load periods to effectively delay or reduce costly grid upgrades needed to meet short-term peak demand.
3. On the user side, commercial and industrial enterprises and household users utilize energy storage systems for arbitraging peak and off-peak electricity price differentials, substantially reducing electricity costs. Furthermore,

they serve as reliable insurance, acting as an uninterrupted backup power source during grid failures to ensure the continuity of critical production processes or household electricity supply.

However, opportunities and challenges coexist. When tens of thousands or even hundreds of thousands of high-energy-density cells are densely integrated into a limited space, their inherent safety risks are dramatically amplified. Thermal runaway—an uncontrollable temperature rise phenomenon triggered by internal chain exothermic reactions—has become the safety crisis hanging over the energy storage industry. Mechanical abuse, electrical abuse, or thermal abuse can all trigger thermal runaway in a single cell. If the substantial heat and flammable gases released cannot be effectively controlled, they can cause a catastrophic thermal runaway propagation within the battery pack, potentially engulfing the entire energy storage unit in flames (Figure 2.5), leading to enormous economic losses and social impacts.

Consequently, global industry–academia–research efforts are continuously invested in research and development, aiming to build robust safety defenses while persistently enhancing battery performance and full life-cycle economics. This demands coordinated innovation across three levels: intrinsic safety (e.g., developing flame-retardant electrolytes and thermally stable electrode materials), intelligent early warning (e.g., big data- and AI-based early prediction of thermal runaway), and systemic protection (e.g., efficient thermal management, fire suppression, explosion mitigation, and isolation design). Advancing lithium-ion battery energy storage steadily towards a new phase characterized by enhanced safety, improved economics, and extended longevity is not only an inevitable requirement of technological progress but also a

FIGURE 2.5 Explosion accident at an energy storage power station in California.

fundamental safeguard for achieving the global transition to clean energy and sustainable development.

2.2.2 Overview of Lithium-Ion Batteries

2.2.2.1 Composition and Classification of Lithium-Ion Batteries

The main components of a lithium-ion battery are the metal casing, the cell, and the electrolyte. The cell comprises metal foil current collectors, coating materials, a separator, and other elements. Although structurally different in outer packaging formats, different types of lithium-ion batteries share a similar internal core structure.

The cell is a multilayer structure, entirely immersed in electrolyte, and is assembled inside the battery by stacking or winding in the sequence "separator-positive electrode-separator-negative electrode". The positive and negative electrodes use metal current collectors as substrates: aluminum foil for the positive electrode and copper foil for the negative electrode. Coated on both sides of the current collectors are the coating materials, formulated by mixing electrochemically active material particles with a binder in specific proportions. This coating is where the electrochemical reactions occur and constitutes the most crucial part of the battery [1].

The separator, typically made from polymer materials like polyethylene (PE) and polypropylene (PP), serves to separate each layer of the positive and negative electrodes. Additionally, lithium-ion batteries include other components such as terminals, tabs, and safety vents.

Lithium-ion batteries can be classified based on their cathode materials into types including $LiCoO_2$ (LCO), $LiMnO_2$ (LMO), lithium nickel cobalt manganese oxide (NCM), lithium nickel cobalt aluminum oxide (NCA), $LiFePO_4$ (LFP), and among others.

LCO batteries offer the advantages of stable structure and high capacity. As the most mature cathode material, LCO batteries feature simple manufacturing processes, good cycle performance, and high electrical conductivity. They are commonly used in portable electronic devices like laptops and mobile phones. However, their disadvantages include high production costs, the relative scarcity of cobalt, environmental concerns, and poor thermal stability.

LMO batteries provide high safety and stability. Their energy density is slightly lower than that of LFP batteries, and they exhibit good cycle performance at low temperatures. However, their disadvantages are quite pronounced: the cathode material often requires blending with other materials during manufacturing, leading to significant variations in electrochemical performance.

NCM batteries represent a common type of ternary lithium-ion battery. This ternary cathode material offers higher specific capacity and lower cost compared to single-component cathode materials. Relative to traditional LCO batteries, NCM batteries rely less on cobalt, reducing both cost and resource dependency. They possess high energy density and excellent cycle stability. The drawbacks include poorer performance under overcharge and over-discharge conditions, inferior thermal stability under extreme conditions, and degraded cycle performance.

NCA is another common ternary lithium-ion battery cathode material. NCA is considered a second-generation, greener alternative to LCO, offering high energy density and good low-temperature performance. However, it suffers from poorer safety performance and higher technical requirements for production.

LFP batteries use lithium iron phosphate as the cathode material. LFP batteries hold significant advantages in terms of safety performance and cycle life. However, LFP also has the following disadvantages: relatively poor low-temperature performance, low tap density, and larger battery volume for the same capacity compared to LCO batteries, thus offering no advantage in miniature battery applications.

The characteristics of these several battery types are summarized in Table 2.1.

As evident, because the cathode material of LFP batteries has a higher thermal decomposition temperature compared to other battery types, they are less prone to violent thermal runaway during mechanical abuse tests conducted in a fully charged state. This allows for a clearer observation of intermediate electrode damage and overcharge, facilitating the determination of critical thresholds for various failure stages. Furthermore, the cycle life of LFP batteries is significantly longer than that of other battery types. Therefore, the research work in this chapter also primarily focuses on LFP batteries.

Based on their external form, lithium-ion batteries can be classified into four main types: cylindrical, prismatic, pouch, and blade batteries. A schematic diagram of these four lithium-ion battery types is shown in Figure 2.6.

The external metal casing of cylindrical batteries is typically encapsulated using nickel-plated steel shells or an aluminum alloy. The interior consists of a cylindrical cell body formed by winding the positive and negative electrodes together with a separator. An insulating gasket is placed at the bottom of the battery, and the electrolyte is filled and sealed within the casing. The use of steel or aluminum shells gives

TABLE 2.1
Advantages, Disadvantages, and Application Scenarios of Common Lithium-Ion Batteries

Type	Advantages	Disadvantages	Application Scenarios
LCO	Simple manufacturing process and high voltage plateau	High cost, limited cycle life, and poor safety	Smartphones, laptops, etc.
LMO	Abundant manganese resources, low cost, and good safety performance	Poor electrolyte compatibility and slightly lower energy density	Power tools, low-speed electric vehicles, etc.
NCM	Stable electrochemical performance and high energy density	Poor overcharge/over-discharge performance, and poor thermal stability	Electric vehicles, etc.
NCA	High energy density and good low-temperature performance	Poor safety and high production technology requirements	Electric vehicles, aerospace energy, etc.
LFP	Good safety performance and long cycle life	Poor low-temperature performance and relatively low tap density	Electric vehicles, energy storage systems, etc.

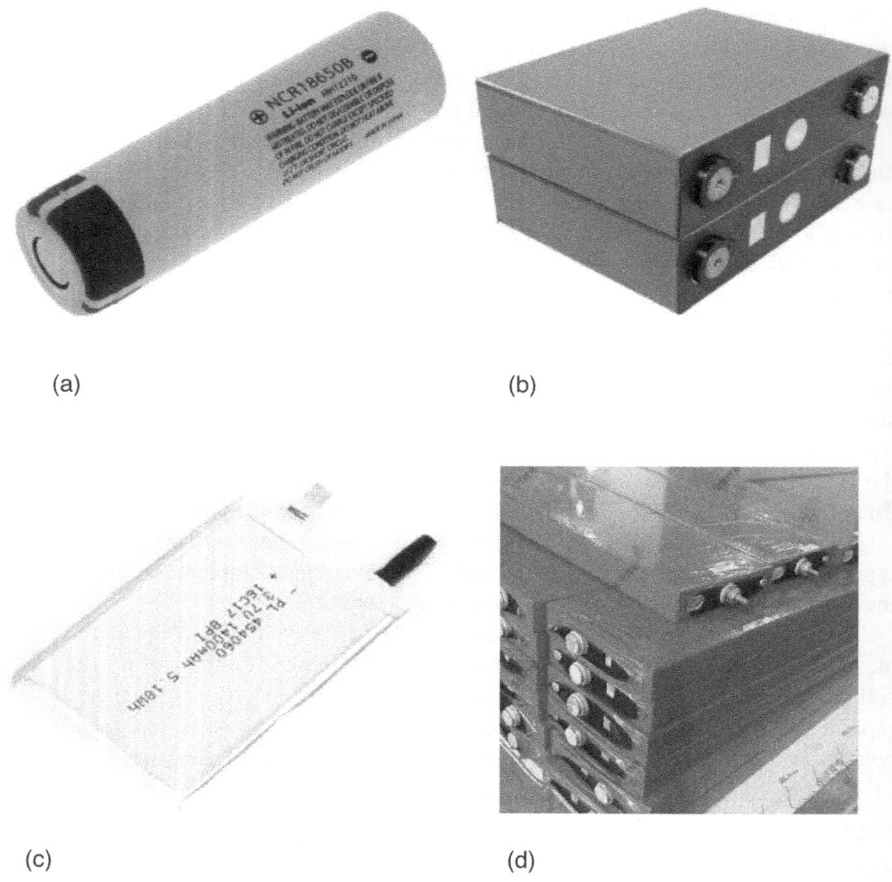

FIGURE 2.6 Different types of lithium-ion batteries. (a) Cylindrical type. (b) Prismatic type. (c) Pouch type. (d) Blade type.

cylindrical lithium-ion batteries high casing rigidity; however, their radial crush resistance is relatively weak. Plastic deformation of the shell can lead to misalignment of the internally wound cell and separator tearing [2].

Prismatic batteries utilize a rigid aluminum metal shell for encapsulation, containing either wound or stacked (laminated) cells. A bursting disc is often located on the side of the casing, and the space between the inner wall of the shell and the cell is filled with cushioning material. The rectangular rigid shell of a prismatic battery can distribute external impact loads. When stacked cells are used internally, they avoid the risk of interlayer slippage present in wound cells, leading to a more uniform distribution of mechanical stress [3]. However, stress concentration is prone to occur at the corners of the rectangular structure, which, during severe collisions, may lead to cracking and electrolyte leakage.

Pouch batteries use a polymer composite film as the external casing, usually containing stacked cells, and are sealed via a heat-sealing process. There are no fixed gaps within the casing structure, and the battery weight is significantly lower than that of prismatic hard-case batteries. Due to the polymer film casing, pouch batteries cannot withstand concentrated loads well, making them susceptible to separator damage that triggers internal short circuits (ISCs). They exhibit strong deformation adaptability and a lower risk of violent failure. Under mechanical abuse, electrolyte leakage is a common occurrence.

A blade battery is essentially a long, thin prismatic hard-case battery. It employs an aluminum alloy casing and contains stacked cells, presenting an overall long, strip-like shape. In practical applications, it can serve as a structural component within the battery pack, replacing traditional longitudinal and transverse beams [4]. Due to their long, thin geometry characterized by a high aspect ratio, blade batteries distribute crushing loads uniformly along their length. As a structural component, blade batteries possess higher structural strength than standard hard-case batteries and exhibit a lower risk of thermal runaway.

2.2.2.2 Working Principle of Lithium-Ion Batteries

The cathode material of a lithium-ion battery is a lithium metal compound, while the anode material is typically carbon. The separator, situated between the cathode and anode, is generally a porous polymer material that physically separates the electrodes to prevent short circuits while allowing lithium ions to pass through. During

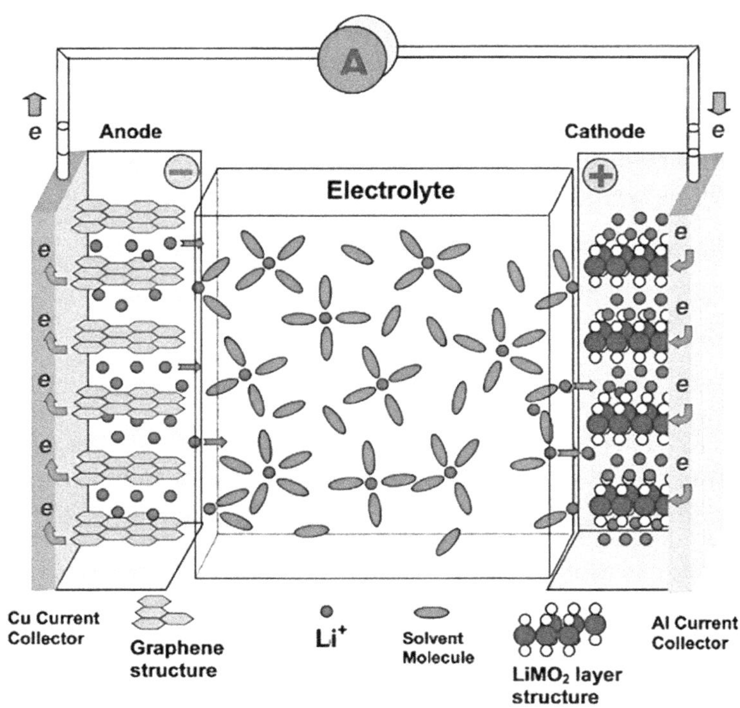

FIGURE 2.7 Schematic of lithium-ion battery operation [4].

operation, the battery involves internal and external circuits [4]. Figure 2.7 illustrates the working principle of a lithium-ion battery.

During charging, electrons move from the positive electrode to the negative electrode through the external circuit. Simultaneously, within the internal circuit, lithium ions are deintercalated from the positive electrode, travel through the electrolyte and separator, and reach the negative electrode, where they combine with the electrons that have arrived via the external circuit. The discharge process is the reverse of the charging process. During discharge, electrons move from the negative electrode to the positive electrode through the external circuit; lithium ions are deintercalated from the negative electrode and travel back through the internal circuit to be re-embedded into the positive electrode. The cyclical movement of electrons and lithium ions through the external and internal circuits enables the rechargeable nature of the battery.

Using a LFP battery as an example: when the battery discharges, the anode active material undergoes an oxidation reaction driven by the potential difference, causing lithium ions to be deintercalated and release electrons. The released electrons cannot travel through the electrolyte and must be conducted via the copper foil current collector to the external circuit, then flow through the load towards the positive electrode, providing electrical energy to the load. Meanwhile, the deintercalated lithium ions from the anode migrate directionally within the electrolyte. These ions move through the separator, composed of porous material, from the electrolyte on the anode side to the electrolyte on the cathode side, eventually reaching the surface of the positive electrode. At the cathode surface, the lithium ions and the electrons conducted to the cathode active material via the aluminum foil current collector participate in a reduction reaction at the active sites of the cathode. The lithium ions embed into the crystal structure of the cathode material, forming a stable lithium-containing compound.

The cathode reaction during discharge:

$$\mathrm{Li}_x\mathrm{C}_6 \rightarrow x\mathrm{Li}^+ + x\mathrm{e}^- + 6\mathrm{C} \tag{2.1}$$

The anode reaction during discharge:

$$\mathrm{Li}_{1-x}\mathrm{FePO}_4 + x\mathrm{Li}^+ + x\mathrm{e}^- \rightarrow \mathrm{LiFePO}_4 \tag{2.2}$$

When the battery is charged, an external electric field forces lithium ions to deintercalate from the positive electrode and migrate back to embed into the negative electrode. During charging, the battery's positive terminal is connected to the positive terminal of the power source, and the negative terminal to the negative terminal. The potential difference provided by the external power source exceeds the battery's discharge potential, initiating the charging process. Driven by the external electric field, lithium iron phosphate at the cathode undergoes an oxidation reaction, deintercalating lithium ions and accepting electrons. These accepted electrons originate from the positive terminal of the power source and are conducted to the cathode active material via the aluminum foil current collector. The deintercalated lithium ions migrate directionally within the electrolyte, moving through the pores of the separator towards the anode side. When the lithium ions deintercalated from the cathode reach the anode, they pass through the SEI membrane under the influence of the external field and embed into the layered structure of the graphite anode. Simultaneously, they combine with electrons conducted via the power source's negative terminal and the copper foil current collector, undergoing a reduction reaction to form lithium-carbon compounds. As the charging process continues, the number of lithium ions embedded in the anode increases, continually raising the battery's energy density. Charging concludes when lithium ions are almost completely deintercalated from the cathode and the embedding of lithium ions into the anode reaches saturation.

The cathode reaction during charging:

$$LiFePO_4 \rightarrow Li_{1-x}FePO_4 + xLi^+ + xe^- \tag{2.3}$$

The anode reaction during charging:

$$xLi^+ + xe^- + 6C \rightarrow Li_xC_6 \tag{2.4}$$

The overall cell reaction:

$$LiFePO_4 + 6C \underset{\text{discharge}}{\overset{\text{charge}}{\rightleftarrows}} \rightarrow Li_{1-x}FePO_4 + Li_xC_6 \tag{2.5}$$

2.2.3 Heat Generation Mechanisms in Lithium-Ion Batteries during Charge and Discharge

Thermal analysis of lithium-ion batteries relies on accurate thermal models. However, the electrochemical, electrical, and temperature fields interact with each other, forming a complex coupling effect during the charge/discharge process. The electrochemical–thermal coupling model describes how electrochemical reactions interact with heat transfer in lithium-ion batteries. During operation, lithium ions move through the electrolyte between the cathode and anode, enabling charge transfer and energy storage (Figure 2.8).

Reaction and diffusion rates depend on temperature, charge transfer resistance, and concentration gradients [5]. Heat is generated as these reactions release or absorb energy [6]. This heat distribution and conduction critically affect the battery's stability and lifespan. By combining electrochemical and thermal dynamics, the model analyzes key performance metrics: efficiency, temperature profiles, and heat dissipation.

2.2.3.1 Electrochemical Model

The electrochemical behavior of lithium-ion batteries is modeled through coupled multi-physics equations under porous electrode theory, comprising five governing partial differential equation systems: solid/liquid phase lithium-ion diffusion equations (Fick's law), solid/liquid phase charge conservation equations (Ohm's law), and the electrode reaction kinetics equation (Butler–Volmer equation) [7].

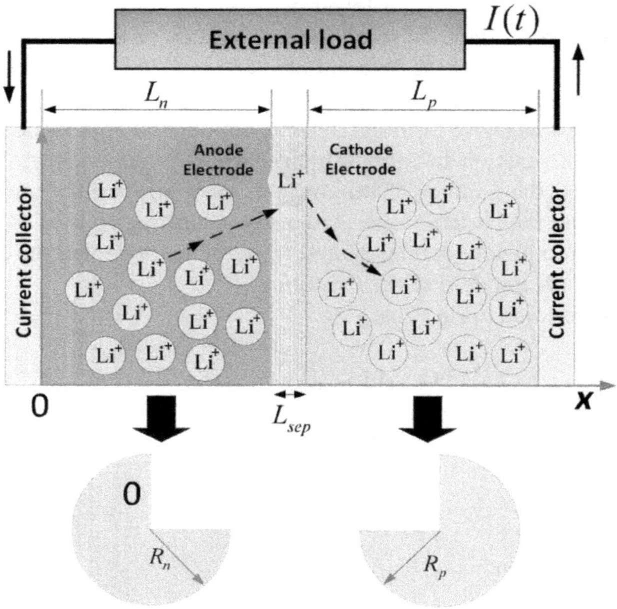

FIGURE 2.8 Schematic diagram of the electrochemical mechanism of lithium-ion batteries.

2.2.3.1.1 Solid-Phase Lithium-Ion Diffusion Equation

The solid-phase lithium-ion diffusion equation describes the lithium-ion diffusion process within electrode active material particles, governed by Fick's second law

$$\frac{\partial c_s}{\partial t} = \frac{D_s}{r^2} \bullet \frac{\partial}{\partial r}\left(r^2 \bullet \frac{\partial c_s}{\partial r}\right) \tag{2.6}$$

where c_s is solid Li^+ concentration, D_s is solid diffusion coefficient, r is particle radius, and t is time.

Assuming the lithium-ion concentration at the center of the active material particles remains constant and does not change over time, then the boundary condition is

$$\left.\frac{\partial c_s}{\partial r}\right|_{r=0} = 0 \tag{2.7}$$

The ion flux passing through the particle surface equals the ion flux produced by the electrochemical reaction. Then

$$-D_s \left.\frac{\partial c_s}{\partial r}\right|_{r=R} = j \tag{2.8}$$

where j is the local current density due to electrochemical reactions.

2.2.3.1.2 Solid-Phase Charge Conservation Equation

According to Ohm's law, the solid-phase potential in the electrode region could be expressed as

$$\frac{\partial}{\partial_x}(\sigma_s^{\text{eff}}\frac{\partial\phi_s}{\partial_x}) = j \tag{2.9}$$

$$\sigma_s^{\text{eff}} = \varepsilon_s^{1.5}\sigma_s \tag{2.10}$$

where σ_s^{eff} is the effective solid-phase conductivity, ϕ_s is the solid-phase potential, ε_s is the solid-phase volume fraction, and σ_s is solid-phase conductivity.

The solid-phase potentials on both sides of the separator are the same, and the solid-phase potentials of the two electrodes are also the same, which can be expressed as

$$\left.\frac{\partial\varphi_s}{\partial x}\right|_{x=L_n} = \left.\frac{\partial\varphi_s}{\partial x}\right|_{x=L_n+L_{\text{sep}}} \tag{2.11}$$

$$-\left.\frac{\partial\varphi_s}{\partial x}\right|_{x=0} = \left.\frac{\partial\varphi_s}{\partial x}\right|_{x=L_n+L_p+L_{\text{sep}}} \tag{2.12}$$

2.2.3.1.3 Liquid-Phase Lithium-Ion Diffusion Equation

The liquid-phase lithium-ion diffusion equation describes the diffusion process of lithium ions in the electrolyte. It is typically formulated by the Nernst–Planck equation

$$\varepsilon_e\frac{\partial c_e}{\partial_t} = \frac{\partial}{\partial_x}\left(D_e^{\text{eff}}\bullet\frac{\partial c_e}{\partial_x}\right) + \frac{1-t^+}{F}j \tag{2.13}$$

where c_e is the lithium-ion concentration in the electrolyte, ε_e is the volume fraction occupied by the electrolyte, D_e^{eff} is the effective diffusion coefficient of lithium ions in the electrolyte, t^+ is the transference number of lithium ions, and F is the Faraday constant.

2.2.3.1.4 Liquid-Phase Charge Conservation Equation

The current density in the electrolyte phase consists of two components, including the migration current and the diffusion current, which satisfy

$$j_l = -\kappa_e^{\text{eff}}\frac{\partial\varphi_e}{\partial_x} + \frac{2RT\kappa_e^{\text{eff}}}{F}\left(1-t^+\right)\frac{\partial\ln c_e}{\partial x} \tag{2.14}$$

where j_l is the liquid-phase current density, κ_e^{eff} is the effective ionic conductivity of the electrolyte, φ_e is the liquid-phase potential, and T is the thermodynamic temperature.

2.2.3.1.5 Electrode Reaction Kinetics Equation

The electrochemical reaction process of lithium ions at the electrode–electrolyte interface is typically described by the Butler–Volmer equation:

$$j = j_0\left(\exp\left(\frac{\alpha_a F}{RT}\eta\right) - \exp\left(\frac{\alpha_c F}{RT}\eta\right)\right) \tag{2.15}$$

$$j_0 = FK(c_{s,\max} - c_{s,\text{surf}})^{\alpha_c}(c_{s,surf})^{\alpha_a}(c_e)^{\alpha_a} \tag{2.16}$$

where j_0 is the exchange current density, α_a is the anode transfer coefficient, α_c is the cathode transfer coefficient, K is the reaction rate coefficient, $c_{s,\max}$ is the maximum intercalation concentration in solid particles, $c_{s,\text{surf}}$ is the surface concentration of solid particles, and η is the overpotential.

η and j_0 are the key parameters determining the progress of the electrochemical reaction, where η is defined as

$$\eta = \varphi_s - \varphi_e - U_{\text{eq}} \tag{2.17}$$

where U_{eq} is the thermodynamic equilibrium potential.

2.2.3.2 Thermal Model

To quantitatively analyze the thermal characteristics of lithium-ion batteries, it is necessary to identify the fundamental principles of heat generation [6]. The total heat generated by a lithium-ion battery consists of four components, including electrochemical reaction heat (Q_r), polarization internal resistance heat (Q_p), Ohmic internal resistance heat (Q_j), and side reaction heat (Q_s). Thus, the total heat generation of the battery can be expressed as

$$Q = Q_r + Q_p + Q_j + Q_s \tag{2.18}$$

2.2.3.2.1 Electrochemical Reaction Heat (Q_r)

Electrochemical heat (Q_r) results from entropy changes during lithium-ion battery reactions. It is negative during discharge (heat release) and positive during charge (heat absorption). The heat is calculated as

$$Q_r = Q / 3{,}600 \cdot F \cdot I \tag{2.19}$$

where Q is the total heat generated, and I is the current during charge/discharge operation.

2.2.3.2.2 Polarization Internal Resistance Heat (Q_p)

Q_p represents the heat generated by polarization resistance during lithium-ion battery operation. This heat is always positive, as polarization releases energy during charge/discharge. The heat generation is calculated as

$$Q_p = I^2 R_p \tag{2.20}$$

where R_p is the polarization resistance.

2.2.3.2.3 *Ohmic Internal Resistance Heat* (Q_j)

Q_j is the irreversible heat generated by internal resistance in lithium-ion batteries. It occurs in conductive layers during operation, which is calculated as

$$Q_j = I^2 R_j \tag{2.21}$$

where R_j is the Ohmic resistance.

2.2.3.2.4 *Side Reaction Heat* (Q_s)

Q_s refers to heat generated during abnormal charge/discharge conditions, and is typically negligible in total heat calculations.

Then, the total internal heat generation (Q) in lithium-ion batteries is given by

$$Q = Q_r + Q_p + Q_j = Q / 3{,}600 \cdot F \cdot I + I^2 R_p + I^2 R_j \tag{2.22}$$

2.2.3.3 Electrochemical–Thermal Coupling Model

Lithium-ion batteries exhibit coupling between electrochemical reactions and thermal effects during operation. First, the heat generated by electrochemical reactions induces temperature variations [7]. Conversely, these temperature changes significantly influence electrochemical parameters and reaction kinetics. To accurately characterize this interaction, a coupling framework integrating electrochemical and thermal models must be established.

2.2.3.3.1 *Temperature Dependent Parameter Modeling*

To account for the temperature sensitivity of electrochemical parameters, we establish temperature-dependent relationships using the Arrhenius equation

$$X(T) = X_{\text{ref}} \bullet \exp\left(\frac{E_a}{R}\left(\frac{1}{T_{\text{ref}}} - \frac{1}{T}\right)\right) \tag{2.23}$$

where $X(T)$ is the temperature-dependent parameter (e.g., solid-phase diffusion coefficient D_s, reaction rate coefficient K), X_{ref} is the reference value at T_{ref}, and E_a is the activation energy.

Electrolyte properties (e.g., liquid-phase diffusion coefficient D_e and ionic conductivity κ_e) follow analogous temperature-correction rules. These relationships enable bidirectional coupling between electrochemical and thermal models.

2.2.3.3.2 *Electrochemical–Thermal Model Coupling Mechanism*

The coupling between electrochemical and thermal models is achieved through temperature-dependent parameter interactions, as illustrated in Figure 2.9.

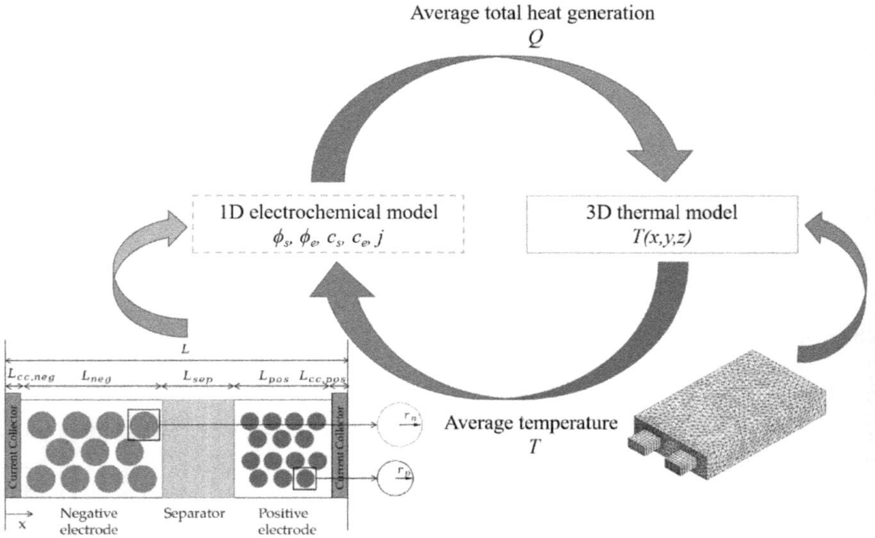

FIGURE 2.9 The coupling mechanism between electrochemical and thermal models.

By solving a 1D electrochemical model, the solid/liquid phase potentials (ϕ_s, ϕ_e), lithium-ion concentrations (c_s,c_e), and current density (j) could be obtained [8]. The average total heat generation Q, which includes contributions from Ohmic, polarization, and electrochemical reaction heating, is then calculated. The total heat generation Q is transferred to the 3D thermal model. The energy conservation equation is expressed as

$$\rho C_p \frac{\partial T}{\partial t} - \lambda \nabla^2 T = Q \tag{2.24}$$

where ρ is the bulk density of the battery, C_p is the bulk specific heat capacity, λ is the bulk thermal conductivity, and T is the battery temperature. $\nabla^2 T$ represents the three-dimensional Laplacian operator of the temperature field, which can be explicitly expanded as

$$\nabla^2 T = \frac{\partial^2 T}{\partial x^2} + \frac{\partial^2 T}{\partial y^2} + \frac{\partial^2 T}{\partial z^2} \tag{2.25}$$

According to Newton's law of cooling, ignoring the radiative heat transfer, the relationship between the battery temperature and the ambient temperature is expressed as

$$-\lambda \nabla T = h(T - T_{\text{amb}}) \tag{2.26}$$

where T_{amb} is the ambient temperature, and h is the convective heat transfer coefficient.

The temperature field $T(x,y,z)$ is solved using the thermal model. The resulting 3D temperature distribution is fed back to the electrochemical model to determine the average temperature [9]. This temperature-dependent feedback updates critical parameters (e.g., solid-phase diffusion coefficient D_s and reaction rate coefficient K) in the 1D electrochemical model, thereby establishing bidirectional coupling between the electrochemical and thermal frameworks.

2.3 ROOT CAUSES OF THERMAL RUNAWAY: BATTERY BEHAVIOR UNDER ABUSE CONDITIONS

Thermal runaway does not occur spontaneously; it is typically triggered by external abuse conditions. These conditions include mechanical impact, electrical overload, or high-temperature environments. Each type of abuse condition can initiate specific failure modes within the battery, such as separator damage, electrode material decomposition, or electrolyte reactions. Identifying these triggering conditions and their corresponding internal battery responses constitutes the first line of defense against thermal runaway and is a prerequisite for developing effective prevention and control strategies [10].

2.3.1 Mechanical Abuse: Separator Rupture and Internal Short Circuit

The safety of lithium-ion batteries is closely tied to their internal structure. Inside a typical lithium-ion battery, the positive and negative electrodes are physically separated by an extremely thin separator. This separator is a polymeric membrane (e.g., PE or PP) with a microporous structure. Its core functions are: to allow lithium ions to pass freely, enabling the necessary electrochemical reactions for charge and discharge, while absolutely preventing direct electron conduction between the positive and negative electrodes, that is, preventing direct contact between them [10]. Any breach of this physical isolation leads to catastrophic consequences—an ISC. Mechanical abuse is one of the most direct and violent ways to compromise this isolation.

Mechanical abuse typically refers to the battery being subjected to external mechanical forces, such as crushing, nail penetration, or impact by sharp objects. These forces are not uniformly distributed across the entire battery but often create significant stress concentration in localized areas. When this localized stress exceeds the mechanical strength limit of the battery's internal materials (particularly the separator), it triggers a series of irreversible structural damages (Figure 2.10).

Separator rupture is the critical turning point in this process. The separator itself is designed with extreme precision, typically having a thickness ranging from just over ten to twenty-some micrometers, with pore sizes far smaller than one micrometer. Under normal mechanical stresses, it effectively maintains its structural integrity. However, when external forces like crushing or nail penetration act upon the battery surface, the force is transmitted through the battery casing and the internal jellyroll (wound) or stacked electrode assembly to the separator. If the point of application is a small-area, sharp object (e.g., nail penetration), the pressure borne by the separator at this point increases drastically, far exceeding its puncture strength, resulting in the separator being directly punctured. In the case of large-area crushing, it may cause relative displacement and deformation of the internal jellyroll or electrode sheets, leading to the separator being overstretched until it tears.

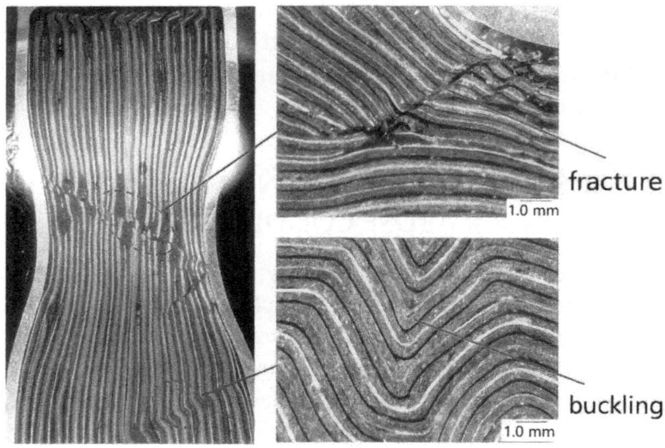

FIGURE 2.10 Structural damage of lithium-ion battery under mechanical abuse.

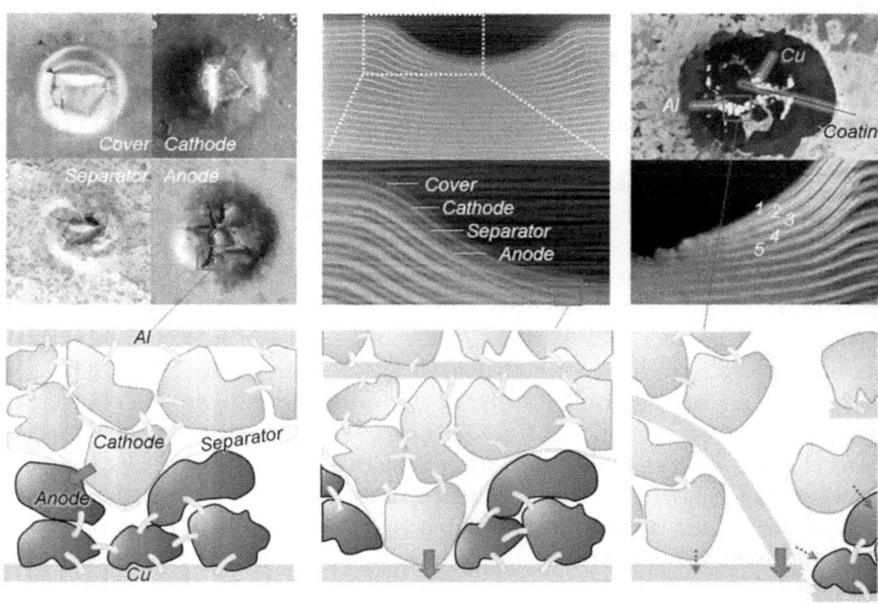

FIGURE 2.11 Formation of internal short circuits under mechanical abuse [4].

Once a physical breach of the separator occurs, whether it's a tiny hole or a crack, the meticulously maintained internal balance of the battery is broken (Figure 2.11). The positive and negative active materials lose their physical barrier at the breach point, achieving direct physical contact. From an electrochemical perspective, this is equivalent to establishing a conductive pathway directly connecting the positive and negative electrodes internally, precisely at the location of the breach.

2.3.1.1 Comprehensive Review of Thermal Management Strategies for Lithium-Ion Batteries: From Heat Generation Mechanism to Advanced Cooling Solutions

This newly formed conductive pathway allows electrons inside the battery to bypass the intended external circuit and instead take an internal shortcut. Electrons flow directly from the negative electrode (lower potential) to the positive electrode (higher potential), creating a substantial localized current. This process is known as an ISC (Figure 2.11). Since this short-circuit point exists within the battery, it cannot be interrupted by any external circuit [11].

The current magnitude at the internal short-circuit point can be estimated using Ohm's law ($I = V/R$). Here, V is the battery's terminal voltage at that moment, which for a fully charged cell could exceed 4.2 V. R is the resistance of the internal short-circuit path, a value typically very small, dependent on the contact area, contact pressure at the short point, and the conductivity of the active materials themselves. Even assuming a short-circuit resistance as high as 100 milliohms (a relatively large estimate), in a battery with a voltage of 4 V, the instantaneous heat generation power at the short point, according to Joule's law ($P = I^2R$), would be as high as 160 W. This represents an extremely significant power density, all released instantaneously within a very tiny localized area (the short point).

The accumulation of this localized Joule heating is the starting point for all subsequent thermal issues. The heat causes the temperature at the short point and its surrounding area to rise rapidly. Elevated temperature further degrades the separator's mechanical properties and chemical stability. For instance, the melting point of common separator materials like PE is typically around 130°C–140°C. When the local temperature exceeds this threshold, the separator begins to shrink and melt. This does not repair the initial breach but rather expands the short-circuit area, creating a lower-resistance path. This, in turn, leads to a larger short-circuit current and more intense heat generation, establishing a localized thermal-electrical positive feedback loop.

Thus, mechanical abuse, by applying external stress, directly compromises the most critical structure ensuring internal electrical isolation—the separator. The rupture of the separator creates the condition for direct contact between the positive and negative electrodes, thereby triggering an ISC. The significant accumulation of Joule heat generated by the ISC locally sows the seeds for potentially subsequent, more intense chain exothermic reactions—namely, thermal runaway. Therefore, understanding the mechanism of separator rupture leading to an ISC is the cornerstone for comprehending the entire process of thermal runaway initiated by mechanical abuse.

2.3.2 Electrical Abuse: Cascading Side Reactions Induced by Overcharge and Overdischarge

The normal operation of lithium-ion batteries relies on a precisely balanced electrochemical system. The fundamental processes of charging and discharging involve the reversible intercalation and deintercalation of lithium ions between the positive and negative electrodes. Each material within this system—the cathode, anode,

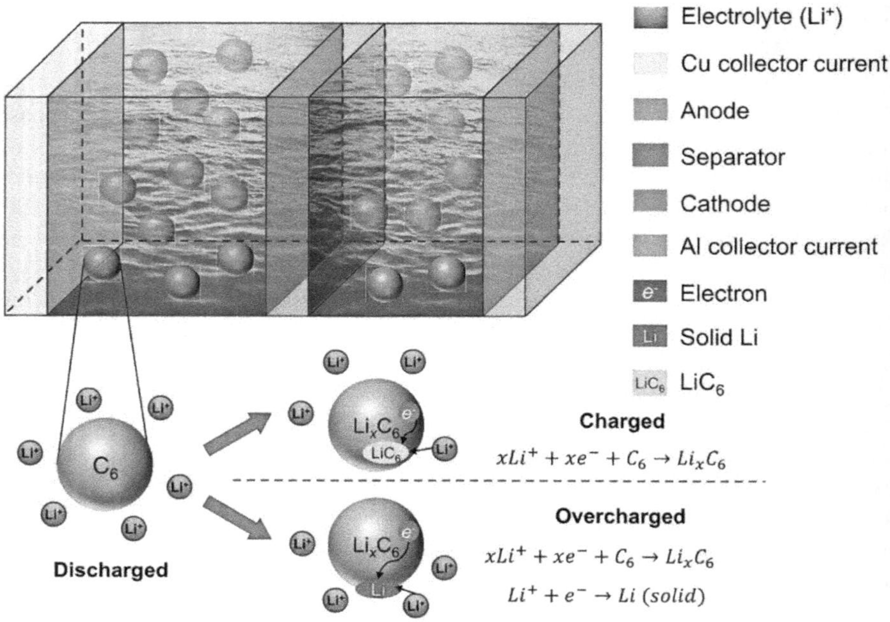

FIGURE 2.12 Overcharged lithium-ion battery.

and electrolyte—remains stable only within specific voltage and chemical potential windows [12]. Electrical abuse, particularly overcharge and overdischarge, forcibly disrupts this equilibrium, pushing the battery beyond its designed safety limits. This triggers a series of uncontrolled, cascading side reactions, ultimately laying the groundwork for thermal runaway.

Overcharge refers to the continued application of charging current after the battery has reached its fully charged state. At this point, the normal lithium intercalation reaction approaches saturation, and the excess energy begins to drive detrimental side reactions.

During normal charging, lithium ions deintercalated from the cathode travel through the electrolyte and embed into the layered structure of the graphite anode—a relatively safe process. However, during overcharge, the anode potential is driven even lower (vs. Li/Li^+), making it extremely difficult for lithium ions to intercalate into the graphite. A large number of lithium ions are transported to the anode surface but cannot be intercalated in time. This causes the lithium ions to directly accept electrons at the anode surface, reducing to metallic lithium and forming a layer of lithium plating. This process is known as lithium plating or lithium dendrite growth (Figure 2.12).

This plated metallic lithium poses several critical hazards.

1. High reactivity: Metallic lithium is highly reactive and directly reacts with the electrolyte, continuously consuming the electrolyte and generating heat and flammable gases.

2. Dendrite formation: The plated lithium often grows in dendritic or mossy forms. These lithium dendrites are needle-like and sharp.
3. Separator penetration: The growing lithium dendrites carry a high risk of penetrating the already extremely thin separator, causing an ISC between the positive and negative electrodes. This generates a large short-circuit current and localized high temperatures.

During overcharging, the cathode material is excessively stripped of lithium ions, causing its crystal structure to become severely unstable. This over-delithiation causes the cathode potential to rise sharply.

1. Structural collapse: When too many lithium ions are removed, the lattice framework of the cathode material may collapse as it can no longer be maintained. This damage is irreversible and leads directly to permanent capacity loss.
2. Oxidation and oxygen release: More dangerously, at high potentials, the transition metal oxides (e.g., oxides of cobalt, nickel, and manganese) in the cathode material become highly unstable and can decompose, releasing oxygen. For instance, a $LiCoO_2$ cathode during overcharge undergoes the reaction: $LiCoO_2 \rightarrow Li_{1-x}CoO_{2+}xLi^{++}xe^{-}$. When x becomes too large, $Li_{1-x}CoO_2$ further decomposes into Co_3O_4, releasing oxygen. The release of oxygen is an extremely dangerous signal. It not only increases the rise in internal battery pressure but also provides an oxidizer for the combustion of the electrolyte, significantly increasing the risk of thermal runaway.

Overdischarge typically occurs in series-connected battery packs. When the cell with the lowest capacity and fastest voltage drop in a pack is forcibly discharged by the other cells in the pack, its voltage is pulled below 0 V, resulting in voltage reversal or polarity reversal.

Under normal conditions, the anode potential is low, and its current collector (copper foil) remains stable. When over-discharge drives the cell voltage to a very low or even negative value, the anode potential is forcibly raised to a very high level [4]. When this potential exceeds the dissolution potential of copper (approximately 3.5 V vs. Li/Li^{+}), the copper foil, which normally serves as the stable current collector, becomes unstable and begins to undergo an oxidative dissolution reaction: $Cu \rightarrow Cu^{2+} + 2e^{-}$ (Figure 2.13).

The generation of copper ions creates two serious problems:

1. Structural damage: Corrosion of the copper foil destroys the anode's conductive network, leading to permanent degradation of battery performance.
2. Metallic dendrites: The dissolved Cu^{2+} ions migrate through the electrolyte to the cathode side. During subsequent charging cycles, these copper ions can be reduced to metallic copper on the cathode surface, forming copper dendrites. Similar to lithium dendrites, copper dendrites are also hard and sharp, possessing the ability to penetrate the separator and initiate ISCs.

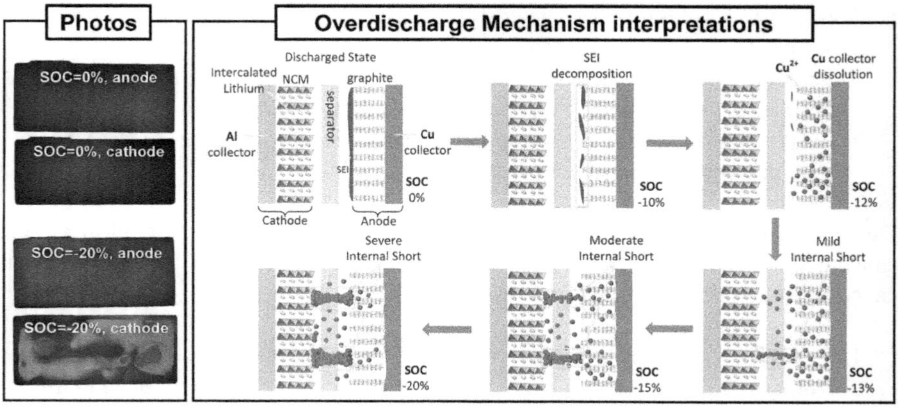

FIGURE 2.13 Overdischarge mechanism schematic [4].

Electrical abuse disrupts the internal chemical equilibrium of the battery through two distinct pathways. Overcharging assaults both electrodes simultaneously: it generates highly reactive lithium dendrites and consumes electrolyte at the anode, while it destroys the crystal structure and releases supportive oxygen at the cathode. Overdischarge, through the unique mechanism of polarity reversal, corrodes the normally stable copper current collector and creates conditions for subsequent copper dendrite formation and ISCs. A common characteristic of these cascading side reactions is that they all irreversibly consume active materials, damage battery structure, generate additional heat and flammable gases, and significantly increase the probability of ISCs, collectively pushing the battery toward the brink of thermal runaway [4]. Understanding these internal side reactions directly triggered by electrical abuse forms the theoretical foundation for constructing battery safety early-warning and management strategies.

2.3.3 Thermal Abuse: Material Decomposition and Reaction Runaway Triggered by Elevated Temperatures

The interior of a lithium-ion battery constitutes a sophisticated electrochemical system, wherein each component material remains stable only within specific temperature ranges. When the ambient temperature surpasses the battery's normal operational upper limit, or when internal heat generation leads to an abnormal temperature rise, it triggers a cascade of material-level decomposition and runaway reactions. This process is termed thermal abuse. Distinct from mechanical and electrical abuse, thermal abuse directly compromises the chemical stability of the battery's internal materials through the physical parameter of temperature, subsequently initiating chain reactions (Figure 2.14).

Within the lithium-ion battery, an increase in temperature first impacts its most sensitive component—the SEI film. The SEI film is a passivation layer formed on the anode surface during the initial formation cycles. Its primary function is to permit the passage of lithium ions while preventing direct contact between the electrolyte

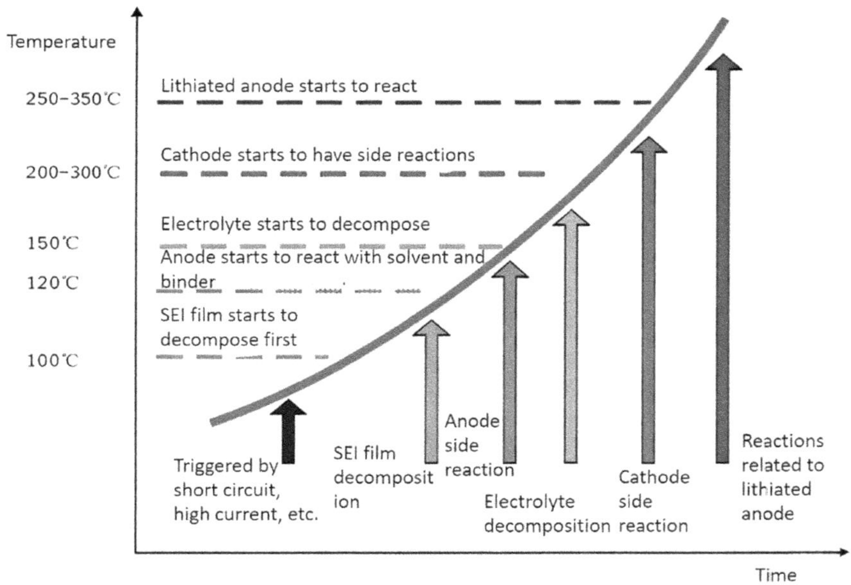

FIGURE 2.14 The corresponding reactions at the temperature under heat abuse.

and the highly reactive anode material. As the temperature rises to approximately 90°C–120°C, the SEI film begins to decompose. The breakdown of this protective layer exposes the reactive graphite anode surface to electrolyte solvent molecules, leading to reduction reactions. These reactions not only consume the electrolyte and generate flammable gases (such as ethane and methane) but also release heat, further elevating the battery temperature [13].

As the temperature continues to increase to around 120°C–150°C, the internal separator begins to undergo a phase change. Commercial polyolefin separators (e.g., PE and PP) typically have melting points between 130°C and 165°C. As the temperature approaches the melting point, the separator shrinks and may melt, causing its microporous structure to collapse or be completely destroyed. The failure of the separator eliminates the physical isolation between the cathode and anode, making a large-scale ISC highly probable. This generates substantial Joule heat, accelerating the temperature rise at an even faster rate.

When the temperature reaches the range of approximately 150°C–200°C, the cathode material starts to decompose. Taking the common $LiCoO_2$ cathode as an example, it decomposes under thermal stress, releasing oxygen. This process can be represented as: $LiCoO_2 \rightarrow Li_xCoO_2 + O_2$. The release of oxygen not only increases the internal pressure of the battery, but more critically, provides a potent oxidizer for the combustion of flammable substances, like the electrolyte, significantly intensifying the reaction severity.

Simultaneously, the lithium salt in the electrolyte (e.g., $LiPF_6$) also decomposes at high temperatures. $LiPF_6$ can start decomposing to produce PF_5 at temperatures as low as ~80°C. PF_5 can further react with trace moisture in the electrolyte to generate

HF. HF is a highly corrosive substance that attacks internal battery components and damages electrode materials, and these reactions are also accompanied by heat and gas generation.

The electrolyte solvent itself decomposes directly at high temperatures. Common carbonate solvents (e.g., EC, DEC, and DMC) undergo oxidation or reduction reactions at elevated temperatures, producing gases like CO and CO_2, and releasing significant heat. These side reactions, along with SEI film decomposition and cathode decomposition, mutually reinforce each other, forming a self-accelerating thermo-chemical reaction cycle.

The aforementioned reactions do not occur strictly in sequence according to fixed temperature thresholds, but rather overlap and mutually reinforce each other as the temperature rises. For instance, the heat generated from SEI film decomposition accelerates the melting of the separator, while the ISC caused by separator failure rapidly increases the battery temperature, subsequently promoting cathode material decomposition. This positive feedback mechanism, where multiple exothermic reactions promote and intensify one another, means that once the battery enters a state of thermal abuse, its internal temperature rises in a self-accelerating manner.

The essence of thermal abuse lies in using temperature as the key parameter to sequentially destabilize the SEI film, compromise the integrity of the separator, disrupt the structure of the cathode material, and impair the chemical stability of the electrolyte. The decomposition reactions of these materials at high temperatures not only directly generate heat and gas but, more critically, the positive feedback relationships among them create a self-accelerating network of chemical reactions. Understanding this mechanism of material decomposition and reaction runaway triggered by high temperatures is of significant importance for establishing effective thermal management strategies and early warning systems.

The lithium-ion battery is a complex electrochemical system whose internal component stability is closely related to temperature. Each key material—including the SEI film on the anode surface, the electrolyte, the cathode active material, and the separator—can only maintain its chemical and physical stability within specific temperature ranges. When the external ambient temperature remains excessively high, or when internal heat generation cannot be dissipated promptly, the battery temperature gradually increases. Once the temperature exceeds the tolerance limits of these key materials, a series of exothermic chemical side reactions are triggered, forming a self-accelerating vicious cycle. This process is referred to as thermal abuse.

2.4 KEY CHARACTERISTIC SIGNALS OF THERMAL RUNAWAY: THE PHYSICOCHEMICAL BASIS FOR EARLY WARNING

During the evolution of thermal runaway, a battery releases multiple detectable physical and chemical signals. These signals, such as abnormal changes in temperature, voltage, current, and gas composition, are direct precursors to imminent thermal runaway. Accurately identifying and interpreting these characteristic signals is central to achieving early warning. Establishing the correlation between these signals and the internal state of the battery is the foundation for developing efficient and reliable early warning systems.

2.4.1 Temperature Characteristics: The Loss of Thermal Stability and Triggering Temperature

Temperature, as one of the most intuitive characteristic parameters during the thermal runaway process of lithium-ion batteries, holds irreplaceable importance in prediction. The essence of thermal runaway is the internal heat accumulation surpassing the rate of dissipation, leading to a sharp local temperature increase that triggers chain reactions. Therefore, by precisely monitoring the battery's temperature changes, two characteristic temperature points with significant warning implications can be identified: the loss of thermal stability and the triggering temperature. These two points demarcate the critical phase of the battery's transition from a safe to a dangerous state.

The loss of thermal stability temperature is the critical point at which the heat generation rate from internal chemical reactions begins to exceed the heat dissipation rate to the surrounding environment, causing the battery temperature to rise spontaneously. This temperature point is typically measured using precision equipment like Accelerating Rate Calorimeters. When a battery is placed in an adiabatic or nearly adiabatic environment and heated at a constant rate, its temperature initially rises in sync with the environment. Once weak exothermic side reactions begin inside the battery, and the heat produced exceeds the system's heat loss, the battery temperature detaches from the control of the ambient temperature, showing a detectable temperature rise. The corresponding temperature at this point is the loss of thermal stability temperature.

From a materials science perspective, the loss of thermal stability temperature is closely related to the initiation of various exothermic reactions within the battery. Typically, the first to occur is the decomposition of the SEI film. The SEI film is a passivation layer covering the graphite anode surface, which begins to decompose at temperatures around 90°C, releasing a small amount of heat. As the temperature continues to rise, the decomposed SEI film can no longer effectively isolate the anode from the electrolyte, leading to direct contact between the electrolyte and the highly reactive lithiated graphite anode, causing reduction reactions that generate more heat and flammable gases. The combined effect of these initial reactions causes the overall heat generation capability of the battery system to surpass its heat dissipation capacity, thereby triggering the onset of self-heating.

The specific value of the loss of thermal stability temperature is influenced by several factors. Different cathode material systems show significant variations. LFP batteries typically have a higher onset temperature, around 130°C–150°C, whereas NMC batteries have a relatively lower onset temperature, approximately in the range of 110°C–130°C. Furthermore, the battery's state of charge (SOC) also affects this temperature point. A fully charged battery, with its electrodes in a high-energy state, often has a lower loss of thermal stability temperature compared to a battery at a low SOC.

The triggering temperature is the tipping point at which internal reactions accelerate drastically, and the temperature surges rapidly within a very short time. The stage from the loss of thermal stability temperature to the triggering temperature is known as the latency or induction period of thermal runaway. During this stage, a series of exothermic reactions occur sequentially inside the battery, including separator melting and collapse, binder decomposition, and reactions between the cathode material and the electrolyte. These reactions mutually reinforce each other, forming a positive feedback loop until the system reaches the triggering temperature.

When the battery temperature reaches the triggering temperature, the most critical change is the large-scale failure of the separator. Commercial PE separators typically have melting points between 130°C and 160°C. When the temperature reaches this range, the separator undergoes large-area melting, losing its electronic insulation property, leading to a large-scale ISC between the positive and negative electrodes. This ISC releases a massive amount of Joule heat in an extremely short time. Simultaneously, the cathode material begins to decompose violently, releasing oxygen, and the electrolyte enters a stage of rapid decomposition. The synergistic effect of these multiple intense exothermic reactions causes the battery temperature to rise sharply at a rate of tens of degrees Celsius per second.

As shown in Figure 2.15 illustrating the characteristic temperature pattern of thermal runaway, it is recognized that thermal runaway has three characteristic temperatures [13]: T1, T2, and T3. T1 is the loss of thermal stability temperature. From this temperature onward, internal active materials begin to exhibit significant exothermic reactions. In this stage, various reactions overlap and have a long duration. T2 is the thermal runaway triggering temperature. This temperature represents the occurrence of an ISC within the battery. Thermal runaway occurs at this point, characterized by an instantaneous temperature spike, rapid gas generation and accumulation, and a high potential for jetting flames. T3 is the maximum temperature of thermal runaway, representing the highest temperature the battery can reach when its thermal and electrochemical energy is fully released. This temperature generally corresponds to the peak heat release rate and occurs almost simultaneously with the triggering temperature T2.

Based on these, thermal runaway can be divided into three periods: T1 to T2: The early stage of thermal runaway. T2 to T3: The occurrence stage of thermal runaway. After T3: The initial fire stage.

The characteristic temperature pattern of thermal runaway can provide a reference for prevention and control technologies and measures. Specifically, if the signals of

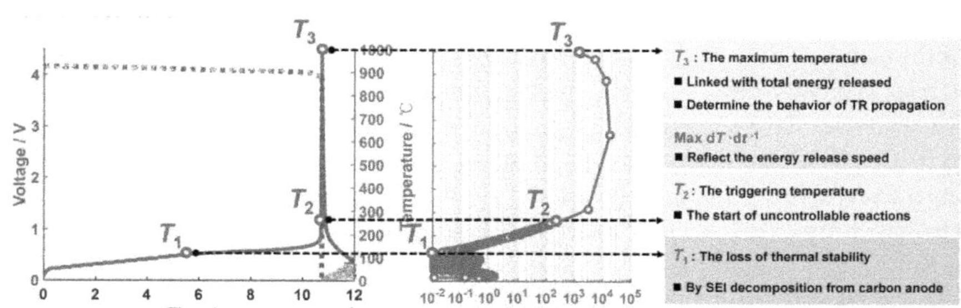

FIGURE 2.15 Temperature–time curve of thermal characteristics under thermal runaway of lithium battery [14].

evolving thermal runaway can be identified during the early stage (T1 to T2), fire accidents can be prevented.

Risk assessment and inclusive design research on thermal runaway of aviation lithium-ion batteries.

The temperature interval between the loss of thermal stability and the triggering temperature provides a critical time window for battery thermal safety warning and intervention. Monitoring whether the battery temperature begins to rise autonomously, decoupling from the ambient temperature, allows for the earliest detection of battery abnormality. Conversely, monitoring if the battery temperature approaches the known thermal runaway triggering temperature can serve as a key criterion for initiating emergency measures (such as activating fire suppression systems and forced cooling).

Understanding the physical significance and determination methods of the loss of thermal stability and the triggering temperature is crucial for establishing effective battery thermal runaway warning systems and formulating rational thermal management strategies. By monitoring the occurrence of these two characteristic temperature points, corresponding prevention and control measures can be implemented at different stages, providing key information to prevent or delay the onset of thermal runaway.

2.4.2 Voltage/Current Anomalies: Manifestations of Internal Short Circuit

Among the precursor characteristics of thermal runaway in lithium-ion batteries, abnormal changes in voltage and current are the key signals that can be captured earliest by electrical monitoring means. These anomalies directly reflect alterations in the battery's internal microstructure, particularly the initiation and progression of an ISC. Understanding the evolution patterns of these electrical characteristics is of paramount importance for the early identification of thermal runaway risk.

2.4.2.1 Voltage Anomaly: From Gradual Decline to Precipitous Plunge

Abnormal battery voltage typically appears earlier than a noticeable temperature rise and is the most direct electrical manifestation of an ISC. An ISC can be understood as the formation of an unintended conductive pathway between the positive and negative electrodes inside the battery. This pathway shunts a portion of the operating current, leading to a drop in the effective output voltage. During different stages of thermal runaway development, voltage anomalies exhibit distinct phase characteristics (Figure 2.16).

In the initial stage, the ISC point might be microscopic and localized, with a relatively high short-circuit resistance. At this point, the voltage may only show minor fluctuations or a slow decline. These changes are very subtle and might require high-precision acquisition equipment for detection. As the internal temperature rises or the mechanical structure deteriorates further, the short-circuit point may expand or multiply, decreasing the short-circuit resistance and gradually accelerating the voltage drop rate.

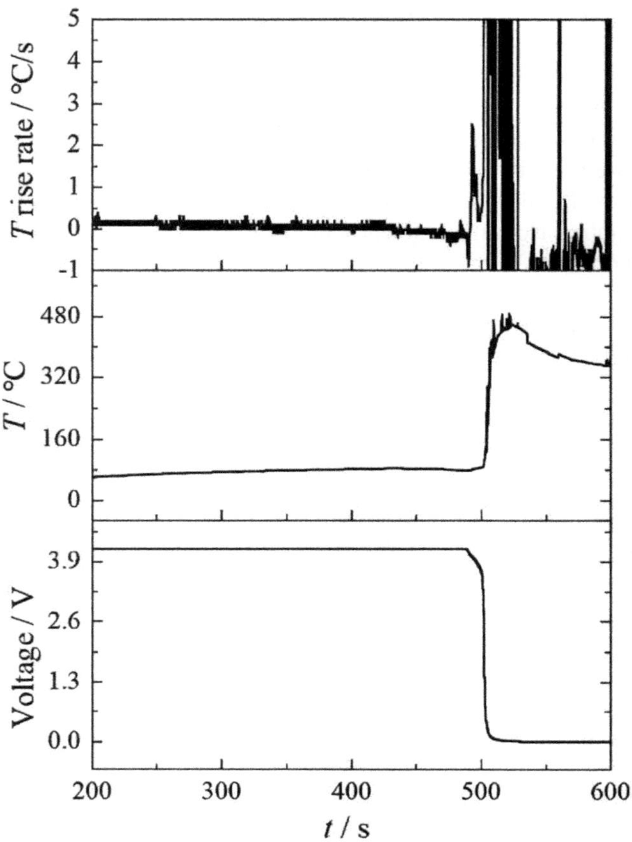

FIGURE 2.16 Voltage, temperature, and temperature-rise-rate curves during heating.

A critical turning point in voltage change occurs when the temperature reaches the separator melting point (typically 130°C–160°C). Large-scale melting of the separator causes a massive ISC between the electrodes, leading to a rapid, cliff-like plunge of the voltage from its normal plateau to near zero volts within a very short time. This precipitous voltage drop is a strong signal of imminent thermal runaway.

Notably, different battery types exhibit variations in voltage anomaly behavior. For instance, LFP batteries have a very flat voltage plateau, making early minor voltage drops potentially less noticeable than in NMC batteries, which complicates early diagnosis.

2.4.2.2 Current Anomaly: From Shunting to Sudden Change

Under normal operating conditions, the battery current is determined by the external load or charger. However, when an ISC occurs, a portion of the current is shunted through the ISC path, causing a discrepancy between the externally measured current and the actual internal operating current. The manifestation of this current anomaly varies depending on the battery's state.

During charge or discharge cycles, an ISC manifests as an additional self-discharge current. During charging, part of the charging current is consumed by the ISC path, prolonging the time required to reach a fully charged state. During discharging, the ISC path additionally drains energy from the battery, leading to reduced discharge capacity and shortened discharge time. Although these changes are subtle, they can be detected through precise coulombic efficiency analysis.

When the battery is in an idle state, an ISC leads to abnormal self-discharge. Normal lithium-ion batteries have a very low self-discharge rate during storage, typically only 2%–5% capacity loss per month. In contrast, a battery with an ISC will exhibit a significantly increased self-discharge rate and markedly reduced voltage retention capability.

As thermal runaway approaches, coinciding with separator melting and the formation of a large-scale ISC, the current can change abruptly. If the battery is at a high SOC at this time, the ISC can induce a very high short-circuit current, generating substantial Joule heat and further accelerating the thermal runaway process.

2.4.2.3 Electrical Characteristics and Detection Methods for Internal Short Circuits

ISC exhibits several typical characteristics in electrical signals, which provide the basis for early diagnosis.

First is the inconsistency between voltage and temperature changes. In the early stages of thermal runaway, voltage anomalies or fluctuations may appear before any significant temperature rise is observed. This time lag between electrical and thermal signals offers a possibility for early warning.

Second is the abnormality in relaxation voltage. A normal battery exhibits predictable voltage relaxation characteristics during idle periods or intervals between charge/discharge cycles. In contrast, a battery with an ISC shows distorted voltage relaxation curves, with anomalies in both the relaxation speed and the stabilized voltage value.

Based on these characteristics, researchers have developed various ISC detection methods. Model-based methods involve establishing a battery equivalent circuit model and identifying changes in model parameters in real time. Specifically, the emergence and decrease of a short-circuit resistance parameter can sensitively detect early-stage ISC. Inconsistency-based methods leverage voltage differences among individual cells within a battery pack. When the voltage of a particular cell significantly deviates from the others, it may indicate the occurrence of an ISC in that cell.

It is important to note that the electrical characteristics of an ISC are influenced by several factors, including the short-circuit type (e.g., anode-cathode short and current collector short), severity, location, and the battery's SOC. These factors collectively determine the specific manifestation of voltage and current anomalies, making accurate diagnosis of an ISC reliant on the comprehensive judgment of multiple parameters.

Monitoring voltage and current anomalies provides the earliest warning window for battery thermal runaway. By establishing sensitive detection algorithms and reasonable warning thresholds, potential risks can be identified minutes or even earlier before thermal runaway occurs, buying valuable time for implementing safety measures. This early warning method based on electrical signals holds significant engineering application value due to its fast response and low implementation cost.

2.4.3 Gas Generation Characteristics: Analysis of Gas Composition and Release Rate

The generation of gaseous products is a critical early warning indicator during the thermal runaway process of lithium-ion batteries. Analyzing the gas composition and release rate provides deep insights into the internal chemical reactions and serves as a vital basis for the early diagnosis of thermal runaway [14].

2.4.3.1 Chemical Mechanisms of Gas Generation

The gases produced during the thermal runaway of lithium-ion batteries primarily originate from the following chemical reactions:

First is the decomposition of the SEI film. When the temperature rises to the range of 90°C–120°C, the SEI film begins to thermally decompose, generating gas products such as CO and CO_2. Key components of the SEI film, like lithium carbonate and alkyl lithium carbonate, become unstable at these temperatures and decompose, releasing gases.

As the temperature continues to increase to 120°C–150°C, reactions between the electrolyte and the anode material commence. The lithiated graphite anode is highly reactive and undergoes reduction reactions with carbonate solvents in the electrolyte, producing hydrocarbon gases such as C_2H_4 and C_2H_6, along with H_2 and CO_2.

When the temperature reaches 150°C–200°C, the cathode material begins to participate in the gas generation process. O_2 released from the decomposition of the cathode material reacts with the electrolyte solvents in oxidation reactions, producing large amounts of CO_2 and water vapor. Simultaneously, the decomposition of the lithium salt $LiPF_6$ generates harmful gases like HF and phosphorus fluorides (PF_3, PF_5).

Under even higher temperature conditions (above 200°C), thermal decomposition of the electrolyte solvents becomes the primary source of gases. Solvent molecules like EC and DMC undergo bond breaking and recombination reactions, producing a series of small-molecule gaseous products.

2.4.3.2 Characteristic Analysis of Gas Composition

The sequence of appearance and concentration changes of different gas components reflect specific internal chemical reaction processes:

CO and CO_2 are typically the earliest detected gas components. Their emergence often marks the initial decomposition of the SEI film. As the temperature rises, the concentrations of hydrocarbon gases like C_2H_4 and C_2H_6 begin to increase significantly, indicating that reactions between the electrolyte and the anode have started. The generation of H_2 primarily results from the decomposition reactions of the electrolyte solvents, and its concentration change reflects the intensity of the thermal decomposition reactions.

The detection of fluorine-containing gases like HF is particularly significant. The appearance of these gases indicates that the lithium salt $LiPF_6$ has begun to

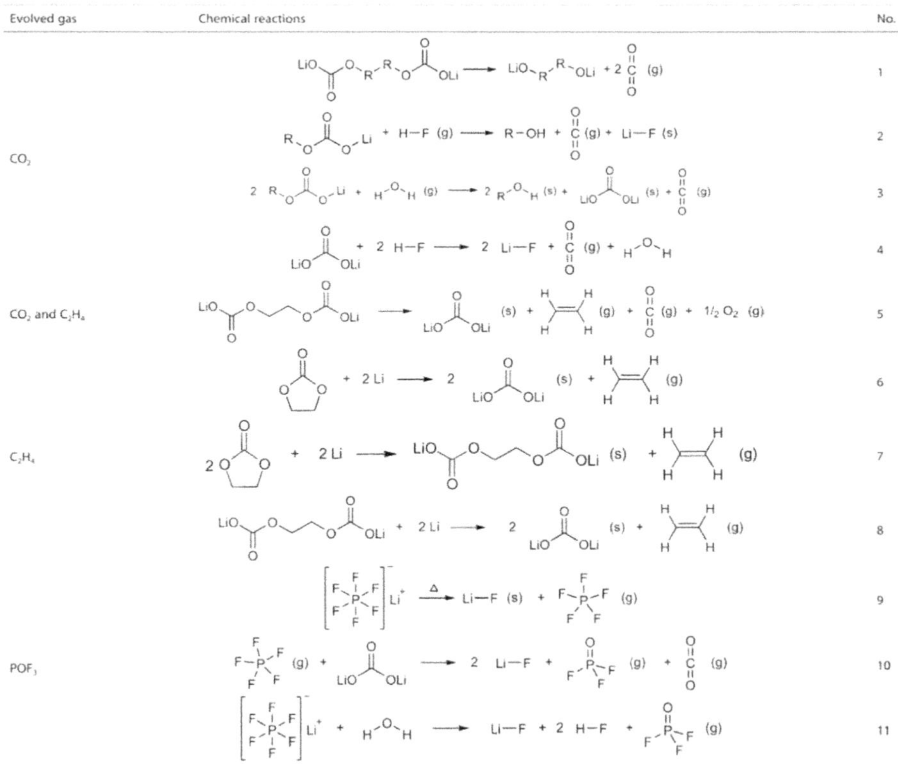

FIGURE 2.17 Formula diagram of various gas generation.

decompose, signifying that the internal battery temperature has reached a considerably high level. Since HF is highly corrosive, its generation not only reflects the progression of thermal runaway but also causes further damage to the internal battery structure.

Gas generation characteristics differ noticeably among battery chemistries. For instance, LFP batteries primarily produce CO, CO_2, and H_2 during thermal runaway, whereas NMC batteries, due to the oxygen release characteristic of their cathode material, also generate significant amounts of hydrocarbon gases. These differences make it possible to identify the thermal runaway state of different battery types through gas analysis. The formation reactions corresponding to various gases are shown in Figure 2.17.

2.4.3.3 Kinetic Characteristics of Release Rate

The variation pattern of the gas release rate also contains crucial early warning information. During the initial stage of thermal runaway, gas release is typically gradual. As the temperature rises and reactions accelerate, the release rate exhibits an exponential increase trend.

Using analytical techniques such as online mass spectrometry or Fourier Transform Infrared Spectroscopy (FTIR), the change in gas release rate can be monitored in real time. Research has found that the gas release rate curve typically exhibits a three-stage characteristic: an initial slow release stage, a middle accelerated release stage, and a final explosive release stage. This phased characteristic is closely related to the cascading amplification process of internal chemical reactions within the battery.

The mutation point (or abrupt change point) in the gas release rate often corresponds to the critical moment of failure for key internal battery materials. For instance, when the separator begins to melt, the temperature spike caused by the resulting ISC triggers a sharp increase in the gas release rate. This mutation point occurs earlier than the precipitous drop in voltage and the rapid rise in temperature, providing an even earlier time window for warning.

2.5 THERMAL RUNAWAY PREDICTION TECHNOLOGIES: FROM MODELS TO INTELLIGENT DIAGNOSIS

To predict thermal runaway earlier and more accurately, advanced technologies that go beyond traditional threshold-based judgments are required. This includes simulating the multi-physics coupled processes inside the battery through physical models, as well as utilizing data-driven algorithms to learn failure patterns from historical data. Prediction methods that integrate physical mechanisms with artificial intelligence can significantly improve the accuracy and timeliness of early warnings, representing a major direction for current technological development.

2.5.1 Thermal Runaway Physical Models

Thermal runaway in lithium-ion batteries is an intense process driven by chain exothermic reactions. When the internal battery temperature reaches the trigger temperature of specific materials, a series of side reactions are activated sequentially. If the substantial heat released by these reactions cannot be dissipated promptly, it causes a further rise in battery temperature, forming an uncontrollable positive feedback loop that ultimately leads to thermal runaway.

Among the numerous side reactions, the decomposition of the SEI film, the reaction between the anode and electrolyte, and the decomposition of the cathode active material are the three most critical initial reactions. Accurately describing the kinetics of these reactions is the theoretical foundation for building thermal runaway prediction models and formulating safety prevention and control strategies. To this end, researchers have established corresponding classic kinetic models to quantify the relationship among reaction rate, temperature, and material quantity.

2.5.1.1 SEI Film Decomposition Reaction Model

The SEI film begins to thermally decompose when the temperature reaches approximately 90°C–120°C. Its decomposition reaction is typically described using a first-order reaction model. The reaction rate is expressed as

$$\frac{dc_{\text{SEI}}}{dt} = -A_{\text{SEI}} c_{\text{SEI}} \exp\left(-\frac{E_{a,\text{SEI}}}{\text{RT}}\right) \tag{2.27}$$

where c_{SEI} represents the remaining amount of the SEI film, a time-dependent variable, and t is the reaction time. A_{SEI} is the pre-exponential factor, representing the frequency of reaction occurrences, and its value is determined by fitting experimental data. $E_{a,\text{SEI}}$ is the activation energy for the SEI decomposition reaction, reflecting the energy barrier for the reaction.

The physical meaning of this model is that the decomposition rate of the SEI film is proportional to its current remaining amount (c_{SEI}) and accelerates exponentially with increasing temperature. This reaction is exothermic, and its heat generation rate Q_{SEI} can be calculated by the product of the reaction rate and the molar reaction enthalpy ΔH_{SEI}, which can be expressed as

$$Q_{\text{SEI}} = -\Delta H_{\text{SEI}} \frac{dc_{\text{SEI}}}{dt} \tag{2.28}$$

2.5.1.2 Anode–Electrolyte Reaction Model

After the SEI film decomposes, the exposed highly active lithiated carbon (Li_xC) reacts with the electrolyte. This reaction is often described using a shrinking core model. The reaction rate equation is expressed as

$$\frac{d\delta}{dt} = A_{\text{neg}}(1-\delta)\exp(-\frac{E_{a,\text{neg}}}{\text{RT}}) \tag{2.29}$$

where δ is the conversion rate of the anode active material, ranging from 0 to 1. $\delta = 0$ indicates no reaction, and $\delta = 1$ indicates complete reaction. Aneg is the pre-exponential factor, $E_{a,\text{neg}}$ is the activation energy for the anode-electrolyte reaction, and $(1-\delta)$ represents the relative amount of remaining unreacted lithiated carbon.

This model assumes the reaction starts at the surface of the electrode particles and progresses inward. The reaction rate is controlled by both chemical kinetics and the reaction interface area. This is a highly exothermic reaction, and its heat generation rate Q_{neg} is expressed as

$$Q_{\text{neg}} = -\Delta H_{\text{neg}} c_{\text{neg},0} \frac{d\delta}{dt} \tag{2.30}$$

where $c_{\text{neg},0}$ is the initial molar concentration of the anode active material, and ΔH_{neg} is the molar reaction enthalpy for this reaction.

2.5.1.3 Cathode Decomposition Reaction Model

The cathode material (e.g., Li_xCoO_2) decomposes at high temperatures, releasing oxygen. This reaction is also commonly described by a first-order kinetic model, which can be expressed as

$$\frac{dc_{\text{pos}}}{dt} = -A_{\text{pos}}c_{\text{pos}} \exp(-\frac{E_{\text{a},pos}}{\text{RT}}) \tag{2.31}$$

where c_{pos} represents the remaining amount of cathode active material, A_{pos} is the pre-exponential factor, and $E_{a,\text{pos}}$ is the activation energy for the cathode decomposition reaction.

Cathode decomposition is not only exothermic itself, but the released oxygen can further react with the electrolyte in oxidation reactions, generating significant additional heat and greatly accelerating the thermal runaway process. The heat generation rate Q_{pos} for this reaction is expressed as

$$Q_{\text{pos}} = -\Delta H_{\text{pos}} \frac{dC_{\text{pos}}}{dt} \tag{2.32}$$

where ΔH_{pos} is the molar reaction enthalpy for the cathode decomposition.

These three classic kinetic models, utilizing the Arrhenius equation form, closely link the reaction rate to temperature, accurately capturing the exponential self-accelerating nature of the core side reactions during thermal runaway. When constructing a complete multi-physics coupled model, these reaction models are incorporated as key heat source terms into the system's energy conservation equation. This enables effective prediction of battery temperature evolution and thermal runaway trigger time under different abuse conditions. The parameters of these models (e.g., A and Ea) depend on the specific battery material system and are typically calibrated using experimental data from techniques like differential scanning calorimetry and thermogravimetric analysis.

2.5.2 Early Warning Methods Based on Gas Generation Composition

Lithium-ion batteries undergo complex internal chemical changes before thermal runaway occurs, often accompanied by the generation and release of specific gas components. Early warning methods based on gas composition monitor the emergence and concentration changes of these characteristic gases, enabling the identification of abnormal battery states before significant temperature rise and voltage drop occur. This provides a critical time window for the prevention and intervention of thermal runaway.

2.5.2.1 Relationship between Gas Composition and Thermal Runaway

Lithium-ion batteries produce multiple gases during thermal runaway, and the composition and concentration changes of these gases are closely related to the initiation and progression of thermal runaway. Before thermal runaway occurs, the battery releases low levels of gases such as H_2, CO, CO_2, CH_4, C_2H_4, and among others (Figure 2.18).

The production of these gases results from internal chemical reactions, such as electrolyte decomposition and redox reactions involving cathode materials. Therefore, by detecting the presence and concentration changes of these gases, one can determine if the battery is in the early stages of thermal runaway.

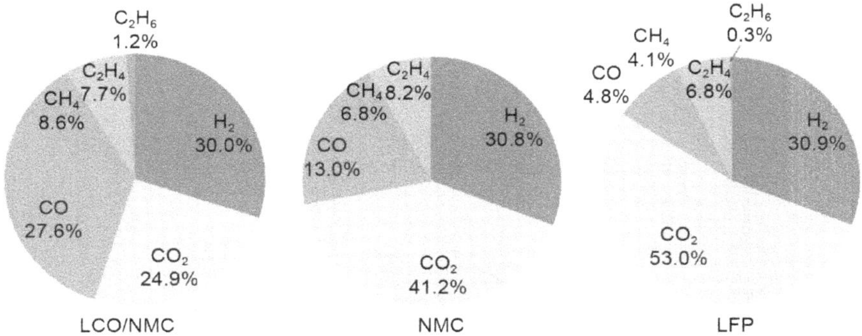

FIGURE 2.18 Gas concentration change.

2.5.2.2 Methods for Implementing Early Warning

1. Gas detection technologies: The implementation of early warning strategies based on gas composition requires the use of highly sensitive gas detection technologies. Common techniques include Gas Chromatography-Mass Spectrometry, FTIR, and various types of gas sensors (Figure 2.19). These technologies enable the real-time monitoring of gas species and concentrations generated by the battery, providing essential data support for early warning systems.
2. Selection of characteristic gases: Among the numerous gas components produced, selecting appropriate characteristic gases is crucial for effective early warning. Research indicates that H_2 and CO are primary gases generated during the early stages of thermal runaway, and their concentration changes are relatively significant. Therefore, these gases can be designated as characteristic indicators for early warning. Monitoring the variation in their concentrations allows for the assessment of whether a battery is entering the early phase of thermal runaway.
3. Establishment of warning models: Once characteristic gases are detected, establishing an effective warning model is necessary to determine if the battery is likely to undergo thermal runaway. By analyzing extensive experimental data, a correlation model between gas concentration and the probability of thermal runaway occurrence can be developed. When the detected gas concentration exceeds a predefined threshold, the warning system will trigger an alarm, prompting relevant personnel to take preventive actions to avert thermal runaway.

Furthermore, early warning methods based on gas composition can be integrated with other monitoring techniques, such as temperature and voltage monitoring, to further enhance the accuracy and reliability of the warning system. The comprehensive application of multiple monitoring methods enables a more holistic understanding of the battery's status and facilitates the timely detection of potential safety hazards.

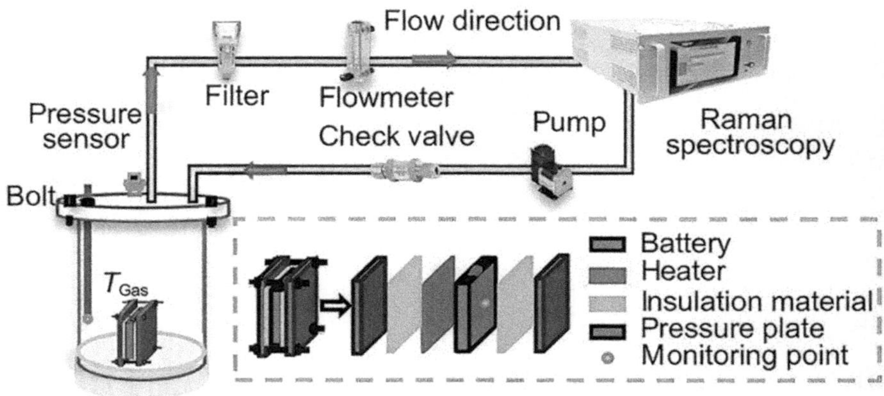

FIGURE 2.19 Real-time detection platform for thermal runaway gas of battery.

2.5.3 Application of Machine Learning in Thermal Runaway Prediction

With the rapid development of data science and artificial intelligence technologies, machine learning methods offer a novel technical pathway for predicting thermal runaway in lithium-ion batteries. Compared to traditional physics-based models, machine learning can automatically learn complex patterns of battery behavior from vast amounts of historical data, enabling the intelligent identification and prediction of early thermal runaway features and bringing new possibilities for safety warning at the individual cell level.

Machine learning is an automated, data-driven learning approach that extracts patterns and regularities from large datasets to build predictive models (Figure 2.20). In battery thermal runaway prediction, machine learning models can learn the characteristics of thermal runaway from various operational data (such as voltage, current, temperature, and gas concentration), thereby achieving early warning. Common machine learning methods include supervised learning, unsupervised learning, and reinforcement learning. For thermal runaway prediction, supervised learning is the most frequently used method. It utilizes labeled data (i.e., data known to represent either thermal runaway or normal operation) to train a model, enabling it to recognize the signatures of thermal runaway.

2.5.3.1 Data Collection and Preprocessing

Before constructing a machine learning model, it is necessary to collect a large amount of battery operational data. This data typically includes parameters such as battery voltage, current, temperature, and gas concentration. Data collection can be achieved through laboratory testing or sensor networks deployed during actual operation. The collected data requires preprocessing, including steps like data cleaning, normalization, and feature extraction, to improve the model's training effectiveness.

For instance, data cleaning can remove noise and outliers from the dataset; normalization can transform data of different units/scales into a uniform scale, facilitating model processing; feature extraction can derive features useful for thermal

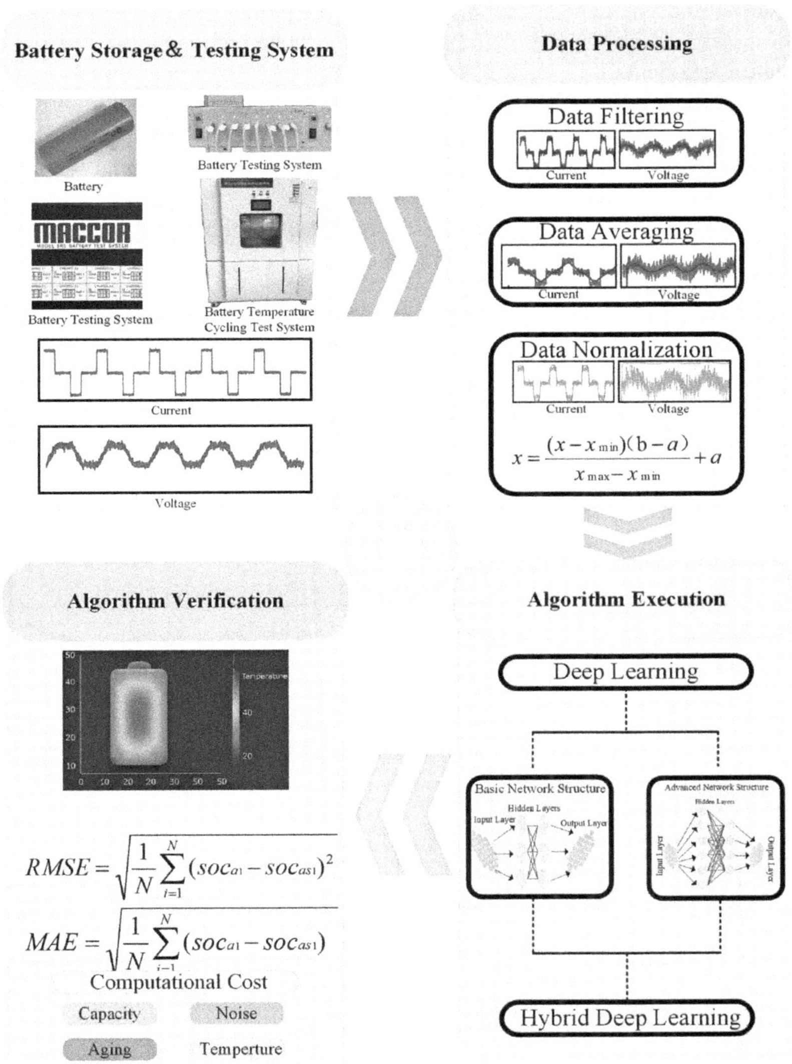

FIGURE 2.20 Procedure of machine learning to study thermal runaway.

runaway prediction from the raw data, such as the rate of voltage change and the rate of temperature change. These preprocessing steps can significantly enhance the model's accuracy and robustness.

2.5.3.2 Feature Selection and Extraction

Feature selection and extraction are critical steps in machine learning. In battery thermal runaway prediction, selecting appropriate features can markedly improve model performance. Common features include parameters like voltage, current, temperature, and gas concentration, as well as their rates of change. Furthermore, more complex features can be engineered, such as voltage fluctuation frequency and temperature gradients.

For example, the rate of voltage change can reflect the rate of internal chemical reactions, while the rate of temperature change can indicate the accumulation of internal heat. These features provide the model with richer information, aiding it in better identifying early signs of thermal runaway. Through feature selection and extraction, raw data is transformed into feature vectors that are more amenable to model processing, thereby enhancing the model's training efficacy.

2.5.3.3 Selection and Construction of Machine Learning Models

In battery thermal runaway prediction, commonly used machine learning models include decision trees, support vector machines (SVMs), random forests, neural networks, and among others (Figure 2.21). Each of these models has its own advantages and disadvantages, making them suitable for different application scenarios.

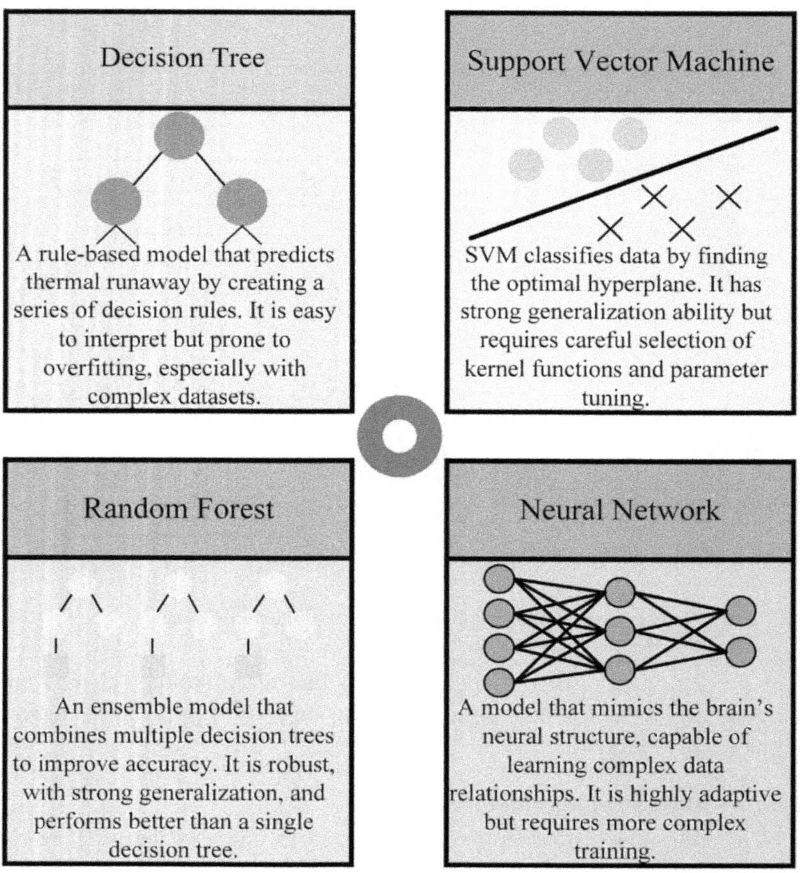

FIGURE 2.21 Schematic diagrams of four commonly used machine learning models.

1. Decision tree: A rule-based model that predicts thermal runaway by constructing a series of decision rules. Its key advantage is strong interpretability, but it is prone to overfitting.
2. SVM: A model based on statistical learning theory, which classifies data by finding an optimal hyperplane. It possesses good generalization ability but is relatively sensitive to the choice of kernel function and parameter tuning.
3. Random forest: An ensemble learning model that improves prediction accuracy by constructing multiple decision trees and aggregating their predictions (e.g., through voting). It offers high robustness and good generalization capability.
4. Neural network: A model based on artificial neurons, which predicts thermal runaway by learning complex relationships within the data. It boasts powerful non-linear fitting capability and adaptive learning capacity, but the training process is relatively complex.

In practical applications, suitable machine learning models can be selected based on specific requirements. For instance, decision tree or random forest models might be chosen for scenarios requiring fast response times, whereas neural network models could be preferred for applications demanding high prediction accuracy.

2.5.3.4 Model Training and Optimization

After selecting a suitable machine learning model, it needs to be trained and optimized. The model training process involves adjusting the model's parameters using labeled data to enable it to better fit the data. The model optimization process involves tuning the model's hyperparameters (such as learning rate and regularization parameters) to improve its performance.

For example, when training a neural network model, the learning rate can be adjusted to control the model's convergence speed, and regularization parameters can be adjusted to prevent overfitting. Furthermore, methods like cross-validation can be used to evaluate the model's performance, ensuring it possesses good generalization capability. Through model training and optimization, a machine learning model capable of accurately predicting thermal runaway can be constructed.

2.5.3.5 Construction of a Real-Time Early Warning System

A machine learning-based thermal runaway early warning system requires the capability for real-time monitoring and alerting. This system typically includes a data acquisition module, a data preprocessing module, a model prediction module, and an alert module.

The data acquisition module is responsible for collecting battery operational data in real time. The data preprocessing module performs cleaning, normalization, and feature extraction on the collected data. The model prediction module utilizes the trained machine learning model to make predictions based on the (processed) data. The alert module issues a warning signal based on the model's prediction results.

For instance, when the model predicts an imminent thermal runaway event, the alert module can trigger an alarm, prompting relevant personnel to take preventive measures to avoid the occurrence of thermal runaway. By constructing a real-time early warning system, early detection of battery thermal runaway can be achieved, thereby ensuring the safety of the battery system.

Machine learning technology holds broad application prospects in battery thermal runaway prediction. By building data-driven models, the characteristics of thermal runaway can be extracted from battery operational data, enabling early warning. In practical applications, it is necessary to select appropriate machine learning models based on specific requirements and enhance model performance through training and optimization. The construction of real-time warning systems enables early detection, thus safeguarding battery system safety. As machine learning technology continues to develop and find new applications, its role in battery thermal runaway prediction will become increasingly important.

2.6 THERMAL RUNAWAY MITIGATION THROUGH DESIGN AND MATERIAL STRATEGIES

The most effective safety strategy is to prevent thermal runaway at its source. This requires innovations at the material and design levels to enhance the intrinsic anti-thermal runaway capability of the battery itself. This includes developing electrode materials with higher thermal stability, flame-retardant electrolytes, and more robust separators, while also integrating passive safety devices within the individual cell. These strategies aim to improve the battery's thermal stability or achieve automatic safety isolation in emergency situations.

2.6.1 Electrode Material Design for Enhanced Intrinsic Safety

The thermal runaway risk of lithium-ion batteries is closely related to the intrinsic thermal stability of their internal materials. Through rational electrode material design, the thermal stability of the battery can be improved at the source, reducing the likelihood and severity of thermal runaway. Current strategies for enhancing the intrinsic safety of electrode materials primarily focus on cathode material modification, anode material optimization, and the development of new material systems.

2.6.1.1 Cathode Material Design

The cathode material is a crucial component of lithium-ion batteries, and its chemical and thermal stability significantly impacts battery safety. By optimizing the composition, structure, and surface properties of cathode materials, their thermal and chemical stability can be markedly improved, thereby reducing the risk of thermal runaway.

2.6.1.1.1 Modification of High-Nickel NCM/NCA Cathode Materials

High-nickel NCM (e.g., NCM811) and NCA cathode materials are extensively researched and applied due to their high energy density. However, these high-nickel materials are prone to thermal runaway at elevated temperatures, primarily because they undergo structural phase transitions that release significant oxygen, thereby triggering combustion reactions inside the battery. To enhance the thermal stability of high-nickel cathode materials, researchers have employed various modification methods.

FIGURE 2.22 SEM images of NCM (a) and NCM-2%Na (b), and EDS images of NCM-2%Na (c).

Surface coating: Applying a coating of a stable material (such as Al_2O_3 or ZrO_2) onto the surface of the cathode particles can effectively suppress oxygen release and improve the material's thermal stability. For example, an Al_2O_3 coating has been shown to significantly enhance the thermal stability of NCA cathode materials, reducing oxygen release at high temperatures by approximately 50%.

Doping modification: Doping the cathode material with small amounts of metal ions (such as Na^+, Mg^{2+}, and Al^{3+}) can improve the structural stability of the material, thereby enhancing its thermal stability. For instance, NCM811 cathode material doped with Na^+ exhibits significantly improved structural stability at high temperatures (Figure 2.22).

2.6.1.1.2 Optimization of LFP Cathode Materials

LFP cathode materials are widely used in electric vehicles and energy storage systems due to their excellent thermal stability and safety. However, the relatively low energy density of LFP materials limits their use in some high-performance applications. To increase the energy density of LFP materials while maintaining their good thermal stability, researchers have adopted various optimization methods.

Nanonization: Fabricating LFP material into nanoparticles can significantly increase its specific surface area, thereby enhancing the material's electrochemical performance. Studies have shown that nanonized LFP materials, while

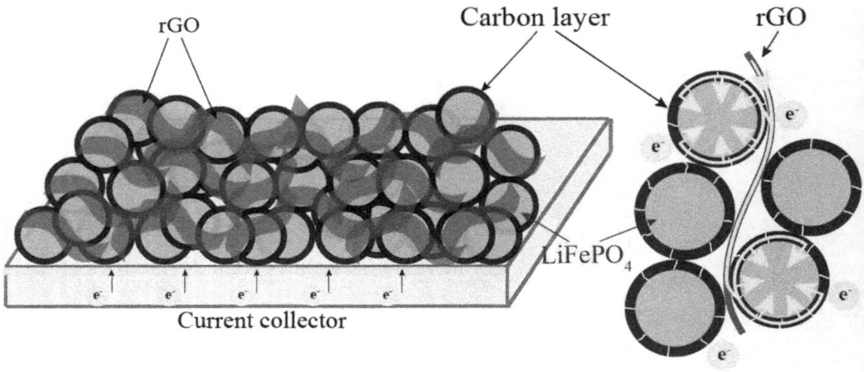

FIGURE 2.23 Schematic diagram of the $LiFePO_4$@C/rGO composite material.

maintaining good thermal stability, can achieve an increase in energy density of approximately 20%.

Composite materials: By compositing LFP material with other materials (such as graphene and carbon nanotubes), the electrical conductivity and structural stability of the material can be improved. For example, LFP/graphene composites exhibit significantly enhanced electrochemical performance while maintaining good thermal stability (Figure 2.23).

2.6.1.2 Anode Material Design

Anode materials also play a significant role in the safety of lithium-ion batteries. The thermal and chemical stability of the anode material directly impacts the battery's thermal runaway risk. By optimizing the composition, structure, and surface properties of anode materials, their thermal and chemical stability can be significantly improved, thereby reducing the risk of thermal runaway.

2.6.1.2.1 Modification of Silicon-Based Anode Materials

Silicon-based anode materials are extensively researched due to their high theoretical specific capacity. However, silicon-based materials undergo severe volume expansion during charge and discharge cycles, which can easily lead to structural degradation and potentially trigger thermal runaway. To enhance the stability and safety of silicon-based anode materials, researchers have adopted various modification methods.

Nanonization: Fabricating silicon into nanoparticles can significantly reduce the volume expansion during charge and discharge cycles. Studies indicate that the volume expansion rate of nano-silicon particles is approximately 60% lower than that of micron-sized silicon particles.

Composite materials: Compositing silicon with other materials (such as carbon and graphene) can improve the material's electrical conductivity and structural stability (Figure 2.24). Silicon/graphene composites demonstrate a significantly reduced volume expansion rate while maintaining good conductivity.

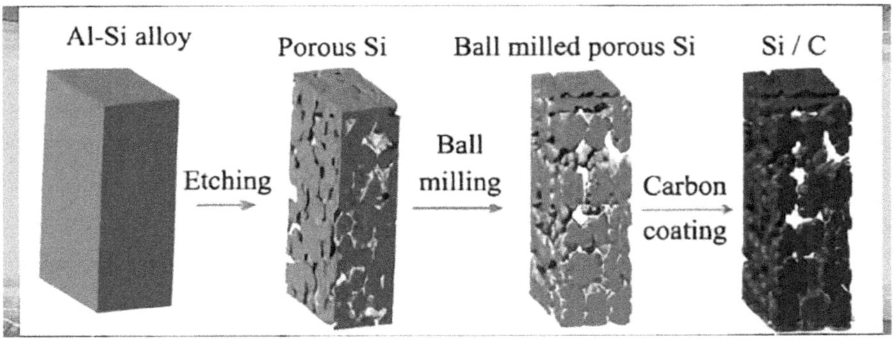

FIGURE 2.24 Si/C composite material.

2.6.1.2.2 Optimization of Graphite Anode Materials

Graphite anode materials are widely used in lithium-ion batteries due to their good electrical conductivity and stability. However, graphite anode materials are prone to side reactions at elevated temperatures, which can trigger thermal runaway. To improve the thermal stability and safety of graphite anode materials, researchers have employed various optimization methods.

Surface coating: Applying a coating of a stable material (such as Al_2O_3 or PVDF) onto the graphite surface can effectively suppress the occurrence of side reactions and enhance the material's thermal stability (Figure 2.25). For instance, an Al_2O_3 coating has been shown to significantly improve the thermal stability of graphite anode materials by reducing their side reactions at high temperatures.

Doping modification: Doping graphite with small amounts of elements (such as B and P) can improve the material's structural stability, thereby enhancing its thermal stability. For example, graphite anode materials doped with B exhibits significantly improved structural stability at high temperatures, accompanied by a reduction in side reactions.

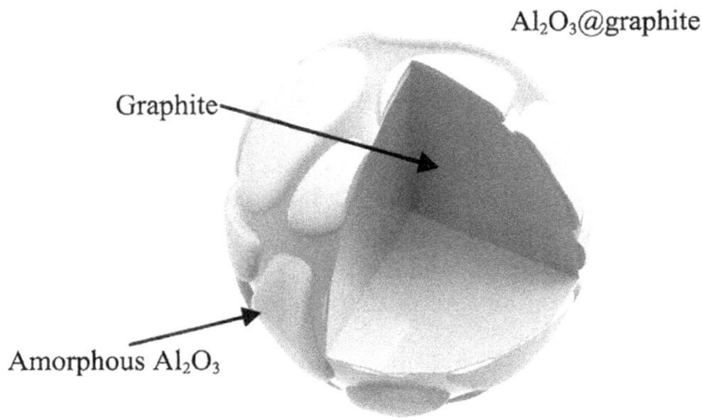

FIGURE 2.25 Schematic illustration of the Al_2O_3 coated on a graphite surface.

2.6.1.3 Integrated Design and Application

By optimizing the composition, structure, and surface properties of both cathode and anode materials, the intrinsic safety of lithium-ion batteries can be significantly enhanced. In practical applications, the design of cathode and anode materials must be considered comprehensively to achieve optimal performance and safety for the battery system.

2.6.1.3.1 Cathode-Anode Material Matching

The matching of cathode and anode materials is a critical aspect of improving battery safety. Through rational selection of these materials, optimal system performance and safety can be achieved. For instance, pairing a high-nickel NCM cathode with a silicon-based anode can increase the battery's energy density, while simultaneously optimizing the composition and structure of both materials to effectively mitigate the risk of thermal runaway.

2.6.1.3.2 Battery System Integration

In the design of the battery system, the cathode and anode materials must be integrated holistically to realize the best possible performance and safety. Further enhancements to battery safety can be achieved by optimizing the system's structure and design. For example, employing a multilayer separator design can effectively prevent ISCs, thereby reducing the risk of thermal runaway.

In summary, optimizing the composition, structure, and surface characteristics of cathode and anode materials significantly improves the intrinsic safety of lithium-ion batteries. For cathodes, surface coating and doping enhance the thermal stability of high-nickel NCM materials, while nanonization and composite design boost the energy density and thermal stability of LFP materials. For anodes, nanonization and composite design improve the stability and safety of silicon-based materials, and surface coating and doping enhance the thermal stability and safety of graphite materials. Integrated design of cathode and anode materials enables optimal battery system performance and safety. The role of ongoing developments in materials science will become increasingly vital for enhancing battery safety.

2.6.2 Application of High-Stability Separators

The separator, a critical component in lithium-ion batteries, serves the dual function of providing electronic insulation between the electrodes while allowing ionic conduction. Traditional polyolefin separators (e.g., PE and PP) are prone to shrinkage and melting at high temperatures, leading to ISCs and potentially triggering thermal runaway. High-stability separators significantly improve heat resistance and safety through material modifications, structural design, and functional treatments, offering an effective solution for preventing thermal runaway.

2.6.2.1 Development of High-Temperature Substrates

Traditional polyolefin separators typically have melting points between 130°C and 165°C. When the internal battery temperature reaches this range, the separator melts

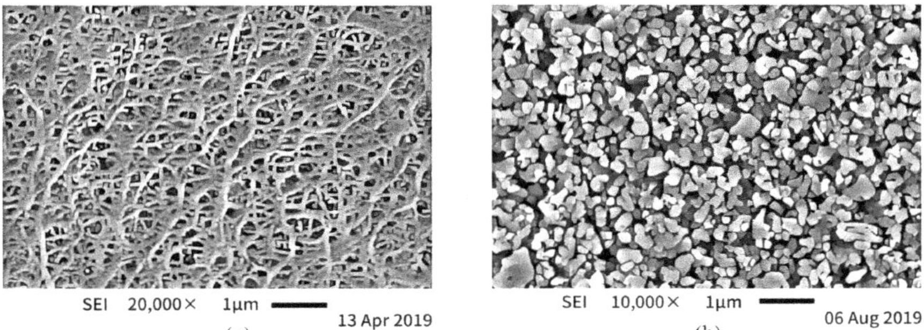

FIGURE 2.26 Ceramic-coated separator. (a) PE SEM. (b) Al_2O_3 Coated SEM.

and shrinks, losing its isolating function. Developing high-temperature-resistant substrates is a fundamental approach to enhancing separator stability.

Ceramic-coated separators are created by coating a layer of inorganic ceramic particles (e.g., Al_2O_3 and SiO_2) onto the surface of a polyolefin base film, forming a composite structure (Figure 2.26). The ceramic layer remains stable at high temperatures. Even when the polymer substrate melts, the ceramic particles maintain physical separation, preventing direct contact between the cathode and the anode. Ceramic-coated separators can increase the heat-resistance temperature to over 200°C and reduce the thermal shrinkage rate to less than 5%, compared to ~80% for traditional separators.

Aramid separators utilize aromatic polyamide materials, such as poly(m-phenylene isophthalamide), which possess excellent thermal stability and mechanical strength. Aramid materials have a melting point exceeding 400°C and do not melt at high temperatures; instead, they undergo carbonization. This characteristic allows aramid separators to maintain structural integrity even under extreme temperature conditions, effectively preventing ISCs.

Nonwoven separators are fabricated from high-temperature-resistant fibers like polyimide or cellulose through processes such as electrospinning or wet-laid forming. Their three-dimensional network structure provides good thermal stability, dimensional stability, and electrolyte wettability. Polyimide separators can withstand temperatures above 300°C and exhibit almost no dimensional change at elevated temperatures.

2.6.2.2 Design of Intelligent Shutdown Functionality

Intelligent separators are designed to automatically shut down ionic transport pathways when abnormal temperature increases occur, halting further battery reactions and thereby preventing the initiation and progression of thermal runaway.

Thermal shutdown separators are functionalized by incorporating thermally sensitive materials onto the surface or within the bulk of the base membrane (Figure 2.27). When the temperature rises to a preset threshold, these thermal-sensitive materials undergo a phase change, blocking the pores of the separator and obstructing ion transport. Commonly used thermal-sensitive materials include polymers such as

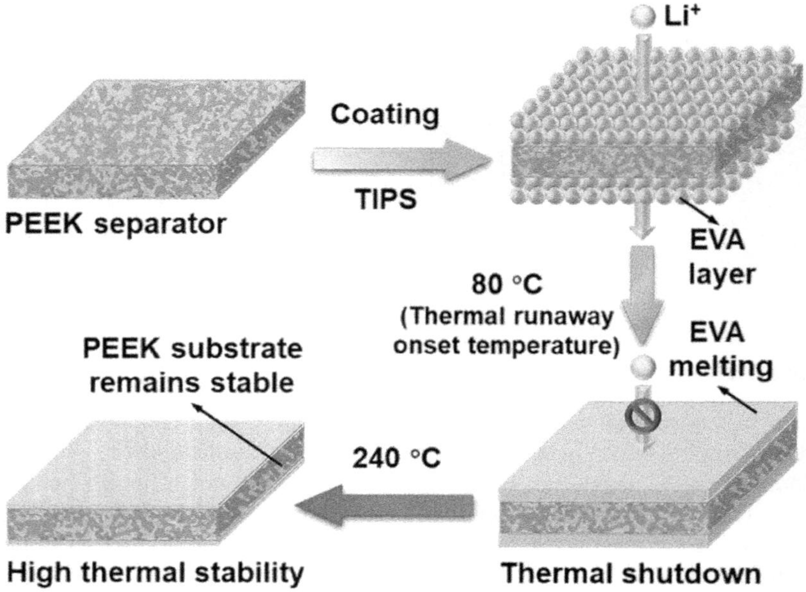

FIGURE 2.27 Thermal shutdown separator.

poly(methyl methacrylate) and poly(vinyl alcohol). The shutdown temperature can be precisely controlled by adjusting the material composition.

Microsphere shutdown separators embed thermally sensitive polymer microspheres (e.g., PMMA microspheres) on the surface or within the bulk of the separator. At normal operating temperatures, the microspheres remain stable, allowing ions to pass freely through the separator pores. When the temperature exceeds the safety threshold, the microspheres melt and expand, effectively blocking the pores and achieving rapid shutdown. This design can achieve a response time on the order of seconds, with a shutdown efficiency exceeding 95%.

2.6.2.3 Multifunctional Composite Structures

Modern high-safety separators often employ composite structural designs that integrate multiple functions to comprehensively enhance battery safety performance. Common configurations include ceramic-coated shutdown separators and multilayer composite separators.

Sandwich-structured separators are composed of multiple layers of materials with different functions laminated together (Figure 2.28). For instance, a PP separator can serve as the mechanical support layer, which is coated on both sides with PVDF polymer layers to create a sandwich-type PVDF/PP solid polymer electrolyte. The high-strength PP separator acts as a mechanical barrier, significantly enhancing the mechanical strength of the solid polymer electrolyte and effectively inhibiting the growth of lithium dendrites during electrochemical cycling.

Gradient-pore structure separators are fabricated using special processes to create a pore structure with a gradient in pore size. The side facing the cathode features smaller

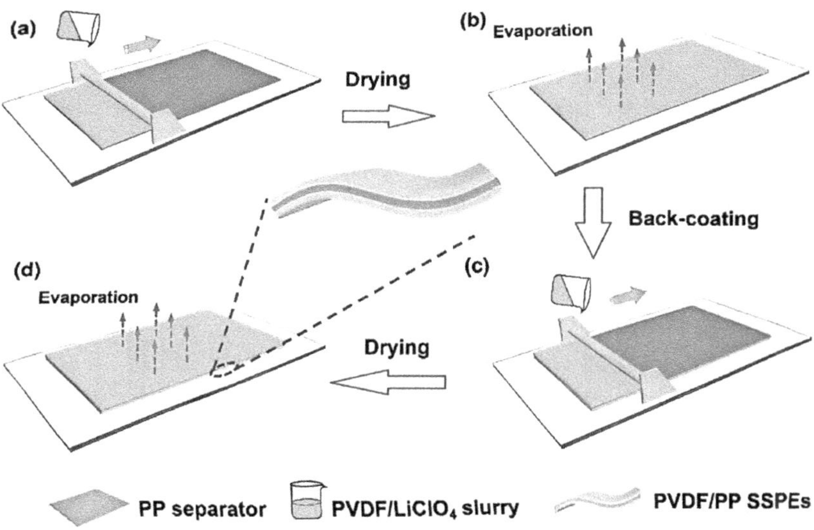

FIGURE 2.28 The preparation process of the sandwich-structure PVDF/PP SSPEs.

pores, effectively blocking the passage of detached cathode particles; the side facing the anode has larger pores, facilitating the rapid transport of lithium ions. This structural design enhances safety while also optimizing electrochemical performance.

2.6.2.4 Surface Modification Treatments

The physical and chemical properties of the separator surface significantly impact its safety and electrochemical performance. Surface modification can further optimize these separator characteristics.

Hydrophilic treatment: Methods such as plasma treatment or graft modification are used to increase the separator surface's affinity for the liquid electrolyte, enhancing its wettability. This helps reduce interfacial impedance, improve rate capability, and mitigate the risk of localized overheating.

Surface charge regulation: Introducing specific functional groups onto the separator surface allows for the regulation of surface charge distribution. A positively charged surface can adsorb anions from the electrolyte, increasing the lithium-ion transference number. A negatively charged surface favors a uniform lithium-ion flux, suppressing dendrite growth.

The application of high-stability separators significantly elevates the safety margin of lithium-ion batteries. Future development in separator technology will focus on several key directions: developing new high-temperature-resistant polymer materials (e.g., polyetheretherketone PEEK and polyphenylene sulfide PPS); designing intelligent, responsive separators capable of shutdown triggered by multiple parameters like temperature and voltage; researching self-healing separators that can automatically repair localized damage; and exploring integrated designs combining solid electrolytes with separators, aiming to fundamentally resolve safety hazards associated with liquid electrolytes.

2.7 FUTURE PERSPECTIVES IN THERMAL RUNAWAY PREVENTION AND CONTROL

Despite the existence of various prevention and control strategies, the challenge of thermal safety management remains severe with the continuously increasing demand for higher battery energy density. Performance degradation over the battery's entire lifecycle also affects its safety margins. Future development must balance energy density with safety, explore new material systems, and integrate embedded intelligent sensing technologies to construct inherently safer, next-generation lithium-ion batteries for energy storage.

2.7.1 Balancing High Energy Density and Safety

With the accelerating global energy transition, the demand for energy storage is experiencing explosive growth. The large-scale integration of renewable energy into grids and the widespread application of distributed energy storage systems impose higher requirements on the energy density of lithium-ion batteries. In the renewable energy sector, wind and solar power generation are intermittent and unstable, requiring energy storage systems to achieve a smooth balance between power supply and demand. Distributed energy storage systems, on the other hand, need to store more energy within limited spaces to meet the peak-shaving demands of local grids. Therefore, enhancing the energy density of lithium-ion batteries is not only an inevitable trend in technological development but also key to addressing the challenges of the energy transition.

Although high-energy-density batteries offer significant advantages in improving energy storage efficiency, their potential safety risks cannot be ignored. High-energy-density batteries possess inherent drawbacks in their material properties. For example, NMC/NCA batteries, due to their high nickel content, exhibit poor thermal stability and are prone to thermal runaway under high-temperature conditions. These materials have lower thermal decomposition temperatures and release substantial heat and hazardous gases during thermal runaway, further exacerbating safety risks. Furthermore, the charge and discharge processes of high-energy-density batteries significantly impact thermal safety. Under high-rate charge and discharge conditions, the rate of internal chemical reactions accelerates, leading to intensified heat accumulation and thereby increasing the likelihood of thermal runaway.

Beyond material properties and charge/discharge processes, the safety risks of high-energy-density batteries are closely related to their complex operating environments. In high-temperature environments, the battery's thermal management system might be unable to dissipate heat effectively, leading to excessively high battery temperatures and potentially triggering thermal runaway. In low-temperature environments, charging efficiency decreases, potentially causing localized overheating. External factors like mechanical shock or short circuits can also induce thermal runaway in high-energy-density batteries. The combined effect of these factors presents severe challenges for the safety management of high-energy-density batteries.

2.7.2 Evolution of Thermal Runaway Risk throughout the Entire Lifecycle

Within the full lifecycle of energy storage lithium-ion batteries, the production stage is a critical link determining their thermal safety performance. Factors such as electrode manufacturing processes, battery assembly precision, and material consistency can profoundly impact the battery's thermal stability. Inhomogeneous slurry coating or thickness variations during electrode coating may lead to uneven internal current distribution, potentially causing localized overheating. Furthermore, poor battery assembly, such as separator misalignment or electrode tab misplacement, can increase the risk of ISCs, further aggravating the potential for thermal runaway. Research indicates that minor defects introduced during manufacturing can gradually amplify during long-term use, ultimately leading to battery failure or even thermal runaway incidents.

As energy storage lithium-ion batteries age, their performance gradually degrades, and their thermal safety margins consequently shrink, planting potential hidden dangers for thermal runaway. During the battery aging process, capacity fade and increased internal resistance are the most prominent manifestations. Capacity fade is typically caused by structural degradation of the cathode material, thickening of the anode SEI film, and decomposition of the electrolyte. Increased internal resistance can lead to greater heat generation during charge and discharge cycles, further exacerbating thermal runaway risk. Moreover, the thermal stability of aged batteries decreases significantly. Especially under high-rate charge/discharge or extreme environmental conditions, the probability of thermal runaway is markedly higher than in new batteries.

Improper handling of retired batteries can also lead to serious safety issues. Residual energy in retired batteries, if not properly discharged, can cause thermal runaway incidents during transportation or storage. Improper disassembly and recycling of retired batteries may lead to leakage of hazardous substances, causing secondary pollution to the environment. To address these issues, a comprehensive management system for retired batteries needs to be established. Scientific assessment methods should classify retired batteries for tiered processing. Those with remaining value can be repurposed for less demanding applications, while unusable batteries should be sent to specialized recycling facilities. Moreover, developing efficient thermal runaway suppression technologies, such as barrier technologies and cooling techniques, can quickly control situations where thermal runaway occurs in retired batteries, preventing greater losses, and strengthening supervision over the retired battery handling process to ensure all steps comply with environmental and safety standards is an important direction for future research.

2.7.3 Integration Pathways for New Materials and Intelligent Sensing Technologies

In recent years, with the widespread application of lithium-ion batteries in energy storage, research on new battery materials has become an important direction for enhancing thermal safety. Solid-state electrolytes, as a highly promising alternative material, have attracted significant attention due to their nonflammability and high thermal stability. Compared to traditional liquid electrolytes, solid-state electrolytes can effectively suppress the intense exothermic reactions triggered by electrolyte decomposition during thermal runaway, thereby significantly reducing the battery's thermal runaway risk. However, the application of new materials still faces numerous challenges. The ionic conductivity of solid-state electrolytes is relatively low, and interface contact issues between the electrode and electrolyte may exist, limiting their application in large-scale energy storage systems. Simultaneously, the research and development costs for new cathode and anode materials are high, and achieving low-cost manufacturing while ensuring performance remains a problem to be solved.

Intelligent sensing technology plays a crucial role in the early warning of thermal runaway in lithium-ion batteries. By monitoring key state parameters of the battery in real time, it provides technical support for the early identification and prevention of thermal runaway. Currently, mainstream sensor technologies include temperature sensors, voltage sensors, and gas sensors. These sensors can detect temperature changes, voltage fluctuations, and gas release during battery operation. In energy storage power stations, multiparameter coupled monitoring technology combining multidimensional data is used for full-process warning of thermal runaway. This technology, based on electrical signals supplemented by temperature and gas signals, can issue alarms in the early stages of thermal runaway, providing a time window for subsequent safety measures. Furthermore, electrochemical impedance spectroscopy and ultrasonic detection technology, as noninvasive detection methods, can assess the internal state of the battery without damaging its structure and enable functions like overcharge warning. Although intelligent sensing technology has made significant progress in thermal runaway warning, its application still has limitations. The warning effectiveness of a single sensor is poor, making it difficult to fully reflect the thermal runaway risk of the battery, while multi-sensor fusion technology faces challenges such as large data volumes and complex algorithms. Therefore, future research needs to further optimize sensor design and data processing algorithms to improve the accuracy and reliability of warning systems.

Effectively integrating new materials with intelligent sensing technologies is a key pathway to constructing inherently safe battery systems. The development of intelligent sensors based on new materials offers the possibility for real-time monitoring and active regulation. Leveraging the excellent thermal stability of solid-state electrolytes, high-temperature-resistant embedded sensors can be designed, enabling precise monitoring of the battery's internal state. Moreover, integrating materials science, electrochemistry, and artificial intelligence technologies can establish highly accurate and widely applicable battery thermal management systems. Combined with big data technology and machine learning algorithms, it becomes possible to predict the thermal runaway risk of batteries based on sensor data and inhibit the occurrence of thermal runaway through active control strategies, such as cooling or power cut-off protection.

The application of distributed prevention and control technologies provides new ideas for building inherently safe battery systems. By installing multiple intelligent sensors within the battery pack and combining them with barrier technologies to limit the propagation range of thermal runaway, the risks of fire and explosion can be effectively reduced. However, the integration of new materials and intelligent sensing technologies still faces many technical bottlenecks, such as compatibility issues between sensors and battery materials and the high complexity of system integration. Therefore, future research needs to further explore integrated material-sensor design methods and optimize system architecture to enhance overall performance, thereby laying a solid foundation for achieving inherent safety in the next generation of energy storage lithium-ion batteries.

REFERENCES

1. Jia Z, Wang Z, Wang Q, et al. Study on thermal runaway mechanism and safety risk control methods for power batteries in new energy vehicles. *Automotive Engineering*, 2022, 44(11):1689–1705.
2. Li Q, Yu J, Liu S, Zhu G. A Comprehensive review of thermal runaway characteristics and hazards of lithium-ion batteries under different influencing factors. *Fire Science and Technology*, 2023, 42(11):1482–1487.
3. Li G. Promotion of practical technology of the thermal management system for cylindrical power battery. *Energy Inform*, 2024, 7:33.
4. Du K, Chen G, Zhang Y, et al. Comprehensive review of thermal management strategies for lithium-ion batteries: From heat generation mechanisms to advanced cooling solutions. *Energy Storage Materials*, 2025, 82:104597.
5. Zhou W, Liu Z, An Y, et al. Thermal behavior analysis of lithium-ion capacitors at transient high discharge rates. *Journal of Energy Storage*, 2022, 53:105208.
6. Zheng X, Liu J. Temperature field analysis and simulation and optimization of flow field in high-low temperature test chamber. *Journal of Refrigeration*, 2024, 45(2):119–126.
7. Bandhauer T, Garimella S. Passive, internal thermal management system for batteries using microscale liquid–vapor phase change. *Applied Thermal Engineering*, 2013, 61(2):756–769.
8. Sim M, Kim D, Yoon Y, et al. Real-time predicting the low-temperature performance of WLTC-based lithium-ion battery using an LSTM-PF sequential ensemble model. *IEEE Access*, 2024, 12:90171–90180.
9. He X, Liu Y, Huang X, et al. A reduced-order thermal runaway network model for predicting thermal propagation of lithium-ion batteries in large-scale power systems. *Applied Energy*, 2024, 373:23955.
10. Duan X, Wang H, Jia Y, et al. A multiphysics understanding of internal short circuit mechanisms in lithium-ion batteries upon mechanical stress abuse. *Energy Storage Materials*, 2022, 45:667–679.
11. Liu J, Wang Z, Bai J, et al. Heat generation and thermal runaway mechanisms induced by overcharging of aged lithium-ion battery. *Applied Thermal Engineering*, 2022, 212:118565.
12. Guo R, Lu L, Feng X, et al. Mechanism of the entire overdischarge process and overdischarge-induced internal short circuit in lithium-ion batteries. *Scientific Reports*, 2016, 6:30248.
13. Feng X, Zheng S, Ren D, et al. Investigating the thermal runaway mechanisms of lithium-ion batteries based on thermal analysis database. *Applied Energy*, 2019, 246:53–64.
14. Abd-El-Latif A, Sichler P, Kasper M, et al. Insights into thermal runaway of Li–ion cells by accelerating rate calorimetry coupled with external sensors and online gas analysis. *Batteries & Supercaps*, 2021, 4(7):1135–1144.

3 Lithium-Air Batteries

Yulin Ma and Chuankai Fu

3.1 OVERVIEW OF LITHIUM-AIR BATTERIES

Li-air batteries are generally regarded as batteries that use metallic lithium as the anode and oxygen from the air as the cathodic reactant [1,2] They function through the dissolution/deposition of lithium metal at the anode and the oxygen reduction reaction (ORR)/oxygen evolution reaction (OER) at the cathode. Therefore, Li-air batteries are also commonly referred to as lithium-oxygen batteries (Li-O_2 batteries) [3,4]. Their structure typically consists of a lithium metal anode (LMA), an electrolyte that provides electronic insulation while allowing ionic conduction, and an air cathode that facilitates oxygen transport and possesses catalytic activity. The air cathode is a composite porous matrix conductive to both electrons and ions, enabling electrochemical contact between the oxygen species in the air and lithium ions in the electrolyte.

3.1.1 Structure and Working Principle of Lithium-Air Batteries

In 1996, Abraham first reported a rechargeable nonaqueous Li-air battery based on polymer electrolytes [5]. Although the initial Li-air battery had a cell-level energy density of only 250–350 Wh kg^{-1} and a cycle life of 2–3 cycles, it was still a milestone discovery that opened up a new field. During the discharge process of a Li-O_2 battery, lithium ions at the anode are conducted through the electrolyte to the cathode, where they react with oxygen transported from the external environment at the three-phase reaction interface to form discharge products. During charging, the discharge products decompose, releasing oxygen back to the environment, and lithium ions conduct

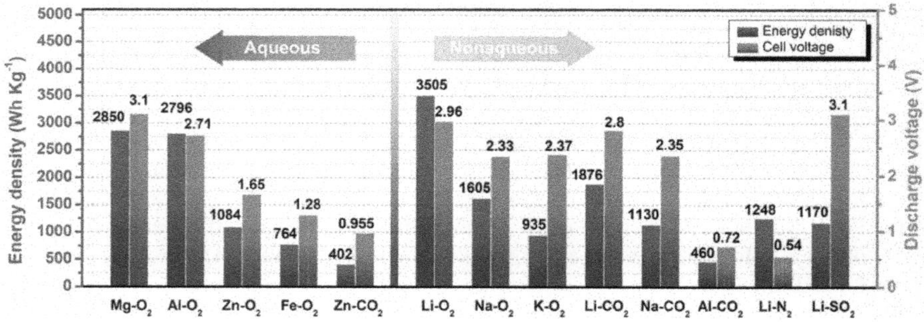

FIGURE 3.1 The performance of various lithium metal-based batteries.

DOI: 10.1201/9781003492511-3

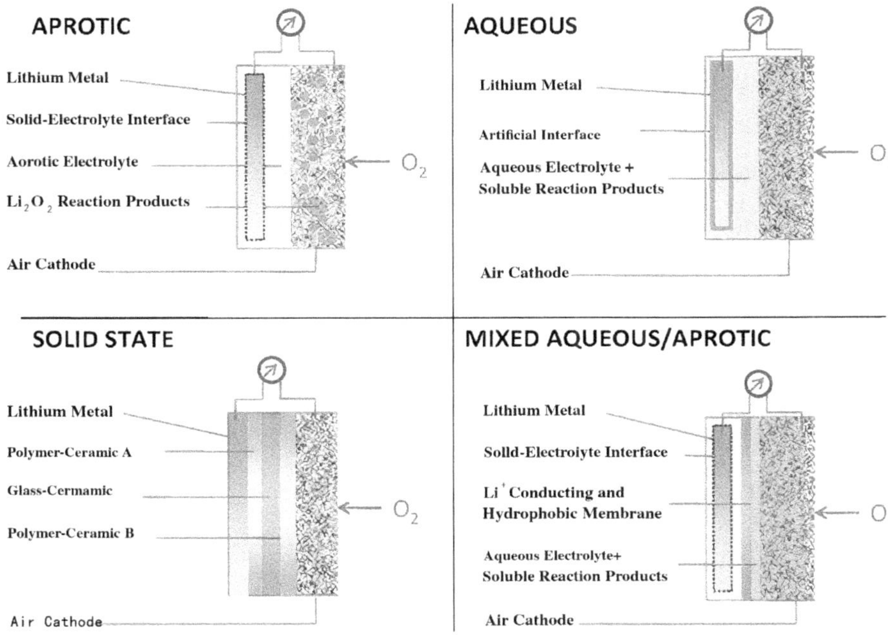

FIGURE 3.2 The structure diagram of Li-air batteries [6].

back to the anode. Overall, the battery continuously "inhales" and "exhales" oxygen, resembling a breathing battery.

As shown in Figure 3.2, based on the type of electrolyte, Li-air batteries are currently mainly divided into four systems: nonaqueous systems, aqueous systems, solid electrolyte systems, and nonaqueous/aqueous hybrid systems [6]. These four types differ in the species of electrolytes involved, and the choice of electrolyte species determines the specific electrochemical reactions during the energy storage and release processes.

3.1.1.1 Nonaqueous Lithium-Oxygen Batteries

A typical nonaqueous Li-O_2 battery consists of a LMA, a nonaqueous organic electrolyte containing organic/inorganic lithium salts, and an air cathode. In 2006, Bruce et al. [7] proposed the reversible electrochemical reaction equation for nonaqueous Li-O_2 batteries, which marked the establishment of Li-O_2 batteries as a new type of secondary rechargeable battery system:

$$2Li^+ + O_2 + 2e^- \rightarrow Li_2O_2, \quad E\theta = 2.96 \text{ V vs. Li / Li}^+ \tag{3.1}$$

This overall reaction can be refined to involve the single-electron transfer reduction of oxygen to form a lithium superoxide intermediate, which then further reacts to produce lithium peroxide.

The single-electron transfer reduction reaction between oxygen and lithium ions is:

$$O_2 + Li^+ + e^- \rightarrow LiO_2 \tag{3.2}$$

The subsequent disproportionation reaction is:

$$2LiO_2 \rightarrow Li_2O_2 + O_2 \tag{3.3}$$

Alternatively, another single-electron transfer electrochemical process is:

$$LiO_2 + Li^+ + e^- \rightarrow Li_2O_2 \tag{3.4}$$

The transition state reactions during discharge are actually more complex, and the discharge mechanisms are continually being refined. Differences in the solvent properties of the electrolyte can critically affect the reaction concentrations of the products at the microscopic level, as well as the conversion energies of various intermediate states, thereby promoting the formation of different microscopic reaction mechanisms [8]. Currently, the industry generally classifies microscopic reaction mechanisms into two main categories: the solution-mediated pathway and the surface-mediated pathway (Figure 3.3) [9].

For the overall reaction, oxygen first transforms into dissolved oxygen;

$$O_2(g) \rightleftharpoons O_2(sol) \tag{3.5}$$

The solution-mediated pathway includes:

$$O_2\ (sol) + e^- \rightarrow O^{2-}(sol) \tag{3.6}$$

$$2Li^+(sol) + 2O^{2-}(sol) \rightarrow Li_2O_2(ads) + O_2(g) \tag{3.7}$$

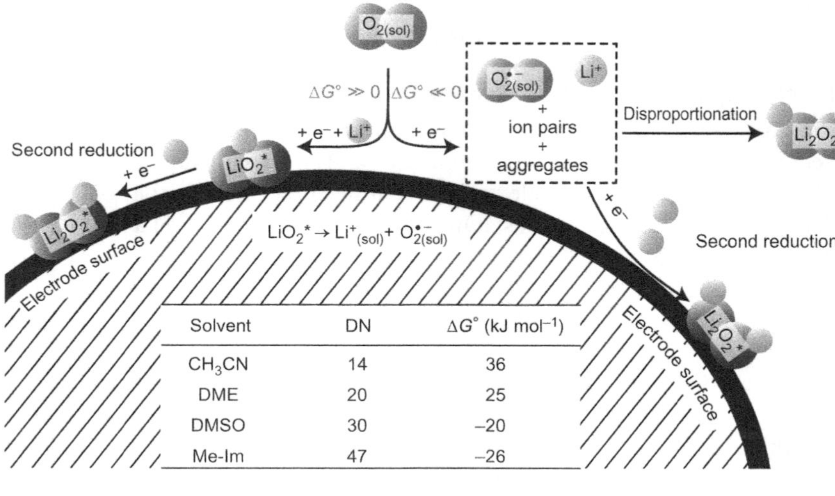

FIGURE 3.3 Surface pathway and solution pathway for discharge product formation [9].

$$2Li^+(sol)+O^{2-}(sol)+e^- \rightarrow Li_2O_2(ads) \quad (3.8)$$

The main characteristic of the solution pathway is the formation of solvated lithium superoxide intermediates; the formation of solvated lithium superoxide requires interaction with suitable solvents to stabilize it in the solution. Then, the lithium superoxide intermediates participate in the subsequent reactions in their solvated state; generally, they undergo a disproportionation reaction in the liquid phase to produce lithium peroxide, while there is also a secondary reduction process that further deposits lithium peroxide on the surface.

The surface pathway includes:

$$Li^+ + O_2(sol)+e^- \rightarrow LiO_2(ads) \quad (3.9)$$

$$Li^+ + LiO_2(ads)+e^- \rightarrow Li_2O_2(ads) \quad (3.10)$$

$$2LiO_2(ads) \rightarrow Li_2O_2(ads)+O_2(g) \quad (3.11)$$

The main characteristic of the surface pathway involves an internal charge transfer process of surface-adsorbed species. Due to the limited solvation effect of the corresponding solvent on lithium superoxide, lithium superoxide intermediates form adsorbed species on the electrode surface, which then undergo rapid secondary reduction, depositing thin-film discharge products on the electrode surface. Disproportionation reactions also occur in the surface reaction pathway; however, overall, the surface pathway is generally considered to be based on secondary reduction reactions. From a computational perspective, it has a relatively low kinetic reaction barrier [10,11].

The traditional view holds that, because the surface adsorption pathway tends to produce thin-film Li_2O_2 growth, and due to the poor conductivity of Li_2O_2, a relatively dense, insulating passivation layer forms on the cathode surface. This passivation layer hinders electron transport at the electrode surface, limiting the electron reduction reactions, and thus negatively impacting the battery's discharge capacity [12]. On the other hand, Li_2O_2 formed through the solution-mediated pathway more readily grows into larger-sized Li_2O_2 particles. Once nucleated, these particles slowly grow on the nuclei to form disc-shaped products, resulting in a larger discharge capacity for the battery, which is mainly constrained by the volume of the porous cathode pores [13].

Based on the above reaction characteristics, electrolyte systems that provide greater solvation effects on intermediate species and electrode materials with lower surface binding energy will promote the solution-mediated discharge pathway. Meanwhile, the local reactant concentration, influenced by current density and ion concentration, also affects the dissociation equilibrium of these reactions. Therefore, in addition to the inherent properties of the solvent, electrolyte additives, electrode surface characteristics, and current density all have significant impacts on the discharge reaction pathways [8]. The combined effect of these complex factors results

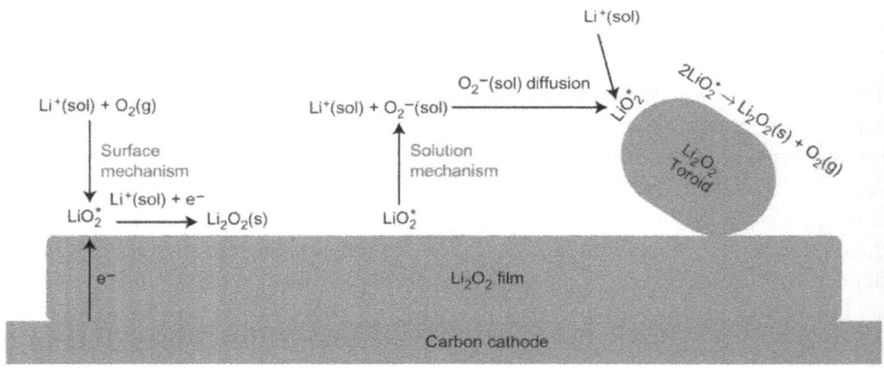

FIGURE 3.4 Dynamic conversion between discharge reaction pathways [14].

in the phenomenon that, in some systems, surface-adsorbed lithium superoxide may also tend to further dissolve to form solvated intermediates. Thus, the surface pathway and solution pathway are not isolated from each other but can interconvert (Figure 3.4) [14].

However, recent studies have challenged the traditional mechanistic models of discharge products. For example, under certain specific current densities and with typical carbon black materials, it has been found that the capacity does not fully follow the positive correlation trend between the electrolyte's solvation strength for lithium superoxide and the discharge capacity. A simple transition from the surface pathway to the solution pathway cannot explain this phenomenon, indicating that LiO_2 solvation may not be the sole factor determining the capacity [15]. In fact, many intrinsic mechanisms still require further in-depth investigation.

3.1.1.2 Solid-State Lithium-Oxygen Battery

Solid-state Li-O_2 batteries use solid electrolytes, which can be further categorized into inorganic solid electrolytes, polymer solid electrolytes, and composite solid electrolytes. Solid electrolytes themselves are highly integrated and suitable for high-specific-energy battery systems. Moreover, due to their inherent advantage of isolating sensitive impurity molecules in the air, they are particularly suitable for open/semi-open Li-O_2 battery systems. In addition, solid electrolytes overall overcome the issues of volatility and leakage common in liquid electrolytes. Their intrinsic flame retardancy and high mechanical strength can significantly enhance safety.

However, the main challenge with solid electrolytes is that an ideal solid electrolyte system has not yet been found that simultaneously meets all practical requirements, including ionic conductivity, high mechanical strength, stable interfacial contact, an intrinsically wide electrochemical stability window, chemical stability, and flame-retardant safety. Nevertheless, some novel solid electrolytes or electrolyte combination systems have met the basic requirements for solid-state lithium-metal batteries in many of these aspects, and the solid-state Li-O_2 battery system is also developing rapidly [16,17].

The surface of solid electrolytes inherently contains various groups that interact with electrolyte salt ions; for example, single lithium-ion conducting solid polymer electrolytes (SPE) can be designed through covalent bonds, or polymer structures and functional groups can be designed at the molecular level to bind anions via intermolecular interactions [18]. Inorganic solid electrolytes such as lithium aluminum germanium phosphate inherently absorb oxygen molecules; thus, they can promote the reduction of oxygen to superoxide and peroxide molecules on the solid electrolyte surface [19]. These characteristics expand the design space for regulating electrochemical mass transport and reaction processes in solid electrolytes. Solid-state batteries can be classified into semisolid (gel electrolytes/solid-liquid hybrid systems), quasi-solid, and all-solid-state battery systems, with the free liquid electrolyte content decreasing sequentially below 10%, 5%, and eventually to none at all. Whether for Li-O_2 batteries or other high-specific-energy battery systems, high-performance batteries based on advanced solid electrolytes represent the future development direction.

3.1.1.3 Aqueous Lithium-Oxygen Battery

Visco et al. [20] proposed an aqueous Li-O_2 battery in 2004. Typically, an aqueous Li-O_2 battery consists of a lithium anode, a porous air cathode, and an aqueous electrolyte. It is well known that water or other strong proton-containing organic solvents readily react violently with highly reactive lithium metal, producing hydrogen gas; excessive local heat generation can even cause combustion or explosion hazards. To protect the lithium anode from the effects of the solvent (water), aqueous Li-O_2 batteries require strict anode protection measures, usually involving a dense inorganic solid-state lithium-ion-conductive membrane that effectively blocks moisture [21,22].

$$4\text{Li} + \text{O}_2 + 6\text{H}_2\text{O} \rightarrow 4(\text{LiOH}\cdot\text{H}_2\text{O}) \tag{3.12}$$

As shown in Equation (3.12), the aqueous electrolyte in Li-O_2 batteries induces the formation of hydrated LiOH as the discharge product. Since the solvent water reacts with the active materials, the theoretical energy density of aqueous Li-O_2 batteries is lower than that of non-proton organic Li-O_2 batteries. However, on one hand, aqueous Li-O_2 batteries alleviate the issue of accumulation and blockage caused by the insolubility of cathode discharge products in organic liquid electrolytes. Therefore, the practically achievable energy density can still approach 500 Wh kg^{-1} (when the concentrations of Li^+ and OH^- reach their solubility limit during deep discharge, LiOH still precipitates and fills the pores of the porous air cathode) [23]. On the other hand, the lithium hydroxide discharge products generated by the aqueous electrolyte to exhibit high reversibility and avoid highly reactive oxygen species, thereby providing better stability for the battery cathode [24].

However, the indispensable inorganic solid electrolyte membrane in aqueous batteries has two main drawbacks. First, insufficient stability: some inorganic solid electrolytes with relatively high ionic conductivity are prone to reduction by lithium metal, resulting in limited stability. Moreover, the solid electrolyte membrane cannot completely suppress lithium dendrite growth, which may pose even greater risks than organic electrolytes in later cycling stages. Therefore, in fundamental research, some researchers use lithium iron phosphate as a pseudo-anode to replace the LMA [25]. Second, there is the issue of mass transport: inorganic solid electrolytes generally possess high Young's modulus and are difficult to deform, leading to poor interfacial contact with lithium metal, which causes mass transport hindrance and large interfacial impedance. To overcome these problems, aqueous Li-O_2 batteries have further evolved to employ hybrid electrolytes.

3.1.1.4 Hybrid Electrolyte Lithium-Oxygen Batteries

As shown above, hybrid (also called mixed) electrolyte Li-O_2 batteries are usually battery systems with complex integrated structures. This is because, to utilize the special properties of strongly polar proton solvents with high reactivity toward lithium metal in the cathode region to achieve more favorable electrochemical reversible reactions and battery performance, another lithium-metal-stable solvent is required as the anode-side electrolyte to simultaneously address the drawbacks of using only inorganic solid electrolyte membranes for anode protection. Generally, a solid electrolyte membrane is still needed between these two electrolytes to isolate the two solvents.

Research on mixed electrolyte systems has been limited in recent years, partly because such complex integrated electrolyte structures cause additional stability issues. Although the lithium-ion migration problem at the anode interface is somewhat alleviated compared to aqueous electrolyte batteries using inorganic solid membranes, different space charge layers at the heterogeneous electrolyte interfaces still cause various electrochemical side reactions [26]. Meanwhile, multiple interfaces also hinder material transfer at the interfaces, increasing the overall battery impedance.

To address these issues, researchers have adopted multiple strategies: First, optimizing the intermediate electrolyte layer; for example, commonly used inorganic solid electrolyte membranes can be improved to composite solid electrolyte layers with better comprehensive performance [27]. Second, avoiding the use of aqueous electrolytes in the cathode region, instead mixing a small amount of water or other proton solvents into organic non-proton solvents combined with more flexible anode protection methods; in such mixed electrolytes, the proton solvent activity is relatively low but still plays a significant regulatory role [28,29]. Third, designing composite electrolyte systems with liquid/liquid immiscibility by utilizing differences in solvent physical properties and different solubility parameters [30]. The advantage of this design is that it does not require an intermediate ceramic electrolyte membrane with large impedance as a phase separator. However, issues of system stability under ion migration and practical conditions such as vibration and impact still need investigation. One approach is to improve system stability through gel solidification.

3.1.2 Research Status of Electrode Reaction Mechanisms in Lithium-Air Batteries

3.1.2.1 Reaction Mechanism of Cathodic ORR/OER Process

3.1.2.1.1 Path of Cathode ORR/OER Reaction Path

It is generally believed that the ORR reaction mechanism during the discharge process of Li-air batteries is as follows [31–33] (* denotes adsorbed species):

$$Li^+ + e^- + O_2^* \rightarrow LiO_2^* \quad (3.13)$$

$$LiO_2^* + Li^+ + e^- \rightarrow Li_2O_2^* \quad (3.14)$$

Peng et al. [34], using surface-enhanced Raman spectroscopy (SERS) combined with electrochemical testing, proposed that reactions (3.13) actually consist of the following two elementary reactions:

$$O_2^* + e^- \rightarrow O_2^- \quad (3.15)$$

$$O_2^- + Li^+ \rightarrow LiO_2^* \quad (3.16)$$

During the formation of $Li_2O_2^*$, LiO_2^* does not undergo the electrochemical reduction reaction described by Equation (3.14), but instead undergoes the following disproportionation reaction:

$$2LiO_2^* \rightarrow Li_2O_2^* + O_2 \quad (3.17)$$

However, since LiO_2^* has not been detected during the charging process, it is believed that the decomposition of Li_2O_2 is a one-step, two-electron oxidation reaction:

$$Li_2O_2^* \rightarrow 2Li^+ + 2e^- + O_2 \quad (3.18)$$

It is noteworthy that the study by Peng et al. employed an acetonitrile (MeCN) electrolyte system and a gold (Au) electrode. For different electrolyte and catalyst systems, the ORR/OER reaction pathways may vary. Bruce et al. [9], using SERS, found that during the ORR process, the specific reaction pathway through which the intermediate LiO_2 converts to Li_2O_2 is closely related to factors such as the solubility of LiO_2 in the electrolyte, electrode potential, and current density. Generally, in solvents with relatively high donor number (DN) and acceptor number (AN), the solubility of LiO_2 is higher, favoring a chemical disproportionation reaction pathway known as the "solution phase pathway." The discharge products in this case mainly present as larger-sized toroidal (ring-shaped) coil structures.

Conversely, in electrolyte systems with low DN and AN values, LiO_2 tends to adsorb onto the electrode surface and undergo electrochemical reduction, referred to as the "surface electrochemical pathway." The discharge products mainly form thin film-like deposits on the air electrode surface. Aetukuri et al. [14], through studying the effect of water as an electrolyte additive, also confirmed the influence of solvation effects on the ORR reaction pathway in Li-air batteries, consistent with the conclusions above. Since the "solution phase pathway" does not involve electron transfer processes and is less affected by electrode surface passivation, it tends to facilitate higher discharge capacities compared to the "surface electrochemical pathway." Additionally, under conditions of high voltage and low current density, the ORR process also tends to follow the "solution phase pathway."

Regarding the OER process, the latest research by Wang et al. [35] found that in electrolyte systems with high DN values, the decomposition of Li_2O_2 tends to produce soluble LiO_2 intermediates, whereas in electrolyte systems with low DN values, LiO_2 is hardly detected. Peng et al.'s study was conducted in a low-DN acetonitrile (CH_3CN) electrolyte system, so their results are not contradictory to the above findings. This indicates that the OER reaction pathway is also closely related to the properties of the electrolyte.

3.1.2.1.2 Oxidative Decomposition Process of Li_2O_2

The typical charge–discharge curve of a Li-air battery is shown in Figure 3.5. From this curve, it can be observed that the polarization degree of the OER process is relatively greater than that of the ORR process [36]. Since Li_2O_2 is insoluble and has poor electrical conductivity, its decomposition is relatively difficult. Therefore, catalysts are generally introduced into the air electrode to accelerate the decomposition kinetics of Li_2O_2. Accordingly, a thorough understanding of the oxidative decomposition process of Li_2O_2 is crucial for developing highly active catalysts and constructing Li-air battery systems with high energy conversion efficiency.

As shown in Figure 3.5, Lu et al. [36] proposed a two-step decomposition and oxidation mechanism for the Li_2O_2 decomposition process. It can be seen that there are two distinct voltage plateau regions during charging. The study found that the first charging region (overpotential < 400 mV) is less affected by charging current and catalysts, whereas the second charging process at higher overpotential is more sensitive to current density and catalysts. Combining Potentiostatic Intermittent Titration Technique testing, Lu et al. suggested that in the initial charging stage, the surface layer of Li_2O_2 first undergoes lithium extraction to form lithium-deficient $Li_{2-x}O_2$ components resembling LiO_2. Subsequently, LiO_2 may undergo disproportionation reactions. The second, more stable charging plateau corresponds to the oxidative

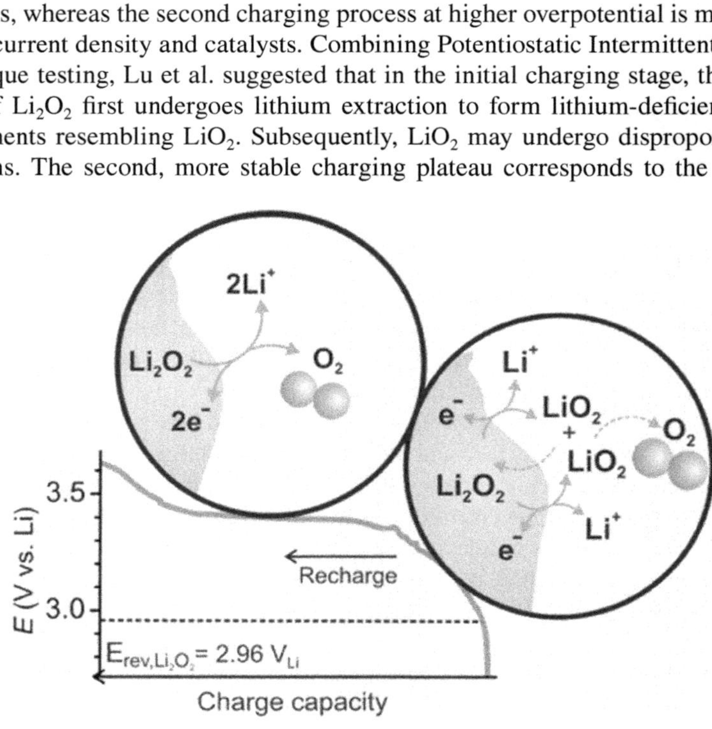

FIGURE 3.5 The discharge/charge curves of Li-O_2 battery and charging mechanism [36].

decomposition of bulk Li_2O_2. The end of the charging process may involve oxidation of by-products such as Li_2CO_3, leading to a significant rise in charging potential.

Ganapathy et al. [37] utilized in situ XRD testing technology to systematically compare the similarities and differences in the oxidative decomposition processes of electrochemically in situ formed Li_2O_2 (E-Li_2O_2) and commercial crystalline Li_2O_2 (C-Li_2O_2). As shown in Figure 3.6, for E-Li_2O_2, two distinct oxidative decomposition stages are clearly observed: the low overpotential region corresponds to the oxidation of amorphous surface layers and Li_2O_2 microcrystals, accompanied by the formation of a small amount of lithium vacancy defect-type $Li_{2-x}O_2$ solid solution; the high overpotential region corresponds to the bulk oxidative decomposition of crystalline Li_2O_2 driven by the $Li_{2-x}O_2$ solid solution intermediate.

Based on the structural and morphological characterization of products at different charging stages, the oxidative decomposition of E-Li_2O_2 starts from the outer surface and proceeds layer by layer. In contrast, for C-Li_2O_2, the oxidative decomposition process occurs in only one form. In the early charging stage, a small amount of $Li_{2-x}O_2$ solid solution is generated, which gradually drives the reduction of Li_2O_2 crystal grains, showing a process generally similar to the oxidative

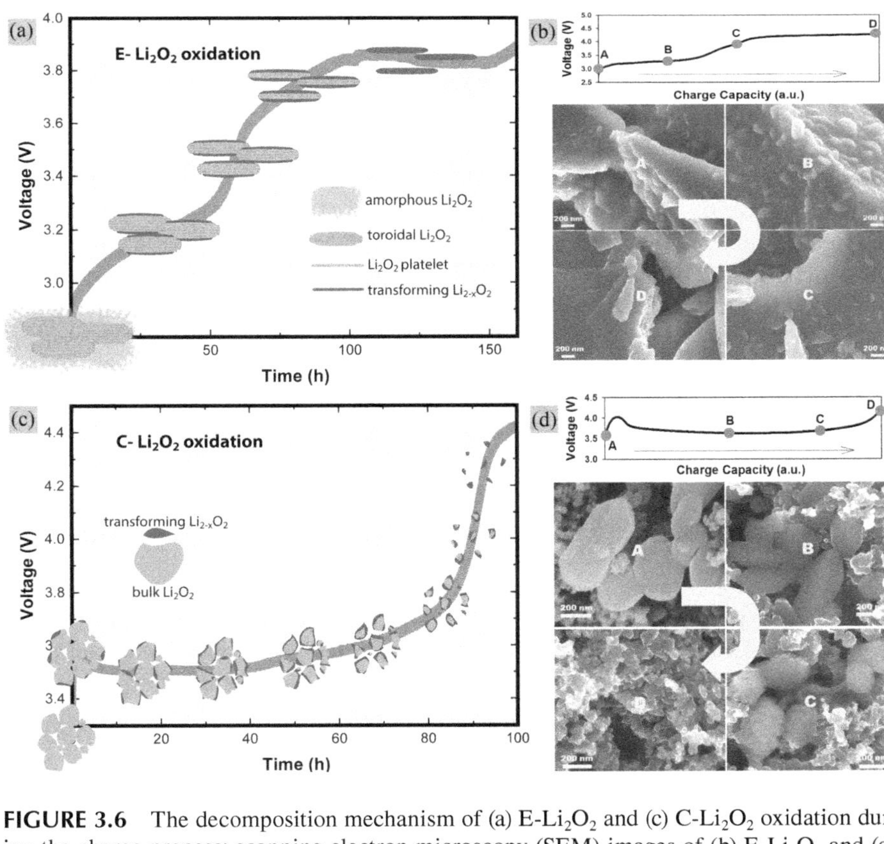

FIGURE 3.6 The decomposition mechanism of (a) E-Li_2O_2 and (c) C-Li_2O_2 oxidation during the charge process; scanning electron microscopy (SEM) images of (b) E-Li_2O_2 and (d) C-Li_2O_2 based cathodes at different charge states [37].

decomposition of E-Li_2O_2. From the above analysis, it is evident that the decomposition reaction of Li_2O_2 is an oxidation process driven by solid-solution reactions. Future research needs to further clarify the rate-controlling steps of the solid-solution reaction process to provide clearer guidance for catalyst design and the study of catalytic mechanisms.

3.1.2.2 Side Reactions of Lithium-Air Battery

Under ideal conditions, the discharge product at the cathode of a Li-air battery should be Li_2O_2. However, under certain conditions, a four-electron oxygen reduction process may occur, resulting in trace amounts of Li_2O. During the charge and discharge processes, reactive radical intermediates such as superoxide radicals and singlet oxygen are generated, making the actual ORR/OER processes far more complex than under ideal conditions. These reactive oxygen-containing radical species during cycling can lead to the decomposition of both the electrolyte and the carbon-based air electrode. Additionally, the relatively high charging overpotential further accelerates the oxidation of the electrolyte, resulting in the accumulation of a large amount of by-products on the electrode surface. Furthermore, since Li-air batteries operate as open systems, air impurities such as CO_2 and H_2O can also cause severe parasitic reactions on the air electrode or lithium anode.

3.1.2.2.1 Side Reactions Associated with Electrolyte Decomposition

Initially, Li-air battery research employed carbonate-based electrolytes, but it was found that the cycle reversibility of the batteries was very poor. Physical characterization revealed that in Li-air batteries using carbonate electrolytes, the main discharge products on the air cathode were lithium carbonate (Li_2CO_3) and small amounts of inorganic lithium salts, rather than Li_2O_2 [38]. Later, it became gradually understood that the ORR process generates intermediate superoxide ions (O_2^-). Superoxide radicals exhibit high nucleophilic reactivity and thus attack the CH_2 groups of carbonate solvent molecules, leading to ring-opening and bond cleavage of the solvent molecules.

Modifying the molecular structure of solvents can improve their tolerance to superoxide radicals to some extent. Li-air batteries based on ether electrolytes have been shown to primarily produce Li_2O_2 as the cathode discharge product. However, superoxide radicals can also abstract hydrogen from ether solvents, resulting in the formation of small amounts of Li_2CO_3 and other alkyl lithium salts. Particularly, with ongoing cycling, the amount of by-products gradually increases, causing gradual passivation and failure of the air electrode [23,24]. Sulfone-based electrolytes, such as dimethyl sulfoxide (DMSO) and sulfolane, are relatively more stable against superoxide radicals, but oxidation decomposition remains inevitable with continued cycling [39]. Additionally, some amide solvents, such as *N,N*-dimethylformamide (DMF) and *N,N*-dimethylacetamide (DMA), are more stable than carbonate and ether solvents, yet side reactions are still unavoidable.

McCloskey et al. [39] investigated a series of electrolyte solvents including diethylene glycol dimethyl ether (DME), DMSO, N-methyl-2-pyrrolidone (NMP), tetrahydrofuran, and MeCN. Differential electrochemical mass spectrometry (DEMS) tests showed that the discharge processes of Li-air batteries based on most organic solvents follow a two-electron/O_2 reduction reaction pathway. However, this does not imply that all the charge is involved in the formation of Li_2O_2. Titration experiments indicated that even in the most stable DME electrolyte system, the yield of Li_2O_2 is only about 91%, suggesting that some electrochemical side reactions occur during the ORR process. It can be said that, up to now, there is no electrolyte system sufficiently stable to truly meet the rechargeable cycling requirements of Li-air batteries under practical conditions. In the future, it is necessary to continue in-depth research on electrolyte decomposition mechanisms. Based on these mechanistic studies, targeted development and design of novel electrolyte solvent molecules or additives are required to improve stability and performance.

3.1.2.2.2 Oxidative Decomposition of Carbon-Based Air Electrodes

Besides the electrolyte, the carbon materials of the air electrode may also undergo oxidative decomposition during the charge–discharge process. This is mainly because commonly used porous carbon materials generally contain abundant defects and functional groups on their surfaces, which lead to high reactivity. Bruce et al. [40] used 13C isotope-labeled carbon materials for the air electrode and analyzed the nuclear magnetic resonance (NMR) spectra of the generated Li_2CO_3 by-products. They found that part of the by-products originated from the labeled carbon cathode. Further studies revealed that hydrophilic carbon materials not only decompose themselves at potentials above 3.5 V but also promote electrolyte decomposition, whereas hydrophobic carbon materials are relatively more stable. If the carbon surface contains oxygen-containing functional groups, its reactivity with superoxide radicals is even higher [41]. Moreover, under the presence of oxygen, the discharge product Li_2O_2 can chemically react with the carbon electrode to form Li_2CO_3 and other alkyl lithium salts, which deposit on the carbon surface [42]. In addition, during charging, the carbon cathode surface may experience growth and decomposition of Li_2CO_3. The formation of Li_2CO_3 is possibly due to reactions between peroxide radicals and carbon, but the exact mechanism remains to be elucidated. In view of this, developing carbon materials with low defect content, modifying and protecting carbon materials, as well as designing non-carbon-based air electrodes are particularly important for improving the reversibility of Li-air batteries.

3.1.2.2.3 Side Reactions Associated with Gas Impurities

Gas impurities introduced during the testing of Li-air batteries, such as H_2O, CO_2, etc., also have a great influence on the structural composition of the charge/discharge products of Li-air batteries. Usually, when the content of H_2O is too high, it not only reacts with Li_2O_2 to produce LiOH, but also spreads to the negative electrode and seriously corrodes the lithium negative electrode, leading to the premature failure of Li-air batteries [43]. Takchi et al. [44] suggested that during the discharge process, CO_2 may react with Li_2O_2 as follows: $O_2+2CO_2+4Li^++4e^- \rightarrow 2Li_2CO_3$. The reaction may involve the following intermediate reaction: $O^{2-}+CO_2 \rightarrow CO^{4-}$, and the composition of the final discharge product will also be affected by the dielectric constant of the electrolyte solvent. In recent years, it has been found that CO_2 may be electrochemically reduced at the cathode to form Li_2CO_3, which is the so-called

"Li-CO_2" battery when CO_2 is utilized as the reactant [45–47]. Although this new metal-air battery system has received much attention, its theoretical mechanism is still controversial. Using in situ surface-enhanced Raman and differential electrochemical mass spectrometry, Peng et al. [48] found that the main discharge product of the Li-CO_2 cell is Li_2CO_3, but it is very likely that trace O_2 plays a key role, and CO_2 itself does not undergo electrochemical reduction. However, some studies have found that trace H_2O or CO_2 may promote the cathodic reaction process to some extent and enhance the stability of the Li-air battery system, but its content must be carefully controlled [49–51].

In addition to this, the lithium anode faces side reactions that should not be ignored as well. The presence of lithium dendrites will exacerbate the reaction between lithium metal and the electrolyte to a certain extent; O_2 and H_2O in the electrolyte diffuse to the negative electrode, which will likewise lead to corrosion and pulverization of the lithium negative electrode. Therefore, the construction of a highly stable SEI protective film layer is crucial for improving the interfacial stability of the lithium anode.

In conclusion, fully clarifying the various possible parasitic reaction mechanisms involved in the air electrode, electrolyte, and lithium anode is crucial for the development of long-life, high-specific-energy Li-air battery systems.

3.1.2.3 Issues and Challenges Facing Lithium-Air Batteries

Given that Li-air batteries are currently the secondary chemical power system with the highest theoretical energy density, many academic teams around the world have carried out related research: Overseas, such as IBM, the U.S. Pacific Northwest National Laboratory and Argonne National Laboratory, the University of Oxford, the Industrial Technology Research Institute of Japan, etc.; domestic, such as Changchun Institute of Applied Chemistry of the Chinese Academy of Sciences, the Shanghai Institute of Silicate Research, Nanjing University, the Shanghai Institute of Space Power Research, Harbin Institute of Technology and so on. Although Li-air batteries in the air electrode, electrolyte, lithium anode, and other key materials and devices have made great progress, the cycle stability, multiplier performance, power density, and so on still cannot meet the actual demand.

At present, Li-air batteries are mainly facing the following challenges:

1. Charging overpotential is too high: Due to Li_2O_2 insoluble and poor conductivity, it is very easy to cause electrode passivation, which leads to a high charging overpotential, seriously reducing the battery energy conversion efficiency.
2. Lithium dendrite and corrosion: The lithium anode is prone to dendrite in the cycle process, which can easily cause internal short circuit of the battery; water and oxygen diffusion to the anode can easily cause lithium anode corrosion and pulverization failure.
3. Electrolyte decomposition: The charging and discharging process will produce superoxide radicals with high nucleophilic reactivity, single-line oxygen, and other intermediates, which will easily cause oxidative decomposition of the electrolyte.

3.1.2.4 Air Electrode for Lithium-Air Batteries

Li-air batteries (Li–O_2 batteries) have emerged as one of the most promising next-generation energy storage systems owing to their exceptionally high theoretical specific energy, potentially exceeding that of conventional Li-ion batteries by several times [35]. In a typical nonaqueous Li–O_2 cell, the air electrode serves as the reaction site for both the ORR during discharge and the OER during charge [52]. The sluggish kinetics of these reactions, along with the insulating nature of the discharge product lithium peroxide (Li_2O_2), lead to large overpotentials, low round-trip efficiency, and limited cycle life. Rational catalyst design for the air electrode is therefore critical for enhancing ORR/OER activity, and the choice and structure of the catalyst profoundly influence the reaction pathways, product distribution, and reversibility of Li_2O_2 oxidation [53]. Presently, Li–O_2 battery cathode catalysts can be broadly categorized into porous carbon materials, precious-metal-based materials, and non-precious transition metal-based materials.

3.1.2.4.1 Carbon Materials

Carbon materials are widely adopted in air electrodes due to their high electrical conductivity, tunable porosity, and chemical versatility. Early studies employed commercial carbons such as Ketjen black, Super P, BP2000, acetylene black, and carbon nanotubes (CNTs) [54,55]. It was found that within certain limits, the discharge capacity of Li–O_2 cells scales positively with the mesopore volume of the carbon, highlighting mesopores as more effective ORR active sites [56]. However, because Li_2O_2 is insoluble in typical electrolytes, it tends to deposit inside the pores, eventually blocking both oxygen and electrolyte transport channels. To address this, hierarchically porous structures have been designed. For example, Xia et al. [57] developed three-dimensional (3D) ordered macroporous/mesoporous carbons, in which the macropores facilitate continuous oxygen and electrolyte transport, while the mesopores provide abundant catalytic sites for ORR/OER. Such architectures provide both the space and access required for controlled Li_2O_2 deposition. Curtiss et al. [58,59] reported that ultrahigh-surface-area activated carbon ($3{,}840\,m^2g^{-1}$) led to the formation of defect-rich LiO_2-like phases ($Li_{2-x}O_2$), which exhibit significantly lower charge potentials (3.2–3.5 V) compared to crystalline Li_2O_2. Kinetic analysis indicated that Li_2O_2 formation proceeds primarily through LiO_2 disproportionation.

Chemical vapor deposition (CVD) has been widely employed to construct advanced carbon architectures for Li-air battery cathodes Kang et al., [60] using a CVD-grown vertically intertwined CNT fiber network, observed beaded Li_2O_2 deposits along the CNT surfaces, minimizing electron transport distances and enabling high-current-density operation. Similarly, Dai et al. [61] produced radially oriented N-doped carbon fibers (CFs), where Li_2O_2 deposits initiated along the fibers and progressively filled the interfiber voids during deep discharge. In Yang's work [62], vertically aligned hollow CFs were grown on a porous alumina substrate via CVD. The fibers possessed an inner diameter of approximately 30 nm and an exceptionally high porosity of up to 90%. Li-air batteries incorporating this material achieved an ultrahigh energy density of 2,500 Wh kg^{-1}. Furthermore, the unique hollow structure of the CFs enabled direct visualization of Li_2O_2 particle morphology evolution during charge–discharge cycles through scanning electron microscopy (SEM), offering valuable insights into the nucleation, growth, and decomposition mechanisms of the discharge products. In another design by Chen [63], a 3D bicontinuous porous graphene oxide (GO) network was in situ grown on the surface of 3D nanoporous metal substrates using a CVD method. This architecture provided a combination of high porosity, excellent electrical conductivity, and stable electrochemical properties, making it a promising cathode framework for high-performance Li-air batteries.

Heteroatom doping has proven to be an effective strategy for enhancing the catalytic activity of carbon-based cathode materials in Li-air batteries. Li [14,64] demonstrated the synthesis of a boron-doped Ketjenblack (B–KB) catalyst, where the introduction of boron atoms with low electronegativity altered the local electronic structure of adjacent carbon atoms. This modification improved the electronic conductivity of carbon and enhanced the overall electrocatalytic activity of the KB support. As a result, Li-air batteries employing the B–KB catalyst delivered a specific capacity of 7,193 mAh g^{-1} at a current density of 0.1 mA cm^{-2}—approximately 2.3 times higher than that of pristine KB. Under capacity-limited cycling, the B–KB-based cell also exhibited a higher discharge plateau and a significantly extended cycle life. Another successful strategy by Qian [65] utilized selenium-doped BP2000 carbon (Se–BP2000) as the cathode catalyst. The incorporation of selenium was found to markedly improve catalytic activity, mitigate polarization during charge–discharge processes, suppress the formation of undesirable by-products, and enhance the overall electrochemical performance of the Li-air battery. Detailed investigations revealed that selenium doping influenced both the growth behavior and spatial distribution of discharge products, shedding light on the underlying mechanisms for performance enhancement. Inspired by the structure of human capillaries, a nitrogen-doped CNT array supported on stainless steel mesh (N–CNTs@SS) was synthesized via a vapor-phase growth method [66]. The resulting electrode material featured a hierarchical porous structure, high electrical conductivity, excellent mechanical stability, and strong hydrophobicity. These combined properties enabled superior electrochemical performance: at a current density of 500 mA g^{-1}, the battery achieved a specific capacity of 9,299 mAh g^{-1}, while under a lower current density of 200 mA g^{-1} with a capacity limit of 500 mAh g^{-1}, stable cycling for 232 cycles was obtained.

Chemical functionalization can further tune performance. Zhang et al. [67] prepared dual-modal porous graphene via PTFE-assisted assembly, achieving ultrahigh discharge capacities. By varying the C/O ratio, they found that oxygen-containing functional groups and lattice defects promoted the formation of nanoscale, isolated discharge products, suppressing electrode passivation. Yin et al. [68] systematically introduced surface functionalities (–COOH, C=O, O–C=O) by controlled acid treatment of CF paper. Electrochemical analysis revealed that C–O and O–C=O moieties enhance both ORR and OER activity, favoring the growth of low-crystallinity, flower-like Li_2O_2.

3.1.2.4.2 Precious Metal-Based Catalysts

The sluggish decomposition kinetics of Li_2O_2 during the charging process often results in a high charge overpotential, leading to substantial capacity loss and poor rate performance in Li-air batteries. Noble metals such as Ru, Au, Pt, and Pd, owing to their outstanding electrocatalytic properties, have been extensively employed to accelerate ORR and OER kinetics in these systems. Early studies by Yang [69] demonstrated that Au/C and Pt/C catalysts could selectively promote ORR and OER, respectively, in nonaqueous Li-air batteries. A subsequent systematic investigation of Ru, Au, Pt, and Pd revealed that ORR activity followed a "volcano-type" dependence on surface oxygen adsorption energy, with the activity trend Pd > Pt > Ru ≈ Au, while Pt exhibited the highest catalytic efficiency for OER [70]. Further advancements were made through the development of an Ir nanoparticle catalyst supported on reduced graphene oxide (Ir–rGO), enabling for the first time a reversible redox conversion between O_2 and LiO_2 [71]. This system achieved over 700 stable cycles under a quasi-air atmosphere. Expanding on this concept, Ir–rGO was later employed to realize reversible Li_2O/Li_2O_2 conversion in a sealed Li-air battery, achieving stable cycling for 2,000 cycles with an ultrahigh energy density of 1,000 Wh kg^{-1} and a Coulombic efficiency (CE) of 99.5% [72]. Notably, confining the solid-phase ORR/OER conversion within an enclosed cathode space addressed the inherent instability of traditional semi-open Li-air battery designs, substantially improving cycle life.

Alloying strategies have also been effective in tuning intermediate adsorption and reaction pathways. A bifunctional Pt/Au alloy catalyst was designed to effectively lower the charge–discharge overpotential and enhance round-trip efficiency [73]. A PdCu alloy catalyst was designed to improve Li-air battery performance by exploiting the electronegativity difference between Pd and Cu, which drives electron transfer from Cu to Pd and thereby modulates LiO_2 intermediate adsorption strength. Incorporation of PdCu into the air cathode significantly reduced ORR/OER overpotentials and enhanced cycling stability [74]. Similarly, a multi-pod PtIr alloy catalyst with low Lewis acidity at the Pt sites was synthesized to improve cathodic reaction kinetics. Because Pt has a higher electronegativity than Ir, electron transfer from Ir to Pt increased Pt's electron density and shifted its d-band center downward, weakening LiO_2 intermediate adsorption and optimizing the ORR/OER pathways. Li-air batteries equipped with PtIr catalysts exhibited a reduced charge–discharge potential gap of 0.44 V and achieved 180 stable cycles at 500 mA g^{-1} under a 1,000 mAh g^{-1} capacity limit [75].

Hybrid novel metal catalysts with novel structures were also developed as high-performance cathode catalysts. Wang et al. [76] fabricated a "sandwich" catalyst composed of graphene, Au nanoparticles, and Au nanosheets. Li_2O_2 grew conformally only on Au nanoparticle surfaces, minimizing contact between Li_2O_2 and the electrolyte, thus enhancing cycling stability. Lu et al. [77] compared Ag nanoclusters of three sizes (Ag_3, Ag_9, Ag_{15}) and found that particle size determined discharge product morphology—ranging from thin films to ring-coil structures—with smaller crystallites and amorphous content correlating with easier decomposition during charge. Wang et al. also demonstrated that Co single atoms dispersed on N-doped carbon promoted uniformly distributed nanoscale Li_2O_2 crystals, achieving a remarkably small charge–discharge voltage gap (0.4 V) and stability over 260 cycles [78]. Zhou et al. [79] showed that Ru/CNT electrodes favored low-crystallinity Li_2O_2, attributed to strong Ru–O_2 interactions that weakened Li–O bonding, promoting defect-rich $Li_{2-x}O_2$ formation. Lu et al. [80] found that Ir-reduced graphene oxide

(Ir-rGO) surfaces generated crystalline lithium superoxide (LiO_2), indicating that ORR was arrested at the single-electron reduction stage. In situ transformation of Ir to Li-containing Ir_3Li solid solution provided a structural match for LiO_2 nucleation.

3.1.2.4.3 Non-Precious Transition Metal-Based Catalysts

Transition metal oxides catalyze ORR/OER through redox cycling of metal cations and by facilitating oxygen adsorption–desorption processes. Common oxides such as MnO_2, NiO, CuO, and Co_3O_4 have shown varying degrees of activity, with Co_3O_4 often exhibiting superior performance [81]. Liu et al. [82] found that Co_3O_4 (111) facets displayed higher activity and stability than (001) facets, as density functional theory calculations revealed that (111) facets facilitated O_2 desorption in OER, accelerating Li_2O_2 oxidation. Oxygen vacancy engineering also proves beneficial. Gao et al. [83] demonstrated that increasing oxygen vacancy concentration in CoO enhanced Li^+/electron transport and provided preferential O_2 adsorption sites. Wang et al. tuned Co_3O_4 oxygen vacancy levels and Co^{3+}/Co^{2+} ratios to control Li_2O_2 morphology and boost ORR/OER kinetics. Adpakpang et al. [84] synthesized low-valence Mn_2O_3 porous nanosheets via thermal treatment, yielding porous Li_2O_2 deposits with lower overpotentials and improved cycling. The influence of exposed crystal facets on the catalytic behavior of β-MnO_2 in Li-air batteries has been systematically investigated by Bruce [85]. β-MnO_2 catalysts were found to significantly reduce the charge–discharge overpotential of the battery. In a low DN tetraethylene glycol dimethyl ether (TEGDME)-based electrolyte, the {111} facets of β-MnO_2 favored a solution-phase growth pathway for Li_2O_2, leading to the formation of large ring-shaped particles. In contrast, the {100} facets promoted a surface-mediated growth pathway, resulting in the formation of thin-film Li_2O_2. These differences were attributed to the variation in LiO_2 intermediate adsorption energy on different crystallographic planes, which in turn dictated the preferred Li_2O_2 growth mechanism.

Beyond single oxides, mixed-metal oxides can deliver synergistic effects. Yan et al. [86] designed a MnO_2/Co_3O_4 composite catalyst via an electrochemical deposition method, leveraging the distinct discharge characteristics of different catalysts in Li-air batteries to achieve synergistic catalytic effects. This strategy facilitated the oriented, embedded growth of large-sized discharge products, thereby enhancing reaction kinetics. As a result, the assembled batteries exhibited a high specific capacity (4,850 mAh g^{-1} at 100 mA g^{-1}), excellent rate capability (3,550 mAh g^{-1} at 600 mA g^{-1}), and extended cycling stability (50 cycles at a capacity limit of 1,000 mAh g^{-1}). Compared with single-metal oxide catalysts, composite oxides such as MCo_2O_4 (M = Mn, Ni, Cu, etc.) demonstrate superior electronic conductivity and higher electrochemical activity. For example, Sun et al. [87] employed Ni foam as a substrate to in situ grow hierarchical porous 3D $MnCo_2O_4$ nanowire networks. The resulting catalyst enabled Li-air batteries to deliver a remarkably high discharge capacity of 12,919 mAh g^{-1} with a low charge–discharge overpotential of only 0.54 V. Similarly, Lu et al. [88] synthesized a bifunctional $CuCo_2O_4$ catalyst via a simple urea-combustion route. The corresponding Li-air batteries exhibited reduced polarization, improved rate performance, and prolonged cycling life. Under simulated air conditions, the battery achieved a discharge capacity of 7,962 mAh g^{-1} while maintaining a charge–discharge voltage gap below 1 V. Zhang et al. [89] first applied

$La_{0.75}Sr_{0.25}MnO_3$ perovskite catalysts in Li–O_2 cells, achieving 124 stable cycles at a fixed capacity. Other perovskites, such as $Sr_{0.95}Ce_{0.05}CoO_{3-\delta}$ and $La_{0.5}Sr_{0.5}CoO_{2.91}$, have also shown promising activity.

3.1.2.5 Liquid Electrolytes for Li-Air Battery

3.1.2.5.1 Ether-Based Electrolytes

Ether-based electrolytes exhibit higher stability than carbonate-based solvents because they are less susceptible to nucleophilic substitution by O_2^-. In addition, ethers have low viscosity and good solubility for oxygen, which promotes the transport of oxygen in Li-O_2 batteries. A high dielectric constant enhances the ability of the solvent to dissolve electrolyte salts. Therefore, ether-based solvents are regarded as strong candidates for electrolytes in Li-O_2 batteries.

In 2006, Read was the first to report the application of ether-based electrolytes in Li-O_2 batteries. They used DOL: DME (with a mass ratio of 1:1) as the solvent, demonstrating high stability and favorable rate performance [90]. Subsequent research by Zhang et al. revealed that the addition of crown ethers, such as 12-crown-4 and 15-crown-5, could enhance the discharge capacity of Li-O_2 batteries [91]. The Abraham team made a significant breakthrough by dissolving $LiPF_6$ in TEGDME and using catalyst-free porous carbon as the cathode, achieving a rechargeable Li-O_2 battery [92]. Building on this work, Jung et al. constructed a highly reversible Li-O_2 battery using a TEGDME-$LiCF_3SO_3$ electrolyte system, achieving nearly 100 charge–discharge cycles. TEGDME has since become a widely used classic solvent for Li-O_2 battery electrolytes [93].

However, studies have demonstrated that ether-based electrolytes are still prone to decomposition during battery operation. Freunberger et al. confirmed that reactive oxygen species (including O_2^-, LiO_2, and singlet oxygen $^1O_2$) generated during the discharge and charging processes of the cathode induce the oxidative decomposition of ether electrolytes, producing by-products such as lithium carbonate. The content of these by-products increases with the number of cycles and eventually becomes the dominant product after a certain number of cycles [94]. Zhao et al. proposed a specific decomposition pathway for TEGDME induced by O_2^- during Li-O_2 battery discharge in Figure 3.7.

In this mechanism, O_2^- first abstracts a hydrogen atom from TEGDME, followed by a series of oxidation reactions forming water (H_2O). The H_2O then reacts with the lithium anode, generating hydrogen gas (H_2), which severely corrodes the lithium metal. Simultaneously, H_2O reacts with the discharge product Li_2O_2 to form lithium hydroxide (LiOH). This pathway highlights the critical role of reactive oxygen species in electrolyte degradation and subsequent battery performance issues [95].

To address this, researchers have developed various strategies to enhance the electrolyte stability. One approach involves substituting active hydrogen atoms with more stable groups or atoms. The Nazar team applied this concept by substituting main-chain protons on DME molecules with methyl (–CH_3) groups, synthesizing 2,3-dimethyl-2,3-butanediol dimethyl ether (DMDMB). Figure 3.8 presents the proposed initial hydrogen abstraction step of glymes by superoxide during cell discharge along with the structures of DME and DMDMB. This modification eliminated

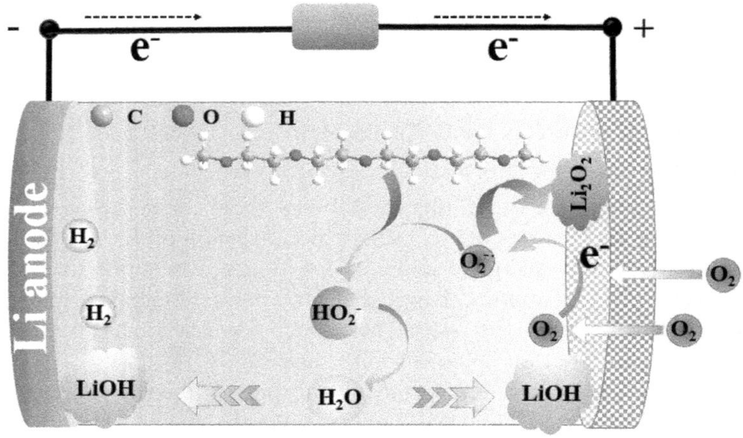

FIGURE 3.7 The possible decomposition pathway of tetraethylene glycol dimethyl ether [95].

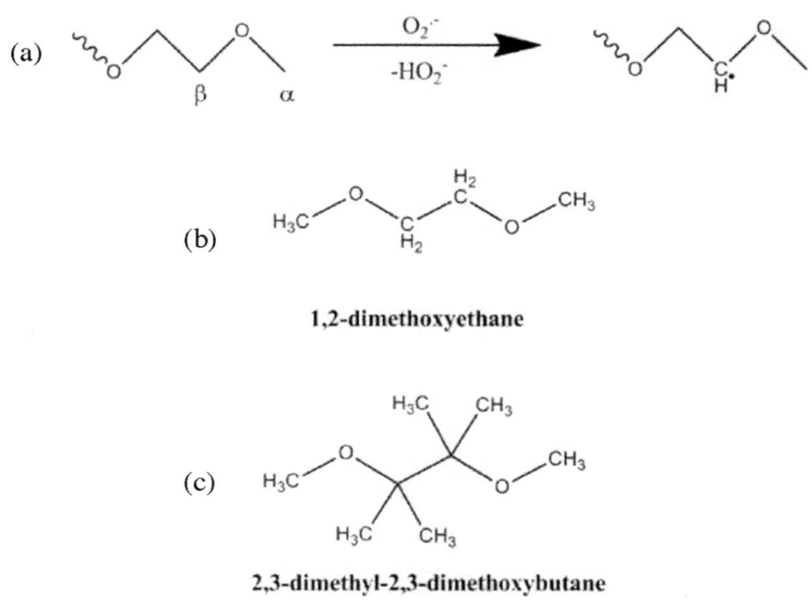

FIGURE 3.8 (a) Proposed initial hydrogen abstraction step of glymes by superoxide during cell discharge and the structures of (b) 1,2-dimethoxyethane (DME) and (c) 2,3-dimethyl-2,3-dimethoxybutane (DMDMB) [96].

methylene dehydrogenation, effectively preventing reactions between superoxide radicals and hydrogen atoms in solvent molecules, and significantly reducing the formation of by-products such as carbonates. Compared to DME, DMDMB usage resulted in a tenfold reduction in CO_2 emission during charging [96]. Similarly, Huang Yunhui et al. developed a new fully methylated cyclic ether as an electrolyte solvent for Li-O_2 batteries. By substituting active α-H positions on ether solvent molecules with methyl groups, they enhanced the solvent's resistance to superoxide radicals and singlet oxygen [97].

In addition to molecular structure modifications, electrolyte concentration regulation has proven effective in suppressing side reactions. The Zhang Jiguang group utilized high-concentration electrolytes, such as 3.0 mol L^{-1} LiTFSI-DME, in Li-O_2 batteries. High-concentration electrolytes reduce the number of free solvent molecules through strong solvation interactions between solvent and salt, thereby enhancing electrolyte stability. Calculations of molecular orbital energies for DME molecules and Li^+-$(DME)_n$ solvates, as well as Gibbs activation energies for C–H bond cleavage in DME and Li^+-$(DME)_n$ when attacked by O_2, confirmed that ultra-high concentration electrolytes enhance stability against oxygen reduction intermediates [98]. The Xie Haiming team from Northeast Normal University used a mixed solvent system of tetraglyme electrolyte (HCG4) and 1,1,2,2-tetrafluoroethyl-2,2,3,3-tetrafluoropropyl ether (TTE) in a 1:1 volume ratio with 3.4 mol L^{-1} LiTFSI, significantly improving ionic conductivity, flame retardancy, and electrochemical stability. TTE, as a cosolvent, contributed to enhanced ionic conductivity and flame retardancy [99].

However, high-concentration electrolytes face challenges like high viscosity, low conductivity and oxygen solubility, and high cost. Localized high-concentration electrolytes with fluoroether diluents retain their solvation structure at lower concentrations, reducing cost and viscosity while improving conductivity and wettability. The Zhang Jiguang group used 1H,2H,5H-octafluoropentyl-1,1,2,2-tetrafluoroethyl ether (OTE) as a diluent with TEGDME to dissolve $LiCF_3SO_3$, forming a localized high-concentration electrolyte. This electrolyte significantly increased Li-O_2 battery discharge capacity and LMA stability while exhibiting resistance to singlet oxygen erosion [100].

At present, the stability of ether-based solvent molecules during the charging and discharging processes of batteries can be effectively improved through methods such as designing and modifying the structure of ether-based solvent molecules, increasing the concentration of electrolytes, and introducing inert diluents. However, the cumbersome design and synthesis of solvent molecules, as well as the expensive high-concentration electrolyte salts and inert diluents, are not conducive to the practical application of batteries. Therefore, finding a stable, easily available, and inexpensive electrolyte remains a difficult problem in the development of metal-air batteries, and there is still a long way to go in the future.

3.1.2.5.2 Sulfone-Based Electrolytes

DMSO exhibits advantageous properties for Li-air batteries, including excellent oxygen diffusion capability, low viscosity, high DN, and high conductivity. It was first employed as a Li-air battery electrolyte by Xu et al. in 2012 [101]. Using 1.0 mol L^{-1} LiTFSI/DMSO as the electrolyte and KB carbon as the cathode, they achieved a remarkable discharge capacity of 9,400 mAh g^{-1}. Subsequent studies with a hierarchically porous graphene oxide-derived carbon cathode demonstrated a significantly reduced charging voltage of 3.7 V, along with an enhanced discharge capacity of 10,600 mAh g^{-1} and improved rate performance.

The Bruce team [102] utilized a nanoporous gold cathode with 0.1 mol L^{-1} $LiClO_4$/DMSO electrolyte, achieving 95% capacity retention after 100 cycles. The nanoporous gold electrode facilitated Li_2O_2 decomposition, with approximately 50% decomposition occurring below 3.3 V. Fourier transform infrared (FTIR) and in situ SERS analyses confirmed that the discharge product was primarily Li_2O_2 (> 99% purity), which fully decomposed upon charging, demonstrating high reversibility (Figure 3.9). Despite these promising results, the practical application of gold cathodes is limited by their high cost, weight, and fabrication complexity. In 2013, Bruce's team [103] developed a TiC cathode with 0.5 mol L^{-1} $LiClO_4$/DMSO electrolyte, forming Li_2O_2 of >99.5% purity and achieving >98% capacity retention after 100 cycles, offering a lighter and more economical alternative.

However, DMSO is chemically unstable with reactive lithium metal. To address this, Sun et al. [104] pioneered the use of $LiNO_3$/DMSO electrolyte, where $LiNO_3$ forms a stable SEI layer on the lithium anode, suppressing DMSO reactivity. This system reduced the charging overpotential by 0.42 V compared to $LiClO_4$/DMSO, exhibiting superior rate capability and cycling stability. Current strategies to mitigate DMSO instability include: (1) Solvation structure modulation: Liu et al. [105] developed a high-concentration LiTFSI/DMSO (1:3 molar ratio) electrolyte, enhancing stability against lithium anodes and DMSO. As shown in Figure 3.10, when the molar ratio of LiTFSI to DMSO is 1:3, the electrolyte contains no free DMSO solvent molecules and exclusively consists of the $(TFSI^-)_a$-Li^+–$(DMSO)_b$ (where $a+b=4$)

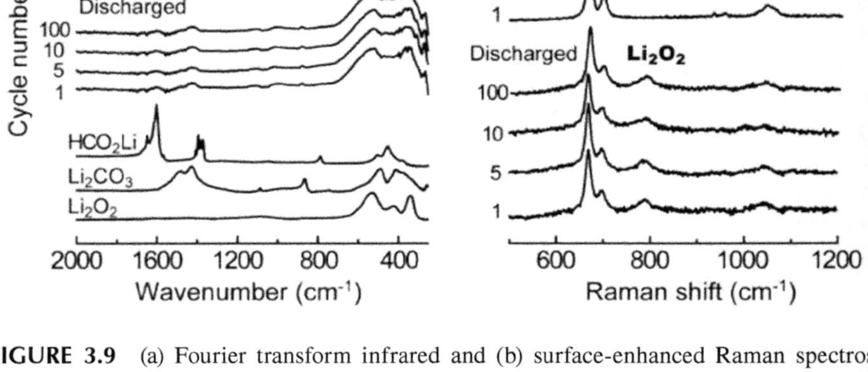

FIGURE 3.9 (a) Fourier transform infrared and (b) surface-enhanced Raman spectroscopy spectra of nanoporous gold cathodes at different cycling stages in lithium-air (Li-air) batteries [102].

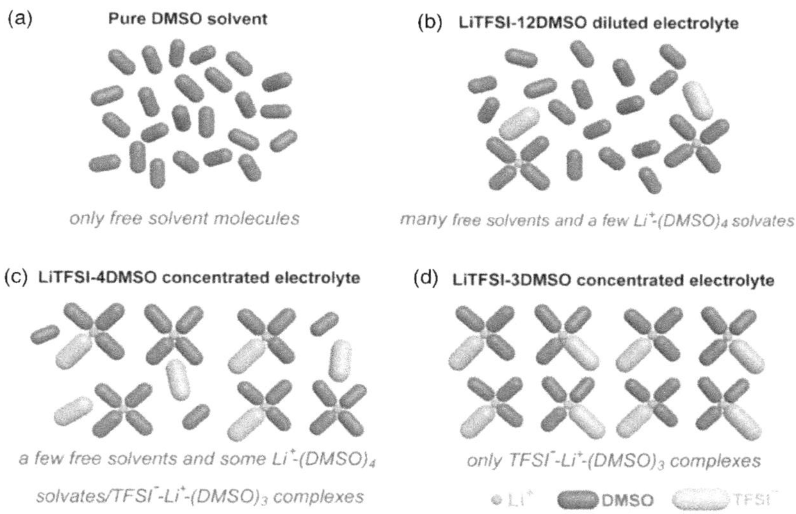

FIGURE 3.10 Solvation structure (a) pure dimethyl sulfoxide (DMSO); LiTFSI/DMSO molar ratios (b) 1:12 (c) 1:4 (d) 1:3 [104].

complex. This enabled 90 stable cycles with CNT air electrodes. (2) Redox mediators (RMs): Studies have demonstrated that incorporating LiI as a RM effectively reduces the charging overpotential in Li-air batteries while accelerating the kinetics of both the ORR and OER. However, Zhang et al. [106] revealed that LiI induces a shuttle effect in DMSO-based electrolytes. As illustrated in Figure 3.11a, during charging, I^- is oxidized to I_3^- at the cathode, whereupon I_3^- diffuses through the electrolyte to the lithium anode and is reduced back to LiI. Owing to LiI's solubility in the electrolyte, a fraction of I^- migrates back to the cathode, perpetuating this cyclic shuttle process. This mechanism not only depletes active LiI but also drastically diminishes the effective I_3^- concentration available for Li_2O_2 oxidation. To address this issue, the authors introduced a self-defense RM, InI_3. Figure 3.11b demonstrates that InI_3 forms a stable metallic In protective layer in situ on the lithium anode surface, which suppresses I_3^-–lithium side reactions and markedly enhances RM utilization, thereby improving Li-air battery performance. (3) Additives and SEI engineering: Zhou and Yoo [107] introduced 2 wt% N, N-dimethyltrifluoroacetamide into $LiNO_3$/DMSO, achieving 92 cycles with multiwalled CNT cathodes. Luo et al. [108] constructed a uniform SEI via low-rate preactivation, and Guo et al. [109] prepared a Li_3PO_4 artificial SEI film on Li metal to inhibit the reactivity of lithium and introduced the RM LiI into the $LiNO_3$/DMSO electrolyte to reduce charging voltages.

Ahn et al. [110] optimized lithium salt/solvent combinations, demonstrating that $LiNO_3$ suppresses side reactions via NO_2^-/NO_2 redox, lowering overpotentials. Among DMA, DMSO, TEGDME, and TMS solvents, TMS/$LiNO_3$ most effectively generated NO_2^-, reducing charging overpotentials and improving oxygen/energy efficiency. The 1 mol L^{-1} $LiNO_3$ TMS system achieved 100 stable cycles. In 2024, Zhong et al. [111] designed a biphasic DMSO/octyltrimethoxysilane (TOOS) electrolyte

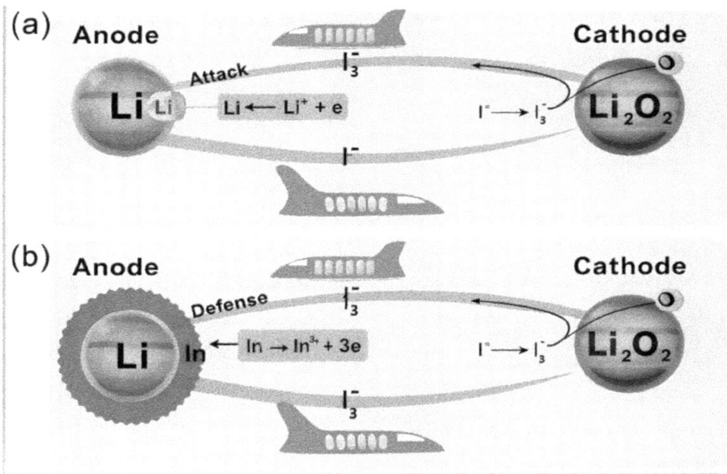

FIGURE 3.11 Schematic illustration of (a) shuttle mechanism and (b) self-defense RM of InI3 [106].

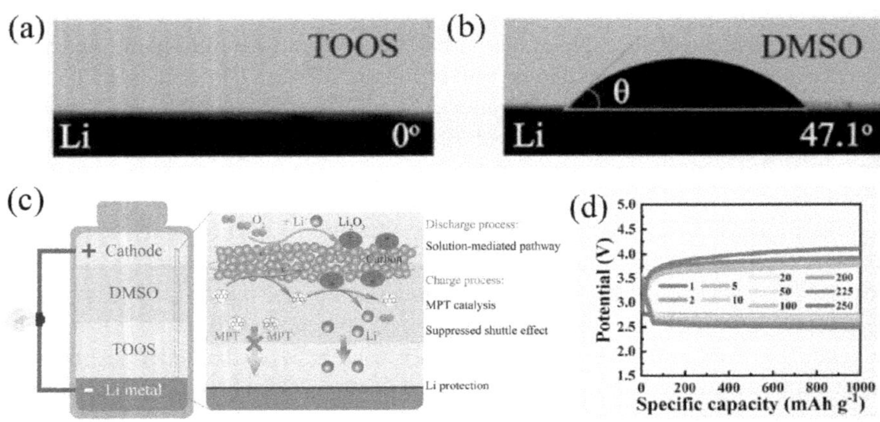

FIGURE 3.12 (a) Contact angle of TOOS and DMSO solvents on lithium foil; (b) Schematic and (c) cycling performance of the biphasic electrolyte Li-air battery [111].

with 10-methylphenothiazine (MPT) in DMSO as the catholyte and TOOS as the anolyte. This system enabled stable cycling for >250 cycles at 3.6 V, leveraging differential wettability on lithium metal (Figure 3.12).

Despite these advances, DMSO reacts with superoxides O_2^- or Li_2O_2, producing by-products like LiOH, $DMSO_2$, Li_2SO_3, and Li_2SO_4. The proposed mechanism involves [39]: O_2 reduction to O_2^-, which reacts with DMSO to form dimsyl anions and hydroperoxide radicals (Figure 3.13). Subsequently, reactions yield LiOH and $DMSO_2$ via nucleophilic attack on DMSO. Alkali-catalyzed dimsyl oxidation produces Li_2SO_3 and Li_2SO_4. The formation mechanism [39] is schematically depicted in Figure 3.14.

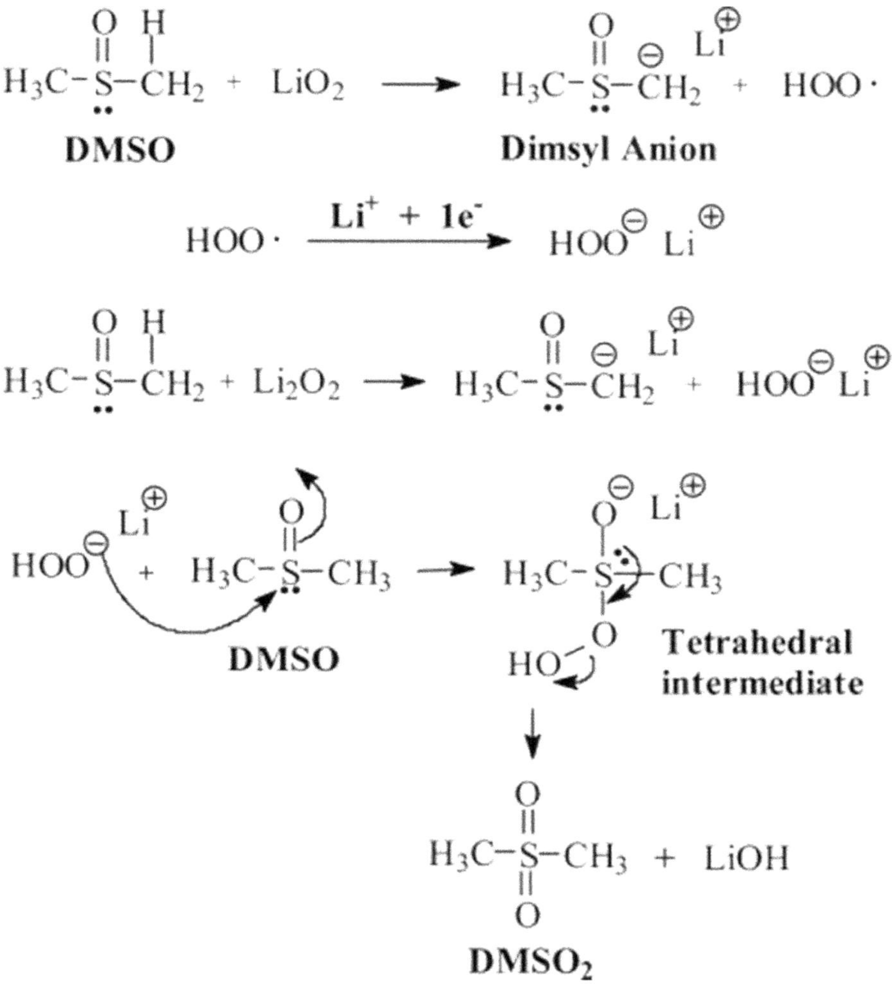

FIGURE 3.13 Degradation mechanism of DMSO during the ORR process [39].

The formation of sulfate and sulfite involves several successive alkali-catalyzed autooxidation steps of dimsyl anions. The LiOH produced from the decomposition of the dimethyl tetrahedral intermediate can generate additional dimsyllithium, which then undergoes further alkali-catalyzed autooxidation to produce lithium methylsulfonate. Under highly alkaline conditions, the remaining methyl groups are oxidized to form Li_2SO_3, which is ultimately converted to Li_2SO_4. Kwabi et al. [112] confirmed that Li_2O_2-DMSO reactivity is independent of carbon electrodes.

3.1.2.5.3 Other Electrolyte Systems

Since the organic carbonates were likely to suffer nucleophilic attack by the reduced O_2 species at the cathode [113] and the ether-based electrolyte had the drawback of

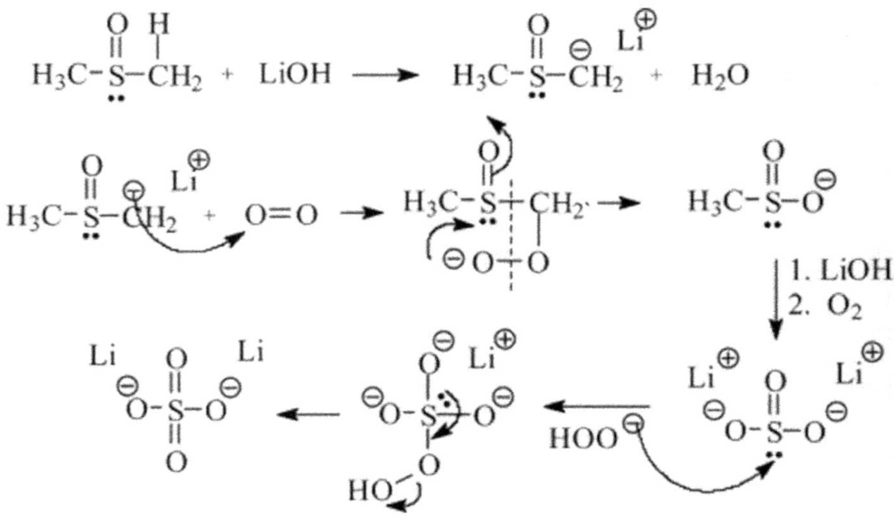

FIGURE 3.14 Base-catalyzed autoxidation of dimsyl ions [39].

decomposition while cycling [114], Bruce's team took advantage of DMF for the first time [115]. Their investigation proved the stability of DMF against the nucleophilic attack of superoxide anions and revealed that the first cycle is based on the reversible Li_2O_2 formation–decomposition with a 0.9 V gap between charge and discharge. Inspired by this method, other amide solvents have also been used in this system, such as N, N-dimethylacetamide (DMA), N-(2-Hydroxypropyl)methacrylamide (HPMA), etc. Walker et al. [116] found that Li anode paired with DMA solvent and lithium nitrate ($LiNO_3$) can stabilize SEI. They eventually manufacture a Li-air cell with this composition cycling for over 2,000 hours (more than 80 cycles) at a current density of 0.1 mA cm^{-2}. Akihiro et al. [117] also used the same composition to obtain an electrolyte with low viscosity, which could accelerate the diffusion of oxygen. Zhou et al. [118] used a strongly solvating HMPA electrolyte coupled with a Li anode surface-protected by LiPON. Li_2O_2, the primary discharge product, could be dissolved in the HPMA. In that case, the kinetics of Li_2O_2 oxidation on recharge can be greatly accelerated. Overall, all aspects of the performance of the cell, such as capacity, rate capability, and cycle numbers, are better than the former ether-based electrolyte Li-air battery.

Nevertheless, the amide-based electrolytes still encountered some barriers on their way to further application. For example, they usually have lower vapor pressure (2.70 mmHg for DMA at 20°C). Compared with the traditional lithium-ion batteries, which are in an enclosed space, the Li-air battery requires an electrolyte with lower vapor pressure to ensure the disappearance of violating while cycling because of its open structure of the cell. Even worse, the decomposition of the electrolyte remains and even increases while cycling. This could be indicated by the 1H NMR and differential electrochemical mass spectrometry in Bruce's work. Specifically, at the 5th cycle, clear evidence was also found by the FTIR at the end of discharge. The relative mechanism of decomposition is presented in Figure 3.15 [115]. The instability was

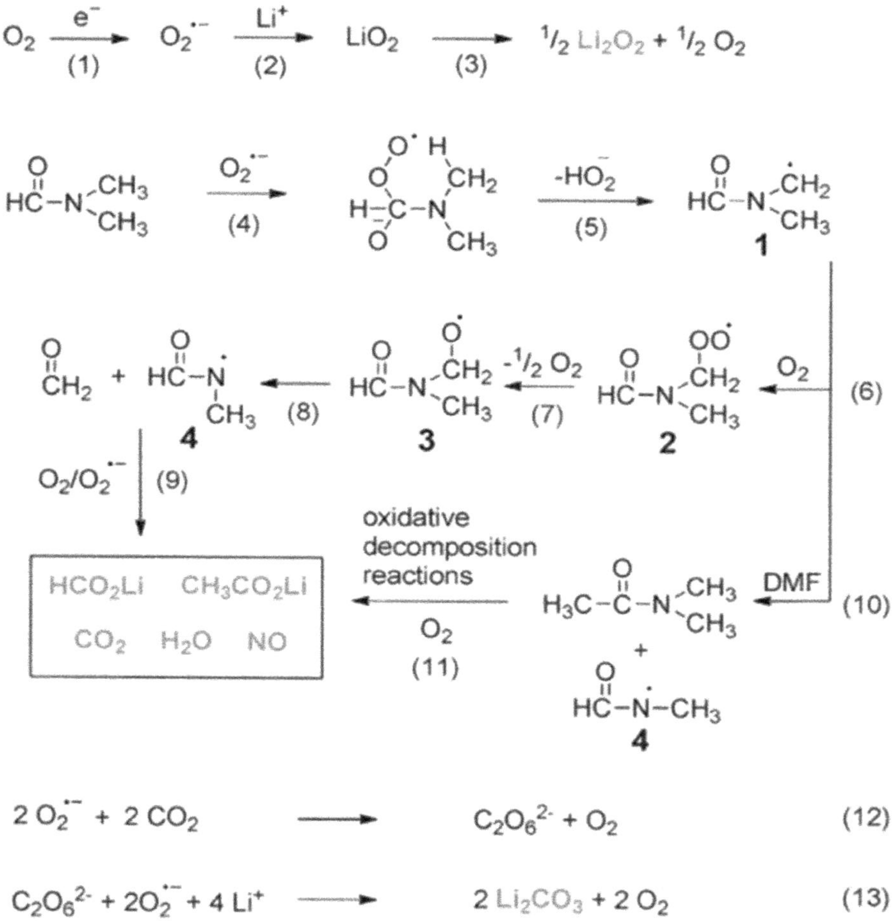

FIGURE 3.15 The possible pathway of the decomposition of the DMF-based electrolyte [115].

not confined to DMF; the DMA and NMP systems were also found to contain the presence of CH_3CO_2Li and HCO_2Li. The main decomposed products at the cathode, such as Li_2O_2, Li_2CO_3, and CH_3CO_2Li, would accumulate while cycling and block the surfaces, thus leading to significant capacity fading.

To cope with the problem of easy evaporation, NMP was selected due to its relatively lower vapor pressure (0.29 mmHg for NMP at 20°C). Additionally, it boasts high stability against superoxide anion. However, Kim et al. [119] found that the NMP electrolyte reacted with oxygen in the presence of lithium metal and that such an oxygen-consuming reaction (i.e., autoxidation) is a possible primary cause for the poor performance of the cell. Fortunately, they also found that $LiNO_3$ could suppress the autoxidation, which is consistent with the earlier discussion above. The overall scheme of this work is shown in Figure 3.16. The NMP with the addition of $LiNO_3$ had an excellent performance in enhancing oxygen efficiency and improving cycling ability.

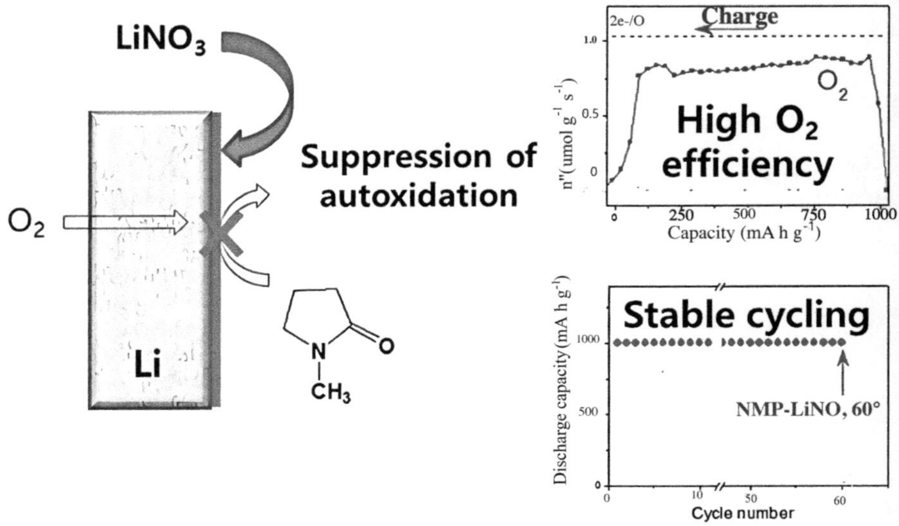

FIGURE 3.16 The overall illustration of the Kim's work [119].

Some other scientists focus on the decomposition and have researched and discovered other newly amide-based electrolytes. Qian [120] proposed that N-methylformamide (NMF) was a suitable electrolyte for the Li-air battery. The system exhibited a charge overpotential of only 0.17 V and a discharge overpotential of 0.32 V, which showed a relatively high reaction activity of ORR/OER. However, the AN value of the NMF solvent was high, which would probably lead to the reaction between the lithium and the electrolyte, accounting for the rising trend of the polarization and cell failure in the end. Considering that concentrated electrolyte was recognized as a strategy to improve the stability between the lithium anode and the electrolyte, he decided to apply it to the experiments. As the electrolyte concentration increases, a solvation structure where the solute encapsulates the solvent would emerge, which would greatly reduce the activity of the solvent molecules and be of great benefit to the compatibility. In the end, the cycle number had been improved from 30 to 108 when the ratio of the LiTFSI and NMF was adjusted from 1:12 to 1:3.

There also exist some different strategies to modify the amide-based electrolyte, such as fluorination. Kiho et al. [121] designed fluorine-containing electrolytes (2,2,2-Trifluoro-N-methyl-N-(1-methylethyl)acetamide($CF_3CON(Me)(iPr)$) and 2,2,2-Trifluoro-N, N-dipropylacetamide($CF_3CON(nPr)_2$)), the formed Li_2O_2 would decompose at a low voltage and the battery was able to maintain a stable plateau of charging voltage at around 3.5 V, as shown in Figure 3.17. The O_2 evolution efficiencies were always higher with the CF_3-containing electrolytes compared to those of the corresponding CF_3-free electrolytes (51.7% for $CH_3CONMe(iPr)$-50, 70.2% for $CF_3CONMe(iPr)$-50, 63.2% for $CH_3CON(nPr)_2$-50 and 80.0% for $CF_3CON(nPr)_2$-50). This work has proven that the molecular design of the electrolyte could regulate the

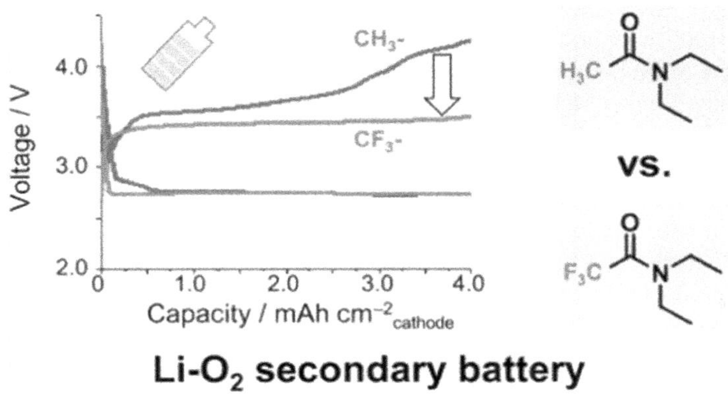

FIGURE 3.17 The overall illustration of Kiho's work [121].

activity of the high-decomposition products and paves a new way to improve the energy density.

In conclusion, the amide-based electrolyte remains a competitive candidate for the Li-air battery due to its strengths, such as oxygen reduction resistance. A couple of innovative methods, such as fluorination, high-concentration electrolytes, have been applied to it to improve the cell performance. The application of the amide-based electrolyte still marks a milestone on the path toward the next generation of energy storage technology.

3.1.2.5.4 Additives

The use of electrolyte additives is primarily aimed at enhancing the electrochemical activity and stability of both the air electrode and the lithium anode in Li-air batteries. Regarding additives for the lithium anode, their design principles are similar to those in lithium metal battery systems like lithium-sulfur (Li-S) batteries. They thus will not be the main focus here. As previously mentioned, during the charge–discharge processes of Li-air batteries, insoluble Li_2O_2 and by-products deposit on the air electrode surface [40,122]. This can lead to the deactivation of catalytically active sites and the blockage of pathways for oxygen and electrolyte transport. In contrast to conventional solid-phase catalysts, soluble molecular catalysts can effectively circumvent the issue of active site passivation faced by solid catalysts, as they are employed as electrolyte additives. Furthermore, compared to traditional solid-phase catalysts, molecular catalysts can more effectively regulate the oxygen electrode reaction pathways and the deposition/dissolution of discharge products. Generally, soluble molecular additives for Li-air batteries fall into the following categories: (1) Reversible oxygen carriers: enhance oxygen transport efficiency; (2) Cationic/anionic additives: regulate the nucleation, growth, and distribution of Li_2O_2; (3) RM molecular catalysts: catalyze the formation and decomposition of Li_2O_2; and (4) Proton shuttles/protonic additives: promote solution-phase pathways during discharge and reduce the charging overpotential.

3.1.2.5.4.1 Reversible Oxygen Carrier Given the low oxygen solubility and transport rate in organic solvents, drawing inspiration from nature by utilizing bioinspired reversible oxygen carriers as electrolyte additives presents an effective strategy. Wan et al. [123] explored a novel perfluorocarbon additive and found that it significantly enhances O_2 mass transfer. This led to a substantial increase in specific discharge capacity from 814 mAh g^{-1} to 16,495 mAh g^{-1}. Hu et al. [124] discovered that cobalt(II) salen (*N,N′*-ethylenebis(salicylideneiminato)cobalt(II), CoII-salen) functions as a reversible oxygen carrier. It promotes the ORR process in Li-air batteries and helps reduce the generation of superoxide radicals and electrode by-products. Inspired by the oxygen transport mechanism of hemoglobin in blood, Samajdar et al. [125] employed human hemoglobin as an electrolyte additive to enhance oxygen transport efficiency. They observed significant improvements in the discharge capacity and cycling performance of Li-air batteries, along with a marked reduction in interfacial impedance after cycling. The design requirements for reversible oxygen carriers necessitate that the carrier molecule possess an appropriate binding strength with oxygen. A binding that is too strong hinders reversible oxygen desorption, while a binding that is too weak offers limited enhancement.

3.1.2.5.4.2 Cationic/Anionic Additives During the discharge process, certain cationic/anionic additives can interact with the intermediate product LiO_2 or specifically adsorb onto the Li_2O_2 surface. This effectively regulates the nucleation, growth, crystallographic orientation, and defect composition of the discharge products. Matsuda et al. [126] demonstrated that introducing Cl^- anions enhances the electronic conductivity of Li_2O_2, thereby delaying air electrode passivation and increasing battery capacity. Similarly, Na^+ metal additives exhibit comparable effects: Na^+-doped $Li_xNa_yO_2$ discharge products display higher charge transfer rates, facilitating easier oxidative decomposition during charging [127]. Further studies by Matsuda et al. [128] revealed that strongly Lewis acidic ions (e.g., NH_4^+) readily interact with LiO_2 intermediates to form relatively stable solvated complex structures, promoting the solution-phase growth pathway of Li_2O_2.

3.1.2.5.4.3 Redox Mediator Another type of electrolyte additive that has been extensively studied is RM-based molecular catalysts. When a RM acts during the discharge process, its oxidized state is preferentially reduced on the electrode surface, and the resulting reduced mediator can effectively promote the reduction of dissolved O_2. As shown in Figure 3.18, during the charging process, the reduced state of the mediator molecule is preferentially oxidized, and then its oxidized state can realize the chemical oxidative decomposition of Li_2O_2 while achieving cyclic regeneration. To achieve the chemical oxidation of Li_2O_2, thermodynamically, the potential of the redox couple corresponding to the mediator molecule needs to be slightly higher than the thermodynamic equilibrium potential for Li_2O_2 formation [129]. Compared with traditional solid-phase catalysts, RMs, as a type of homogeneous molecular catalyst, can effectively avoid the problem of active site passivation during the ORR process and, at the same time, promote the transformation of Li_2O_2 decomposition from point contact to surface contact during the OER process.

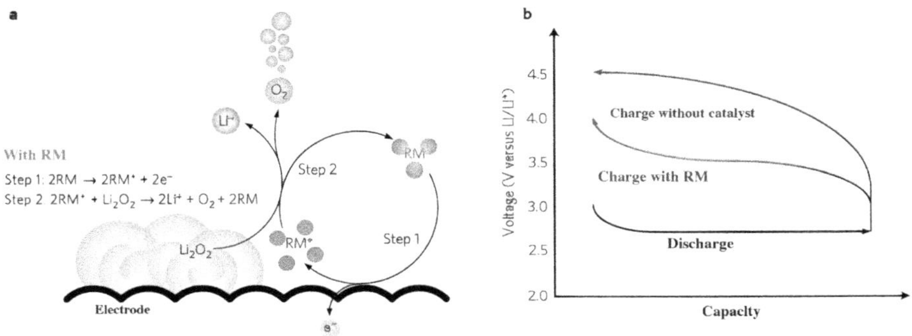

FIGURE 3.18 The functional mechanism of RMs [129].

Bruce et al. [130] were the first to propose the use of tetrathiafulvalene (TTF) as a RM to catalyze the oxidative decomposition of Li_2O_2. It was found that both ORR and OER reactions are close to a $2e^-$ process, and even under the conditions of 1 mA cm^{-2} with a capacity limit of 300 mAh g^{-1}, the charging potential remains below 3.6 V and can stably cycle for more than 100 times. 2,2,6,6-tetramethyl-piperidin-1-oxyl (TEMPO), as another classic organic RM, can also effectively reduce the charging overpotential [131]. TEMPO is a stable free radical scavenger. During the charging process, the N–O group loses one electron and transforms into the stable nitroxyl cation $TEMPO^+$, and TEMPO exhibits better cycle stability than TTF. In addition to TEMPO, other organic compounds containing nitrogen functional groups, such as MPT and DMPZ, can also effectively catalyze the decomposition of Li_2O_2 and improve the energy conversion efficiency of Li-air batteries [25,132].

In addition to organic molecular RMs, some halogen compounds also show excellent homogeneous catalytic activity. Kang et al. [133] used LiI as a RM catalyst, combined with the design of an interwoven CNT air electrode, and the charging potential of the Li-air battery could be maintained below 3.5 V. Grey et al. [134] realized a Li-air battery system based on the reversible formation and decomposition of LiOH using LiI, with an energy conversion efficiency as high as 93.2%. Studies have found that other halogen compounds, such as LiBr, although their redox potential (3.5 V) is slightly higher than that of LiI (3.2 V), can also effectively reduce the charging overpotential of Li-air batteries with almost no by-products generated [135].

In addition to RMs acting on the charging process, Bruce et al. [136] reported a RM molecule that can effectively catalyze the ORR process: 2,5-di-tert-butyl-1,4-benzoquinone (DBBQ). Studies have shown that DBBQ can significantly promote the liquid-phase growth of discharge products and greatly increase the discharge specific capacity of Li-air batteries. During its action, DBBQ first gains electrons on the electrode surface and combines with Li^+, and then combines with O_2 molecules in the solution phase, which can avoid the generation of superoxide anions and possible side reactions. Peng Zhangquan et al. [137] found that coenzyme Q10 has a similar function to DBBQ. When used as a reduction mediator in the ORR process, the discharge specific capacity of the Li-air battery is increased by about 40 times.

Given that the ORR and OER processes have different requirements for the equilibrium electrode potential of the required RMs, conventional RMs are difficult to efficiently catalyze both ORR and OER processes simultaneously. Therefore, designing RMs with a wide potential window and multi-electron transfer capability, or their combined use, should be a key research direction to be focused on in the future. In addition, some studies have found that RM molecules may decompose during charge–discharge processes or shuttle to the surface of the lithium anode for reduction [106], thereby causing passivation of the lithium anode. For this reason, protecting the lithium anode or anchoring molecular catalysts on the surface of the air electrode may be more conducive to exerting the advantages of liquid-phase molecular catalysts [138].

3.1.2.5.4.4 Protonic Additives Water, as a typical proton source, has been extensively studied as an electrolyte additive for Li-air batteries. Studies have shown that a small amount of water, on the one hand, helps to promote the solution-phase ORR reaction pathway, thereby increasing the capacity of Li-air batteries; on the other hand, it also contributes to reducing the charging overpotential and facilitating the decomposition of Li_2O_2. It has been found that H_2O molecules participate in both ORR and OER reactions and promote the conversion of superoxide anions to HO_2^- intermediates, which can effectively inhibit side reactions related to superoxide free radicals [139,140]. Schwenke et al. [43] found that within a certain range, the yield of Li_2O_2 increases with increasing H_2O content; however, the introduction of strong acidic proton sources may lead to a decrease in the yield of Li_2O_2.

Zhou Haoshen et al. [141] introduced an aqueous H_2O_2 solution into the electrolyte, which reduced the charging overpotential of the corresponding Li-air battery by 0.66 V. However, H_2O_2 and H_2O are prone to causing corrosion of the lithium anode. Therefore, researchers chose urea peroxide, an organic H_2O_2 complex, as an alternative additive. This not only achieves a low charging overpotential but also significantly improves the cycle stability of the lithium anode and the Li-air battery. Bruce et al. [142] found that phenol, as a proton additive, has a similar effect to water. It can also promote the liquid-phase growth of Li_2O_2, and the discharge capacity of the corresponding Li-air battery is increased by 35 times.

In conclusion, protonic electrolyte additives can effectively promote the liquid-phase ORR reaction pathway and reduce the charging overpotential. However, when used as additives, their content needs to be carefully controlled to avoid possible side reactions; at the same time, their own chemical and electrochemical stability should also be considered.

3.1.2.6 Lithium Metal Anode Optimization Strategies in Li-Air Batteries

Li-air batteries, owing to their unique electrochemical characteristics, have a theoretically inexhaustible supply of the active material (oxygen from air). Consequently, their capacity is primarily determined by the lithium provided by the negative electrode. Metallic lithium (Li), with its exceptionally high theoretical specific capacity (3,862 mAh g^{-1}) and low electrochemical potential (−3.040 V vs. the standard hydrogen electrode), is considered a viable anode material for lithium-based batteries [143]. However, the practical implementation of Li-air batteries still faces several significant challenges, including lithium dendrite growth, unstable solid electrolyte interphase (SEI) formation, and substantial volume changes [1]. Hostless LMAs fail to maintain uniform lithium-ion regulation during the plating process. This leads to nonuniform lithium deposition during cycling, resulting in the formation of dead lithium, which causes severe volume changes and lithium dendrite growth. Furthermore, continuous volume expansion occurs during repeated stripping/plating cycles, inevitably damaging the SEI. Subsequently, fresh lithium metal is exposed to the electrolyte, triggering excessive SEI growth and uncontrolled dendritic lithium formation, potentially leading to internal short circuits. Additionally, the high electrochemical reactivity of the LMA and the semi-open structure of the air battery further hinder the development of Li-air batteries [144]. Therefore, substantial research efforts have been dedicated to modifying the lithium anode to enable its practical application.

Constructing a protective film on the surface of the LMA is one of the most commonly employed strategies [145]. This includes SEI films and artificial films. For the construction of SEI films, the most common method is to directly add additives to the electrolyte, thereby forming a stable SEI layer on the lithium anode surface. Studies have found that $LiNO_3$ is considered a good choice for forming an SEI layer and significantly enhances the performance of Li-air batteries [146]. Regarding artificial films, applying a film onto the lithium anode surface is a highly popular strategy. Park et al. utilized a layer of poly(vinylidene fluoride-co-hexafluoropropylene) (PVDF-HFP) and Al_2O_3 to suppress electrolyte decomposition on the lithium anode surface. The electrochemical performance and cycle life of Li-air batteries equipped with the PVDF-HFP and Al_2O_3 protective layer were substantially improved compared to the control Li-air batteries [147]. Furthermore, Zhang et al. introduced a simple and effective method in which the Li anode is directly immersed in a solution of 1,4-dioxane (DOA), resulting in the generation of a protective polymer film on its surface [148]. Due to the hostless nature of the lithium anode, the strategies as mentioned earlier do not effectively resolve the fundamental issues inherent to the lithium anode during battery cycling.

Zhang et al. altered the growth direction of dendrites by designing the shape of the host. They assembled a long-life lithium metal battery using a vertically structured lithium anode. This architecture consists of two layers of lithium foil/glass fiber rolled into a coil. The upright shape of the glass fiber membrane promotes horizontal lithium growth, thereby suppressing dendrite formation. Compared to conductive 3D matrices, an insulating matrix prevents lithium from depositing only on the top of the 3D scaffold [149]. Studies have shown that carbon-based materials for 3D matrices, owing to their simple preparation process, versatile morphology, and high electrical conductivity, have become the most commonly used matrix materials for stabilizing LMAs [150]. Consequently, Wang et al. [151] synthesized 3D interconnected lithiophilic CNTs on porous carbon cloth. This significantly reduced the voltage polarization of the lithium metal composite electrode, thereby inducing uniform Li nucleation and ultimately leading to continuous and smooth Li deposition.

In addition, solid-state $Li\text{-}O_2$ ($Li\text{-}O_2$) batteries operate as semi-open systems utilizing lithium metal as the anode, the stability and safety of the LMA are crucial prerequisites for achieving stable cycling performance. However, Li-O_2 batteries also face challenges such as lithium dendrite growth, volume changes of the lithium anode, and high reactivity [152]. Lithium dendrite formation represents a significant challenge for LMAs. When the ion concentration at the anode surface reaches zero, cations and anions in the electrolyte exhibit different behaviors, leading to an excess of lithium ions at the surface. Under these conditions, lithium nucleation occurs, followed by continuous deposition forming lithium dendrites as the current varies. Some dendrites break off, becoming "dead lithium," while others react with the electrolyte. Both phenomena lead to a reduction in battery capacity [153]. As illustrated in Figure 3.19, the semi-open nature of Li-O_2 batteries severely exacerbates the corrosion of the LMA by gases such as H_2O and CO_2 [152].

In summary, a series of strategies for lithium anode protection have been proposed to suppress Li dendrite growth. These include creating artificial SEI layers, introducing additives into the electrolyte, utilizing solid-state electrolytes, and constructing 3D host-based composite anodes. Constructing an artificial SEI layer, particularly one containing a lithium salt interlayer, effectively prevents side reactions between the Li metal and the electrolyte. Furthermore, an artificial SEI containing an inorganic/organic hybrid layer can maintain a highly stable overpotential and extend the cycle life. Introducing additives into the electrolyte helps form a stable passivation layer on the electrode surface and regulates Li deposition behavior, enabling dendrite-free cycling of the LMA with high CE and capacity retention. Notably, replacing liquid electrolytes with solid-state electrolytes is considered the most viable approach to addressing safety concerns in lithium-ion rechargeable batteries. Additionally, carbon-based materials serve as the most commonly used and stable matrix materials for LMAs. The lithiophilicity of the carbon matrix can be further enhanced through methods such as elemental doping.

However, for practical applications, the average CE of the LMA must exceed 99.9%. Thus, realizing truly viable commercial applications requires extensive research and development efforts from the scientific community.

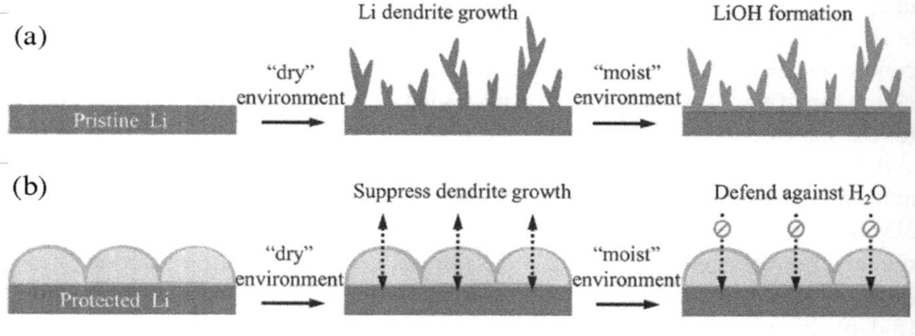

FIGURE 3.19 The lithium diagrams of lithium dendrites in different environments (a) without protective layer; (b) with protective layer [152].

3.1.3 Solid-State Lithium-Air Batteries

3.1.3.1 Working Mechanism of SSLOBs

A typical solid-state lithium–air (oxygen) battery (SSLOB) consists of LMA, solid-state electrolyte, and porous air cathode with well-designed gas diffusion pathways (Figure 3.20) [154]. SSLOBs primarily operate via lithium metal stripping and deposition at the anode during discharge and charge, respectively, while the corresponding ORR and OER occur at the cathode (Figure 3.21) [155]. The electrochemical pathway of ORR is considered to be similar to the widely accepted aprotic lithium–oxygen mechanism proposed by Abraham et al. [5], although notable differences exist. In liquid Li-O_2 batteries, molecular oxygen (O_2) is first reduced via a single-electron transfer to form lithium superoxide (LiO_2), as described by Equation (3.1). However, in solid-state electrolytes, the likelihood of a chemical disproportionation reaction of LiO_2 (Equation 3.2) is considered to be relatively low. Instead, a more plausible mechanism involves the direct electrochemical formation of lithium

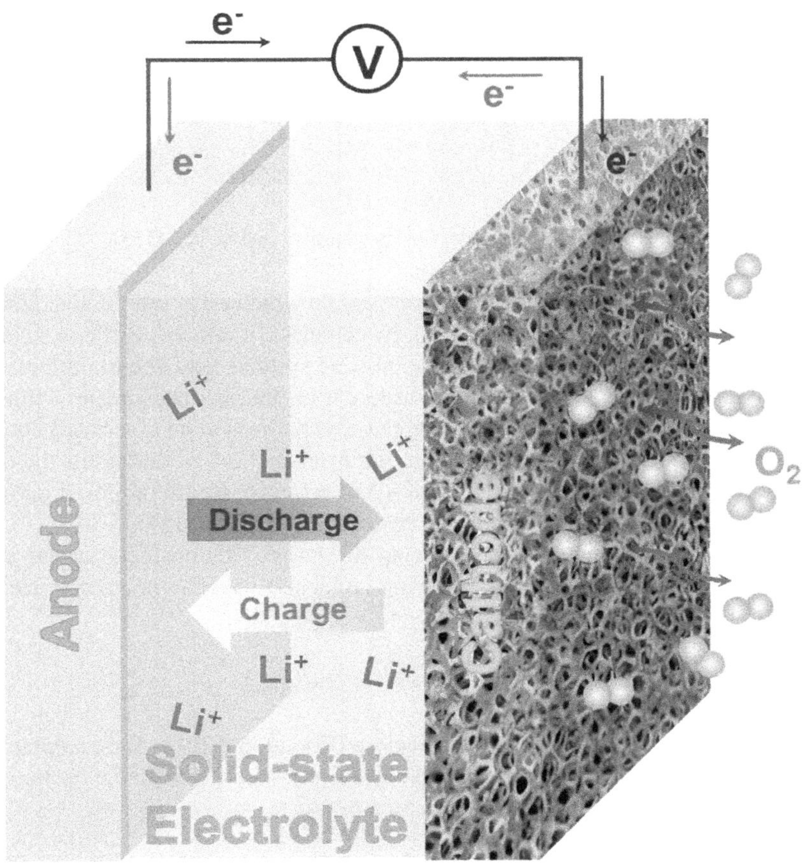

FIGURE 3.20 Schematic diagram of solid-state lithium–air (oxygen) battery (SSLOB) [154].

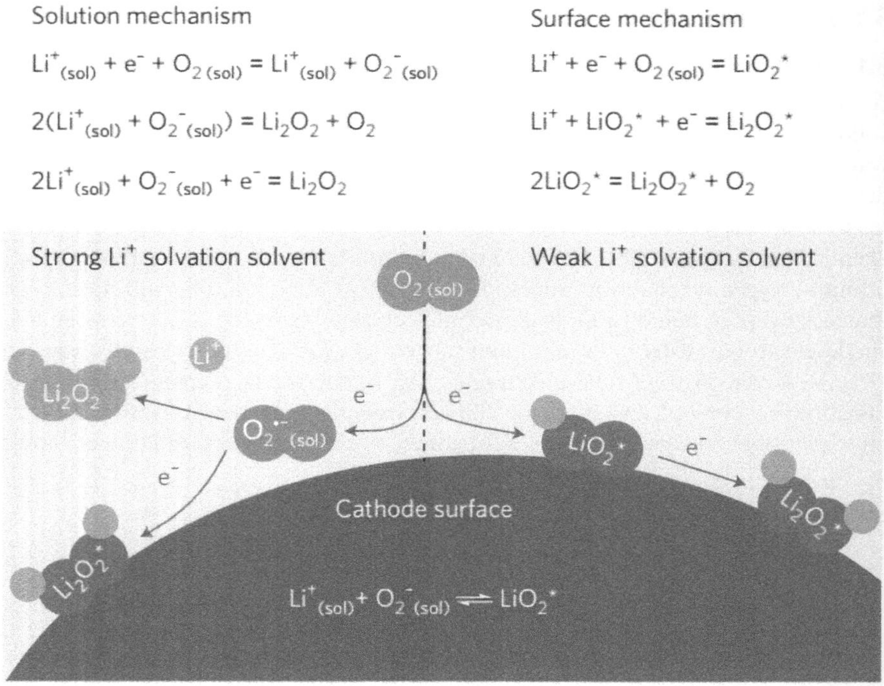

FIGURE 3.21 Schematic of the O_2 reduction mechanism in SSLOBs [155].

peroxide (Li_2O_2) via a surface-mediated process, in which LiO_2 accepts an additional Li^+ ion and electron (Equation 3.3) [156]. Nonetheless, it remains difficult to obtain conclusive evidence for this pathway in solid-state systems, and due to limitations in in situ characterization techniques, the evolution of intermediate species cannot be directly observed. Moreover, the charging reaction pathways and associated intermediates remain undefined and require thorough investigation. Additionally, the direct two-electron transfer mechanism (Equation 3.4), which has been proposed for liquid Li-O_2 batteries, has also been confirmed to occur in SSLOBs [157].

As described by Equation (3.19), during discharge, lithium metal at the anode is oxidized and releases electrons, reducing molecular oxygen via a single-electron transfer to form lithium superoxide (LiO_2).

$$O_2 + Li^+ + e^- \rightarrow LiO_2 \tag{3.19}$$

Through either the disproportionation reaction (Equation 3.20) or the acceptance of a second electron (Equation 3.21), both pathways ultimately lead to the formation of lithium peroxide (Li_2O_2) as the main reaction product.

$$2LiO_2 \rightarrow Li_2O_2 + O_2 \tag{3.20}$$

$$2LiO_2 + Li^+ + e^- \rightarrow Li_2O_2 \tag{3.21}$$

Alternatively, a direct two-electron transfer pathway (Equation 3.22) may also occur, resulting in the formation of lithium peroxide (Li_2O_2) as the primary reaction product.

$$2Li^+ + 2e^- + O_2 \rightarrow Li_2O_2 \tag{3.22}$$

The formation and decomposition of lithium peroxide (Li_2O_2) in the air cathode play a decisive role in determining the energy density and reversibility of SSLOBs. By tuning the electrode architecture, the Li_2O_2 formation pathway can be optimized to provide more active catalytic sites, improved gas diffusion channels, and increased available space. However, one of the most challenging issues is that the decomposition of Li_2O_2 during the charging process is thermodynamically unfavorable. Li_2O_2 has a standard Gibbs free energy of formation (ΔGf) of −649.45 kJ mol^{-1} and, as a wide-bandgap solid insulator, it exhibits limited electronic and ionic conductivity. This inherently slows down the reversibility of SSLOBs [158]. Fundamental studies on the oxidation of Li_2O_2 remain at an early stage, primarily due to the interference from uncontrollable side products such as lithium hydroxide (LiOH) and lithium carbonate (Li_2CO_3), presenting a major challenge. The reaction pathway for Li_2O_2 decomposition is still unclear, particularly regarding the exact location where the reaction takes place. Some studies suggest that both the cathode/Li_2O_2 interface and the electrolyte/Li_2O_2 interface could serve as active sites for Li_2O_2 formation and decomposition [159]. Electron microscopy studies indicate that the cathode/Li_2O_2 interface may be more favorable for Li_2O_2 decomposition. For instance, Zhong's group used in situ transmission electron microscopy (TEM) to observe that in SSLOBs, Li_2O_2 decomposition occurs at the interface between CNTs and Li_2O_2, rather than at the electrolyte/Li_2O_2 interface—a phenomenon consistent with observations in liquid lithium–oxygen systems [160]. They attributed this to the insulating nature of Li_2O_2, which has a bandgap greater than 2 eV, and whose poor electronic conductivity limits

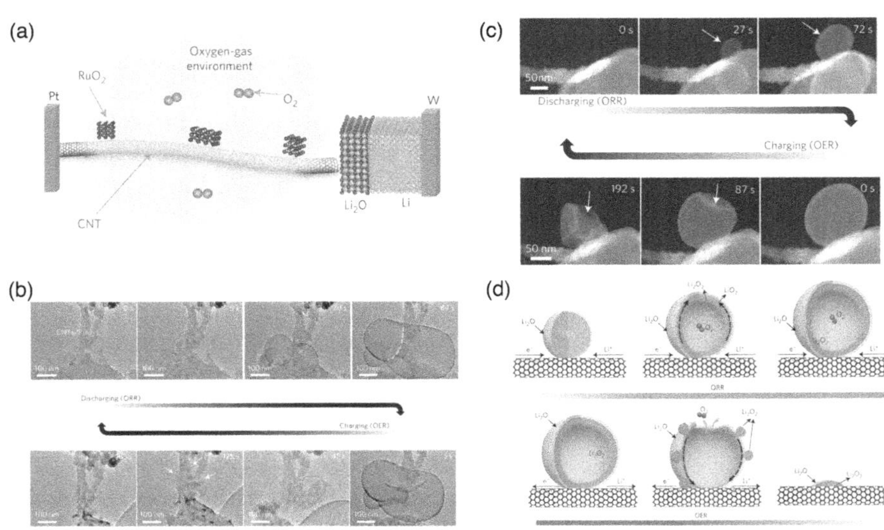

FIGURE 3.22 In situ observation of the evolution of the product in SSLOBs [163].

its oxidation kinetics at high potentials. Conversely, Zheng's research team observed via in situ environmental scanning electron microscopy that Li_2O_2 begins decomposing from the surface of the discharge products rather than from the contact region between the products and either the CNT cathode or the solid electrolyte. This surface-initiated decomposition was attributed to higher electronic conductivity at the Li_2O_2 surface compared to the bulk phase. Under vacuum conditions, the release of O_2 from surface regions further accelerates Li_2O_2 decomposition [161].

Guo's group constructed an SSLOB using a silver nanowire (AgNW) cathode on a gold-coated garnet ceramic electrolyte, with lithium metal as the anode. Thanks to the well-defined contact among Li^+, e^-, and O_2 on AgNWs, they confirmed a surface-mediated reaction pathway. Post-discharge analysis revealed two morphologies of Li_2O_2: a thin-film form around the smooth surfaces of AgNWs and a hollow-disc form located at AgNW junctions and the garnet/AgNW interface [162]. More recently, Luo's team investigated the reaction pathway over the full cycle of a SSLOB system [163]. The battery employed RuO_2-decorated CNTs as the cathode, lithium metal as the anode, and a Li_2O-coated lithium anode as the solid electrolyte (Figure 3.22). Using aberration-corrected environmental TEM, they achieved in situ phase identification during discharge and charge. During discharge, LiO_2 initially formed on CNTs and then disproportionated into Li_2O_2 and O_2, leading to a core-shell structure with Li_2O on the outer layer and Li_2O_2 within. During charging, Li_2O_2 lost a Li^+ and an electron to form LiO_2, which subsequently released O_2 and collapsed its shell structure until fully decomposed. This work established the electrochemical pathway in a specifically designed SSLOB system, providing insights into the reaction mechanisms. However, due to limitations in temporal resolution and the difficulty of capturing transient intermediates or disordered products, in situ TEM cannot fully resolve all aspects of the reaction process. There is a pressing need for advanced in situ characterization techniques, such as in situ X-ray absorption spectroscopy and SERS, to enable real-time observation of electrochemical evolution in SSLOBs.

Although significant progress has been made in understanding the mechanisms of SSLOBs, there remain considerable gaps in both mechanistic insight and reaction kinetics. SSLOBs have attracted widespread attention due to their unique energy characteristics, with their high energy density primarily attributed to the reversible oxygen conversion reactions occurring at the cathode. However, the detailed reaction pathways and the evolution of intermediate species involved in these processes are still not fully understood, leading to uncertainties in both discharge and charge mechanisms. This limited understanding hinders the development of effective strategies at the atomic scale to enhance the thermodynamic and kinetic performance of the cathode, thereby restricting the overall performance improvement of the battery. Currently, many theoretical studies oversimplify the OER by treating it as the direct reverse of the ORR. Such simplification may result in misleading conclusions, as OER does not necessarily proceed via the exact reverse pathway of ORR. This misinterpretation affects not only the accurate comprehension of the reaction mechanisms but also the design of experiments and the development of materials, potentially leading to suboptimal outcomes. Therefore, it is crucial to explore the complex relationship between OER and ORR in greater depth and to clarify their respective mechanisms. This understanding will be key to advancing the performance of SSLOBs. Given the structural and functional similarities between solid-state lithium-ion batteries and SSLOBs, the extensive research conducted on the former can provide valuable insights into the reaction mechanisms of the latter. To address the current bottlenecks faced by SSLOBs, there is an urgent need to develop innovative computational approaches and in situ characterization techniques. These advanced tools can probe reaction mechanisms at the atomic scale and effectively correlate the findings with device-level performance. By enabling real-time monitoring of the formation and transformation of reaction intermediates and combining these observations with state-of-the-art computational simulations, it is expected that the micro-kinetics of battery reactions can be revealed. This will provide a scientific basis for optimizing materials and designing high-performance batteries.

3.1.3.2 Air Electrode Optimization Strategies

In solid-state Li-O_2 batteries, the oxygen electrode reaction interface requires a large number of triple-phase boundaries (TPBs) as active sites, where lithium ions (Li^+), electrons (e^-), and oxygen (O_2) can effectively interact to facilitate electrochemical reactions. To increase the TPB density within the oxygen electrode, many researchers have introduced solid electrolytes directly into the cathode, forming composite-type oxygen electrode structures [164,165]. Currently, common methods for fabricating composite oxygen electrodes include ball milling and sintering, whereby solid-state electrolytes are physically mixed with oxygen electrode catalysts [166–168]. Zhang Xinbo and colleagues at the Changchun Institute of Applied Chemistry [169] developed a thermally induced phase separation technique to successfully introduce a porous plastic-crystal electrolyte in situ into the oxygen electrode, as shown in Figure 3.23a. This approach results in a highly continuous and abundant triple-phase interface within the cathode, enabling rapid Li^+/e^- transport while also facilitating efficient O_2 diffusion [169]. Another advantage of composite oxygen electrodes is the expansion of the active reaction region, as illustrated in Figure 3.23b. Using hot-pressing techniques, a SPE can be uniformly coated onto CNT surfaces. The CNT framework provides sufficient pathways for oxygen transport, creating an extended 3D vertical reaction zone within the cathode [170].

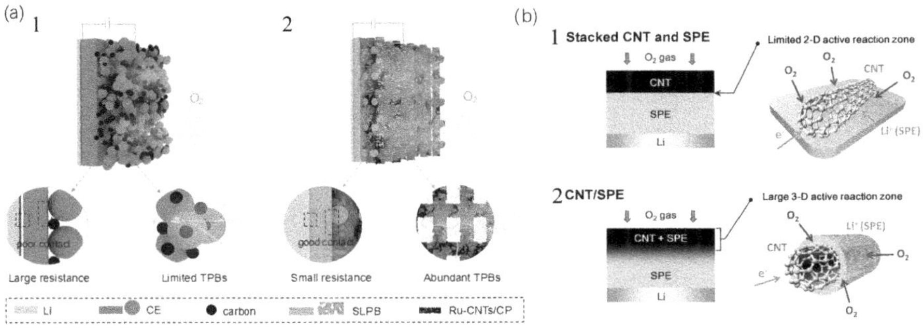

FIGURE 3.23 Structural diagram of the integrated oxygen electrode. (a) Schematic illustration of (1) the conventional SSLOB with CE and (2) the SSLOB with porosity-adjustable SLPB [169]. (b) Schematic illustrations of (1) limited 2-D active reaction zone in conventional carbon nanotube (CNT) and SPE sandwiched structure and (2) enlarged 3-D active reaction zone in 3-D CNT/SPE architecture [170].

Compared to traditional two-dimensional (2D) interfaces, this 3D configuration offers more abundant active sites for Li^+ and e^- transport, significantly increasing the mass transfer pathways at the oxygen electrode interface and thus enhancing the overall reaction kinetics. Despite the increased TPB area provided by integrated oxygen electrodes, one critical issue remains unresolved: the accommodation of solid discharge products. The continuous accumulation of solid-phase products within the cathode gradually reduces its porosity, which leads to a significant deviation between the practical discharge capacity and the theoretical capacity of the Li-O_2 battery.

By tailoring the local electronic environment of oxygen electrode materials, the electrical conductivity of catalysts can be enhanced, thereby accelerating the oxygen electrode reaction kinetics in Li-O_2 batteries [8]. However, in solid-state Li-O_2 batteries, the poor interfacial contact between the solid-state electrolyte and the oxygen electrode impedes the transport of both lithium ions (Li^+) and electrons (e^-) at the electrode–electrolyte interface. This results in increased polarization during charge and discharge processes. In recent years, the strategy of pre-introducing lithium into oxygen electrode catalysts—such as in lithium-transition metal oxides (e.g., Li–Ru–O compounds)—has been demonstrated as an effective way to improve ionic transport kinetics at the oxygen electrode interface. In a recent study [171], as illustrated in Figure 3.24, it was shown that a balanced lithium-ion–electron microenvironment was established within the oxygen electrode. This microenvironment, supported by multidimensional diffusion pathways, simultaneously enhanced the driving force for both Li^+ and e^- transport, thus accelerating the electrochemical formation and decomposition of LiO_2 during the discharge process.

In addition, it is essential to fully consider the limitations imposed by the continuous accumulation of discharge products on electron and ion transport at the electrode reaction interface, which can ultimately prevent the discharge reaction from proceeding further [172]. As illustrated in Figure 3.25, during the charging

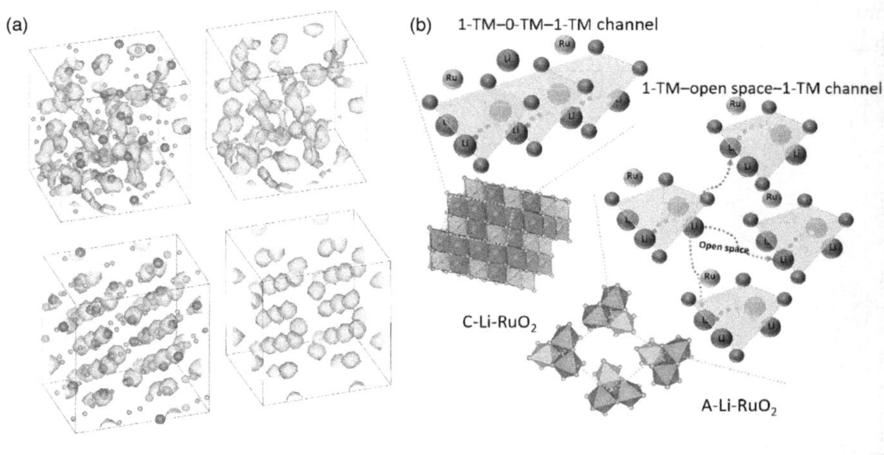

FIGURE 3.24 Local electronic environment of oxygen electrode materials [171]. (a) Balanced lithium-ion–electron microenvironment. (b) Pre-introducing lithium into oxygen electrode catalysts, lithium-transition metal oxides.

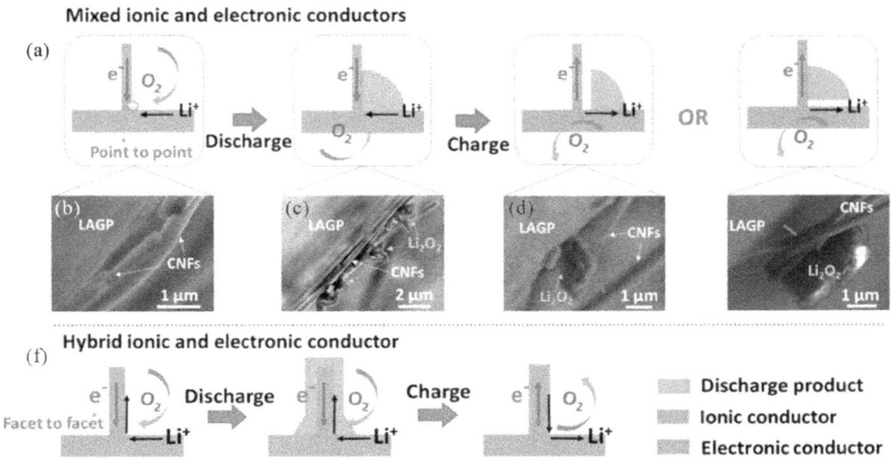

FIGURE 3.25 Illustration and corresponding SEM images to show the discharge–charge process at the interface between ionic and electronic conductors [173]. (a) Bi-boundary and tri-boundary situations. (b) SEM of product corresponding Scenario 1 in image (a). (c) SEM of product corresponding Scenario 2 in image (a). (d) SEM of product corresponding Scenario 3 in image (a). (e) SEM of product corresponding Scenario 4 in image (a). (f) Hybrid ionic and electronic conductor realizing triple boundary.

process, the decomposition of discharge products ceases once they lose contact with either the electronic conductor or the ionic conductor [160], directly confirming the critical role of electron and ion conduction in the formation and decomposition of discharge products. To address this issue, Xueliang Sun and colleagues from Western University in Canada [173] designed and fabricated an ion–electron mixed conductive framework by coating $LiTaO_3$ on nitrogen-doped carbon nanotubes (NCNT), forming a NCNT@$LiTaO_3$ composite structure. This design enabled uniform ion/electron transport throughout the electrode, significantly reduced interfacial resistance, and accelerated reaction kinetics. As a result, the CE improved dramatically from 38.6% to 80.8%.

During the oxygen electrode reaction in Li-O_2 batteries, various reactive intermediates are involved, including ions, electrons, holes, paramagnetic oxygen species, and excitons—species that are highly sensitive to high-energy photons and other external stimuli. External fields such as photoelectric fields, temperature gradients, mechanical stress, ultrasound, microwaves, and magnetic fields have been shown to effectively modulate mass transport within semiconductor catalysts and reduce concentration polarization. In recent years, photo-assisted Li-O_2 batteries have been extensively investigated as a promising strategy to improve the reaction kinetics of the oxygen electrode [174]. However, the photo-assisted oxygen electrode reactions in Li–O_2 batteries are highly complex, involving intertwined mechanisms from photophysics, photochemistry, electrocatalysis, and redox chemistry [174]. As shown in Figure 3.26a, a light-assisted self-powered Li-O_2 battery system has been developed, integrating a flexible solar cell with a Li-O_2 battery [175]. In this system, TiN@TiO_2

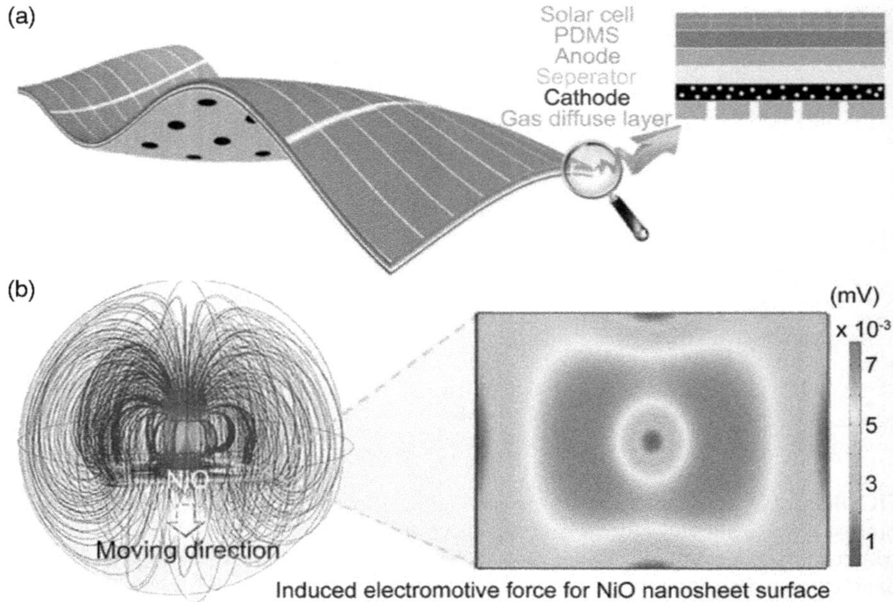

FIGURE 3.26 Schematic diagram of external assisted lithium-oxygen battery [176]. (a) A flexible solar cell with a Li-O_2 battery. (b) A Li-O_2 battery utilizing co-assisted photo–magnetic fields.

is employed as the photo-assisted oxygen electrode catalyst. Under light illumination, the battery exhibited a low charging voltage of 2.94 V (below the standard potential E°(O_2/Li_2O_2) = 2.96 V vs. Li^+/Li), and it can be recharged using solar energy. Professor Xu's group at Jilin University [176] further reported a Li-O_2 battery utilizing co-assisted photo–magnetic fields, as illustrated in Figure 3.26b. They designed a 3D porous nickel oxide nanosheet catalyst (NiO/FNi) supported on nickel foam as the oxygen electrode. Under the synergistic influence of a light field (Xe lamp) and a magnetic field (5 mT), the NiO/FNi-based Li-O_2 battery demonstrated significantly enhanced ORR and OER catalytic activity, as well as improved photoelectric conversion capabilities. At a current density of 0.01 mA cm^{-2}, the battery achieved a low charging voltage of 2.73 V and a high energy efficiency of 96.7%. This work clearly confirms that the synergistic effects of multiple external fields can greatly enhance the overall performance of Li-O_2 batteries.

REFERENCES

1. Lin D, Liu Y, Cui Y. Reviving the lithium metal anode for high-energy batteries. *Nature Nanotechnology*, 2017, 12(3): 194–206.
2. Ue M, Uosaki K. Recent progress in liquid electrolytes for lithium metal batteries. *Current Opinion in Electrochemistry*, 2019, 17: 106–113.

3. Yi J, Guo S, He P, et al. Status and prospects of polymer electrolytes for solid-state $Li-O_2$ (air) batteries. *Energy & Environmental Science*, 2017, 10: 860–884.
4. Li Z, Liu YE, Weng S, et al. Oxygen-permeable and moisture-proof membrane for stable $Li-O_2$/air batteries in humid working environment. *Energy Storage Materials*, 2023, 58: 94–100.
5. Abraham K, Jiang Z. A polymer electrolyte-based rechargeable lithium/oxygen battery. *Journal of The Electrochemical Society*, 1996, 143(1): 1–5.
6. Girishkumar G, McCloskey B, Luntz AC, et al. Lithium-air battery: Promise and challenges. *The Journal of Physical Chemistry Letters*, 2010, 1(14): 2193–2203.
7. Ogasawara T, Debart A, Holzapfel M, et al. Rechargeable Li_2O_2 electrode for lithium batteries. *Journal of the American Chemical Society*, 2006, 128: 1390–1393.
8. Lyu Z, Zhou Y, Dai W, et al. Recent advances in understanding of the mechanism and control of Li_2O_2 formation in aprotic $Li-O_2$ batteries. *Chemical Society Reviews*, 2017, 46: 6046–6072.
9. Johnson L, Li C, Liu Z, et al. The role of LiO_2 solubility in O_2 reduction in aprotic solvents and its consequences for $Li-O_2$ batteries. *Nature Chemistry*, 2014, 6: 1091–1099.
10. Sharon D, Etacheri V, Garsuch A, et al. On the challenge of electrolyte solutions for Li-air batteries: Monitoring oxygen reduction and related reactions in polyether solutions by spectroscopy and EQCM. *Journal of Physical Chemistry Letters*, 2013, 4: 127–131.
11. Kwabi DG, Tulodziecki M, Pour N, et al. Controlling solution mediated reaction mechanisms of oxygen reduction using potential and solvent for aprotic lithium-oxygen batteries. *Journal of Physical Chemistry Letters*, 2016, 7: 1204–1212.
12. Koellisch-Mirbach A, Lohrmann T, Reinsberg PH, et al. The mechanism of Li_2O_2-film formation and reoxidation-influence of electrode roughness and single crystal surface structure. *Journal of Electroanalytical Chemistry*, 2020, 875: 114560.
13. Matsuda S, Kubo Y, Uosaki K, et al. Potassium ions promote solution route Li_2O_2 formation in the positive electrode reaction of $Li-O_2$ batteries. *Journal of Physical Chemistry Letters*, 2017, 8: 1142–1146.
14. Aetukuri NB, Mccloskey BD, Garcia JM, et al. Solvating additives drive solution-mediated electrochemistry and enhance toroid growth in non-aqueous $Li-O_2$ batteries. *Nature Chemistry*, 2015, 7: 50–56.
15. Prehal C, Mondal S, Lovicar L, et al. Exclusive solution discharge in $Li-O_2$ batteries. *ACS Energy Letters*, 2022, 7: 3112–3119.
16. Chi X, Li M, Di J, et al. A highly stable and flexible zeolite electrolyte solid-state Li-air battery. *Nature*, 2021, 592: 551–557.
17. Kondori A, Esmaeilirad M, Harzandi AM, et al. A room temperature rechargeable Li_2O-based lithium-air battery enabled by a solid electrolyte. *Science*, 2023, 379: 499–505.
18. Wang C, Liu H, Liang Y, et al. Molecular-level designed polymer electrolyte for high-voltage lithium-metal solid-state batteries. *Advanced Functional Materials*, 2022, 33: 2209828.
19. Kumar B, Kumar J. Cathodes for solid-state lithium-oxygen cells: Roles of NASICON glass-ceramics. *Journal of The Electrochemical Society*, 2010, 157: A611.
20. Visco SJ, Nimon E, Katz B, et al. Lithium/air semi-fuel cells: high energy density batteries based on lithium metal electrodes. *The 12th International Meeting on Lithium Batteries Abstracts*. Nara, Japan, 2004.
21. Puech L, Cantau C, Vinatier P, et al. Elaboration and characterization of a free standing LiSICON membrane for aqueous lithium-air battery. *Journal of Power Sources*, 2012, 214: 330–336.
22. Shimonishi Y, Zhang T, Johnson P, et al. A study on lithium/air secondary batteries—Stability of NASICON-type glass ceramics in acid solutions. *Journal of Power Sources*, 2010, 195: 6187–6191.

23. Zheng JP, Andrei P, Hendrickson M, et al. The theoretical energy densities of dual-electrolytes rechargeable Li-air and Li-air flow batteries. *Journal of The Electrochemical Society*, 2011, 158: A43–A46.
24. Liu T, Liu Z, Kim G, et al. Understanding LiOH chemistry in a ruthenium catalyzed Li-O_2 battery. *Angewandte Chemie International Edition*, 2017, 56: 16057–16062.
25. Feng N, Mu X, Zhang X, et al. Intensive study on the catalytical behavior of N-methylphenothiazine as a soluble mediator to oxidize the Li_2O_2 cathode of the Li-O_2 battery. *ACS Applied Materials & Interfaces*, 2017, 9: 3733–3739.
26. Huang L, Fu H, Duan J, et al. Negating Li+ transfer barrier at solid-liquid electrolyte interface in hybrid batteries. *Chem*, 2022, 8: 1928–1943.
27. Gu Z, Xin X, Yang J, et al. Bilayer nasicon/polymer hybrid electrolyte for stable solid-state Li-O_2 batteries. *ACS Applied Energy Materials*, 2022, 5: 9149–9157.
28. Lei J, Gao Z, Tang L, et al. Coupling water-proof Li anodes with LiOH-based cathodes enables highly rechargeable lithium-air batteries operating in ambient air. *Advanced Science*, 2022, 9: 2103760.
29. Liu T, Kim G, Jónsson E, et al. Understanding LiOH formation in a Li-O_2 battery with LiI and H_2O additives. *ACS Catalysis*, 2018, 9: 66–77.
30. Ko Y, Kim HI, Cho SJ, et al. Liquid-based janus electrolyte for sustainable redox mediation in lithium-oxygen batteries. *Advanced Energy Materials*, 2021, 11: 2102096.
31. Hummelshøj JS, Blomqvist J, Datta S, et al. Communications: Elementary oxygen electrode reactions in the aprotic Li-air battery. *The Journal of Chemical Physics*, 2010, 132: 071101.
32. McCloskey BD, Scheffler R, Speidel A, et al. On the mechanism of non-aqueous Li-O_2 electrochemistry on C and its kinetic overpotentials some implications for Li- air batteries. *The Journal of Physical Chemistry C*, 2012, 116(45): 23897–23905.
33. Laoire CO, Mukerjee S, Abraham KM, et al. Elucidating the mechanism of oxygen reduction for lithium-air battery applications. *Journal of Physical Chemistry C*, 2009, 113(46): 20127–20134.
34. Peng ZQ, Freunberger SA, Hardwick LJ, et al. Oxygen reactions in a non-aqueous Li+ electrolyte. *Angewandte Chemie-International Edition*, 2011, 50(28): 6351–6355.
35. Wang Y, Lai N, Lu Y, et al. A solvent-controlled oxidation mechanism of Li_2O_2 in lithium-oxygen batteries. *Joule*, 2018, 2(11): 2364–2380.
36. Lu Y-C, Shao-Horn Y. Probing the reaction kinetics of the charge reactions of non-aqueous Li-O_2 batteries. *The Journal of Physical Chemistry Letters*, 2012, 4(1): 93–99.
37. Ganapathy S, Adams BD, Stenou G, et al. Nature of Li_2O_2 oxidation in a LiO_2 battery revealed by operando X-ray diffraction. *Journal of the American Chemical Society*, 2014, 136(46): 16335–16344.
38. Freunberger SA, Chen Y, Peng Z, et al. Reactions in the rechargeable lithium-O_2 battery with alkyl carbonate electrolytes. *Journal of the American Chemical Society*, 2011, 133(20): 8040–8047.
39. Sharon D, Afri M, Noked M, et al. Oxidation of dimethyl sulfoxide solutions by electrochemical reduction of oxygen. *The Journal of Physical Chemistry Letters*, 2013, 4(18): 3115–3119.
40. Ottakam Thotiyl MM, Freunberger SA, Peng Z, et al. The carbon electrode in nonaqueous Li-O_2 cells. *Journal of the American Chemical Society*, 2013, 135(1): 494–500.
41. Itkis DM, Semenenko DA, Kataev EY, et al. Reactivity of carbon in lithium-oxygen battery positive electrodes. *Nano Letters*, 2013, 13(10): 4697–4701.
42. McCloskey BD, Speidel A, Scheffler R, et al. Twin problems of interfacial carbonate formation in nonaqueous Li-O_2 batteries. *The Journal of Physical Chemistry Letters*, 2012, 3(8): 997–1001.
43. Schwenke KU, Metzger M, Restle T, et al. The influence of water and protons on Li_2O_2 crystal growth in aprotic Li-O_2 cells. *Journal of The Electrochemical Society*, 2015, 162(4): A573–A584.

44. Takechi K, Shiga T, Asaoka T. A Li-O_2/CO_2 battery. *Chemical Communications*, 2011, 47(12): 3463–3465.
45. Lim H-K, Lim H-D, Park K-Y, et al. Toward a lithium-air battery: The effect of CO_2 on the chemistry of a lithium-oxygen cell. *Journal of the American Chemical Society*, 2013, 135(26): 9733–9742.
46. Gowda SR, Brunet A, Wallraff GM, et al. Implications of CO_2 contamination in rechargeable nonaqueous Li-O_2 batteries. *Journal of Physical Chemistry Letters*, 2013, 4(2): 276–279.
47. Qiao Y, Yi J, Wu S, et al. Li-CO_2 electrochemistry: A new strategy for CO_2 fixation and energy storage. *Joule*, 2017, 1(2): 359–370.
48. Zhao Z, Su Y, Peng Z. Probing lithium carbonate formation in trace-O_2-assisted aprotic Li-CO_2 batteries using in situ surface-enhanced Raman spectroscopy. *Journal of Physical Chemistry Letters*, 2019, 10(3): 322–328.
49. Li F, Wu S, Li D, et al. The water catalysis at oxygen cathodes of lithium-oxygen cells. *Nature Communications*, 2015, 6: 7843.
50. Yang S, He P, Zhou H. Exploring the electrochemical reaction mechanism of carbonate oxidation in Li-air/CO_2 battery through tracing missing oxygen. *Energy and Environmental Science*, 2016, 9(5): 1650–1654.
51. Qiao Y, Li J, Guo S, et al. Li_2CO_3-Free Li-O_2/CO_2 battery with peroxide discharge product. *Energy and Environmental Science*, 2018, 11 (5): 1211–1217.
52. Lu YC, Shao-Horn Y. Probing the reaction kinetics of the charge reactions of nonaqueous Li-O_2 batteries. *The Journal of Physical Chemistry Letters*, 2013, 4: 93.
53. Song S, Qin X, Ruan Y, et al. Enhanced performance of solid-state lithium-air batteries with continuous 3D garnet network added composite polymer electrolyte. *Journal of Power Sources*, 2020, 461: 228146.
54. Tran C, Yang X-Q, Qu D. Investigation of the gas-diffusion-electrode used as lithium/air cathode in non-aqueous electrolyte and the importance of carbon material porosity. *Journal of Power Sources*, 2010, 195: 2057.
55. Li Y, Wang J, Li X, et al. Nitrogen-doped carbon nanotubes as cathode for lithium-air batteries. *Electrochemistry Communications*, 2011, 13: 668.
56. Xiao J, Wang D, Xu W, et al. Optimization of air electrode for Li/air batteries. *Journal of The Electrochemical Society*, 2010, 157: A487.
57. Guo Z, Zhou D, Dong X, et al. Ordered hierarchical mesoporous/macroporous carbon: A high-performance catalyst for rechargeable Li-O_2 batteries. *Advanced Materials*, 2013, 25: 5668.
58. Yang J, Zhai D, Wang H-H, et al. Evidence for lithium superoxide-like species in the discharge product of a Li-O_2 battery. *Physical Chemistry Chemical Physics*, 2013, 15: 3764.
59. Zhai D, Wang H-H, Yang J, et al. Disproportionation in Li-O_2 batteries based on a large surface area carbon cathode. *Journal of the American Chemical Society*, 2013, 135: 15364.
60. Lim H, Park K, Song H, et al. Enhanced power and rechargeability of a Li-O_2 battery based on a hierarchical-fibril CNT electrode. *Advanced Materials*, 2013, 25: 1348.
61. Shui J, Du F, Xue C, et al. Vertically aligned N-doped coral-like carbon fiber arrays as efficient air electrodes for high-performance nonaqueous Li-O_2 batteries. *ACS Nano*, 2014, 8: 3015.
62. Mitchell RR, Gallant BM, Thompson CV, et al. All-carbon-nanofiber electrodes for high-energy rechargeable LiO_2 batteries. *Energy & Environmental Science*, 2011, 4: 2952.
63. Han J, Guo X, Ito Y, et al. Effect of chemical doping on cathodic performance of bicontinuous nanoporous graphene for Li-O_2 batteries. *Advanced Energy Materials*, 2016, 6: 1501870.

64. Li Y, Wang L, He X, et al. Boron-doped Ketjenblack based high performances cathode for rechargeable Li-O_2 batteries. *Journal of Energy Chemistry*, 2016, 25: 131.
65. Qian Z, Guo R, Ma Y, et al. Se-doped carbon as highly stable cathode material for high energy nonaqueous Li-O_2 batteries. *Chemical Engineering Science*, 2020, 214: 115413.
66. Yang X, Xu J, Chang Z, et al. Blood-capillary-inspired, free-standing, flexible, and low-cost super-hydrophobic N-CNTs@ SS cathodes for high-capacity, high-rate, and stable Li-air batteries. *Advanced Energy Materials*, 2018, 8: 1702242.
67. Xiao J, Mei D, Li X, et al. Hierarchically porous graphene as a lithium-air battery electrode. *Nano Letters*, 2011, 11: 5071.
68. Qian Z, Sun B, Du L, et al. Insights into the role of oxygen functional groups and defects in the rechargeable nonaqueous Li-O_2 batteries. *Electrochimica Acta*, 2018, 292: 838.
69. Lu YC, Gasteiger HA, Parent MC, et al. The influence of catalysts on discharge and charge voltages of rechargeable Li-oxygen batteries. *Electrochemical and Solid-State Letters*, 2010, 13: A69.
70. Lu YC, Gasteiger HA, Shao-Horn Y. Catalytic activity trends of oxygen reduction reaction for nonaqueous Li-air batteries. *Journal of the American Chemical Society*, 2011, 133: 19048.
71. Asadi M, Sayahpour B, Abbasi P, et al. A lithium-oxygen battery with a long cycle life in an air-like atmosphere. *Nature*, 2018, 555: 502.
72. Qiao Y, Jiang K, Deng H, Zhou H. A high-energy-density and long-life lithium-ion battery via reversible oxide–peroxide conversion. *Nature Catalysis*, 2019, 2: 1035.
73. Lu YC, Xu Z, Gasteiger HA, et al. Platinum–gold nanoparticles: A highly active bifunctional electrocatalyst for rechargeable lithium-air batteries. *Journal of the American Chemical Society*, 2010, 132: 12170.
74. Choi R, Jung J, Kim G, et al. Ultra-low overpotential and high rate capability in Li-O_2 batteries through surface atom arrangement of PdCu nanocatalysts. *Energy & Environmental Science*, 2014, 7: 1362.
75. Zhou Y, Yin K, Gu Q, et al. Lewis-acidic PtIr multipods enable high-performance Li-O_2 batteries. *Angewandte Chemie International Edition*, 2021, 60: 26592.
76. Wang G, Tu F, Xie J, et al. High-performance Li-O_2 batteries with controlled Li_2O_2 growth in graphene/Au-nanoparticles/Au-nanosheets sandwich. *Advanced Science*, 2016, 3: 1500339.
77. Lu J, Cheng L, Lau KC, et al. Effect of the size-selective silver clusters on lithium peroxide morphology in lithium-oxygen batteries. *Nature Communications*, 2014, 5: 4895.
78. Wang P, Ren Y, Wang R, et al. Atomically dispersed cobalt catalyst anchored on nitrogen-doped carbon nanosheets for lithium-oxygen batteries. *Nature Communications*, 2020, 11: 1576.
79. Yilmaz E, Yogi C, Yamanaka K, et al. Promoting formation of noncrystalline Li_2O_2 in the Li-O_2 battery with RuO_2 nanoparticles. *Nano Letters*, 2013, 13: 4679.
80. Lu J, Jung Lee Y, Luo X, et al. A lithium-oxygen battery based on lithium superoxide. *Nature*, 2016, 529: 377.
81. Débart A, Bao J, Armstrong G, et al. An O_2 cathode for rechargeable lithium batteries: The effect of a catalyst. *Journal of Power Sources*, 2007, 174: 1177.
82. Gao R, Zhu J, Xiao X, et al. Facet-dependent electrocatalytic performance of Co_3O_4 for rechargeable Li-O_2 battery. *The Journal of Physical Chemistry C*, 2015, 119: 4516.
83. Gao R, Liu L, Hu Z, et al. The role of oxygen vacancies in improving the performance of CoO as a bifunctional cathode catalyst for rechargeable Li-O_2 batteries. *Journal of Materials Chemistry A*, 2015, 3: 17598.
84. Adpakpang K, Oh SM, Agyeman DA, et al. Oxygen evolution reaction: Holey 2D nanosheets of low-valent manganese oxides with an excellent oxygen catalytic activity and a high functionality as a catalyst for Li-O_2 batteries. *Advanced Functional Materials*, 2018, 28: 1707106.

85. Débart A, Paterson AJ, Bao J, et al. α-MnO_2 nanowires: A catalyst for the O_2 electrode in rechargeable lithium batteries. *Angewandte Chemie International Edition*, 2008, 47: 4521.
86. Zhang P, Zhang S, He M, et al. Realizing the embedded growth of large Li_2O_2 aggregations by matching different metal oxides for high-capacity and high-rate lithium oxygen batteries. *Advanced Science*, 2017, 4: 1700172.
87. Wu H, Sun W, Wang Y, et al. Facile synthesis of hierarchical porous three-dimensional free-standing $MnCo_2O_4$ cathodes for long-life Li-O_2 batteries. *ACS Applied Materials & Interfaces*, 2017, 9: 12355.
88. Liu Y, Cao LJ, Cao CW, et al. Facile synthesis of spinel $CuCo_2O_4$ nanocrystals as high-performance cathode catalysts for rechargeable Li-air batteries. *Chemical Communications*, 2014, 50: 14635.
89. Xu J, Xu D, Wang Z, et al. Synthesis of perovskite-based porous $La_{0.75}Sr_{0.25}MnO_3$ nanotubes as a highly efficient electrocatalyst for rechargeable lithium-oxygen batteries. *Angewandte Chemie International Edition*, 2013, 52: 3887.
90. Read J. Ether-based electrolytes for the lithium/oxygen organic electrolyte battery. *Journal of The Electrochemical Society*, 2006, 153(1): A96–A100.
91. Zhang J, Xiao JA, Wang D, et al. Crown ethers in nonaqueous electrolytes for lithium/air batteries. *Electrochemical and Solid-State Letters*, 2010, 14(3): A48.
92. Abraham KM, Cormac ÓL, Edward J, et al. Rechargeable lithium/TEGDME- LiPF6/O_2 battery. *Journal of The Electrochemical Society*, 2011, 158(3): A302.
93. Jung H-G, Hassoun J, Park J-B, et al. An improved high-performance lithium-air battery. *Nature Chemistry*, 2012, 4: 579–585.
94. Mourad E, Petit YK, Spezia R, et al. Singlet oxygen from cation driven superoxide disproportionation and consequences for aprotic metal-O_2 batteries. *Energy and Environmental Science*, 2019, 12(8): 2559–2568.
95. Zhao Y, Liu X, Song X, et al. Decomposition pathway and stabilization of ether-based electrolytes in the discharge process of Li-O_2 battery. *Journal of Energy Chemistry*, 2022, 69: 516–523.
96. Nazar L, Adams B, Black R, et al. Towards a stable organic electrolyte for the lithium oxygen battery. *Advanced Energy Materials*, 2015, 5(1): 1400867.
97. Huang Y, Huang Z, Zheng H, et al. A stable lithium-oxygen battery electrolyte based on fully methylated cyclic ether. *Angewandte Chemie-International Edition*, 2019, 58(8): 2345–2349.
98. Zhang J, Liu B, Xu W, et al. Enhanced cycling stability of rechargeable Li-O_2 batteries using high-concentration electrolytes. *Advanced Functional Materials*, 2016, 26(4): 605–613.
99. Xie H, Zhao Q, Zhang Y, et al. Binary mixtures of highly concentrated tetraglyme and hydrofluoroether as a stable and nonflammable electrolyte for Li-O_2 batteries. *ACS Applied Materials & Interfaces*, 2018, 10(31): 26312–26319.
100. Zhang J, Kwak W, Chae S, et al. Optimized electrolyte with high electrochemical stability and oxygen solubility for lithium-oxygen and lithium-air batteries. *ACS Energy Letters*, 2020, 5(7): 2182–2190.
101. Xu D, Wang Z, Xu J, et al. Novel DMSO-based electrolyte for high performance rechargeable Li-O_2 batteries. *Chemical Communications*, 2012, 48(55): 6948–6950.
102. Peng Z, Freunberger SA, Chen Y, et al. A reversible and higher-rate Li-O_2 battery. *Science*, 2012, 337(6094): 563–566.
103. Ottakam Thotiyl MM, Freunberger SA, Peng Z, et al. A stable cathode for the aprotic Li-O_2 battery. *Nature Materials*, 2013, 12(11): 1050–1056.
104. Sun B, Huang X, Chen S, et al. An optimized $LiNO_3$/DMSO electrolyte for high-performance rechargeable Li-O_2 batteries. *RSC Advances*, 2014, 4(22): 11115–11120.

105. Liu B, Xu W, Yan P, et al. Stabilization of Li metal anode in DMSO-based electrolytes via optimization of salt–solvent coordination for Li-O_2 batteries. *Advanced Energy Materials*, 2017, 7(14): 1602605.
106. Zhang T, Liao K, He P, et al. A self-defense redox mediator for efficient lithium-O_2 batteries. *Energy & Environmental Science*, 2016, 9(3): 1024–1030.
107. Yoo E, Zhou H. Enhanced cycle stability of rechargeable Li-O_2 batteries by the synergy effect of a LiF protective layer on the Li and DMTFA additive. *ACS Applied Materials & Interfaces*, 2017, 9(25): 21307–21313.
108. Luo Z, Zhu G, Guo L, et al. Improving the cyclability and capacity of Li-O_2 batteries via low rate pre-activation. *Chemical Communications*, 2019, 55(14): 2094–2097.
109. Guo H, Hou G, Guo J, et al. Enhanced cycling performance of Li-O_2 battery by using a Li_3PO_4-protected lithium anode in DMSO-based electrolyte. *ACS Applied Energy Materials*, 2018, 1(10): 5511–5517.
110. Ahn SM, Suk J, Kim DY, et al. High-performance lithium-oxygen battery electrolyte derived from optimum combination of solvent and lithium salt. *Advanced Science*, 2017, 4(10): 1700235.
111. Zhong H, Wang J, Mu X, et al. Double-phase electrolyte enabling the long-lasting redox mediation in Li-O_2 batteries. *Energy Storage Materials*, 2024, 67: 103292.
112. Kwabi DG, Batcho TP, Amanchukwu CV, et al. Chemical instability of dimethyl sulfoxide in lithium-air batteries. *The Journal of Physical Chemistry Letters*, 2014, 5(16): 2850–2856.
113. McCloskey BD, Bethune DS, Shelby RM, et al. Solvents' critical role in nonaqueous lithium-oxygen battery electrochemistry. *The Journal of Physical Chemistry Letters*, 2011, 2 (10): 1161–1166.
114. Yang F, Zhang L, Xu J, Liu Q, et al. Progress of cathode material and electrolyte in non-aqueous li-air battery. *Chinese Journal of Inorganic Chemistry*, 2013, 29(8): 1563–1573.
115. Chen Y, Freunberger SA, Peng Z, et al. Li-O_2 battery with a dimethylformamide electrolyte. *Journal of the American Chemical Society*, 2012, 134(18): 7952–7957.
116. Walker W, Giordani V, Uddin J, et al. A rechargeable Li-O_2 battery using a lithium nitrate/n,n-dimethylacetamide electrolyte. *Journal of the American Chemical Society*, 2013, 135(6): 2076–2079.
117. Akihiro N, Shota A, Fumisato O, Morihiro S, Highly porous carbon nanotube air-electrode combined with low-viscosity amide-based electrolyte enabling high-power, high-energy lithium-air batteries. *Journal of Power Sources*, 2025, 633: 236426,
118. Zhou B, Guo L, Zhang Y, et al. A high-performance Li-O_2 battery with a strongly solvating hexamethylphosphoramide electrolyte and a liPON-protected lithium anode. *Advanced Materials*, 2017, 29(30): 1701568.
119. Kim D, Lee D, Ahn S, et al. Autoxidation in amide-based electrolyte and its suppression for enhanced oxygen efficiency and cycle performance in non-aqueous lithium oxygen battery. *Journal of Power Sources*, 2017, 347: 186–192.
120. Qian Z. Preparation and mechanism study of cathode catalysts and electrolytes for lithium air batteries. *Harbin Institute of Technology*, 2021, 2021: 82–112.
121. Kiho N, Mizuki T, Terumi G, et al. Fluorinated amide-based electrolytes induce a sustained low-charging voltage plateau under conditions verifying the feasibility of achieving 500 Wh kg^{-1} Class Li-O_2 batteries. *Applied Materials & Interfaces,* 2024, 16(35): 46259–46269.
122. Freunberger SA, Chen Y, Drewett NE, et al. The lithium-oxygen battery with ether-based electrolytes. *Angewandte Chemie International Edition*, 2011, 50(37): 8609–8613.
123. Wan H, Bai Q, Peng Z, et al. A high power Li-air battery enabled by a fluorocarbon additive. *Journal of Materials Chemistry*, 2017, 5(47): 24617–24620.

124. Hu X, Wang J, Li Z, et al. MCNTs@MnO_2 nanocomposite cathode integrated with soluble O_2-carrier co-salen in electrolyte for high-performance Li-air batteries. *Nano Letters*, 2017, 17(3): 2073–2078.
125. Samajdar RN, George SM, Bhattacharyya AJ. Using hemoglobin as a performance enhancer in rechargeable lithium-oxygen batteries. *Journal of Physical Chemistry C*, 2019, 123(38): 23433–23438.
126. Matsuda S, Kubo Y, Uosaki K, et al. Improved Energy capacity of aprotic Li-O_2 batteries by forming Cl-incorporated Li_2O_2 as the Discharge Product. *Journal of Physical Chemistry C*, 2016, 120(25): 13360–13365.
127. Lyu Z, Wang T, Guo R, et al. Promoting defective-Li_2O_2 formation via Na doping for Li-O_2 batteries with low charge overpotentials. *Journal of Materials Chemistry*, 2019, 7(17): 10389–10396.
128. Matsuda S, Uosaki K, Nakanishi S. Enhanced energy capacity of lithium-oxygen batteries with ionic liquid electrolytes by addition of ammonium ions. *Journal of Power Sources*, 2017, 356: 12–17.
129. Lim H, Lee B, Zheng Y, et al. Rational design of redox mediators for advanced Li-O_2 batteries. *Nature Energy*, 2016, 1(6): 16066.
130. Chen Y, Freunberger SA, Peng Z, et al. Charging a Li-O_2 battery using a redox mediator. *Nature Chemistry*, 2013, 5(6):489–494.
131. Bergner BJ, Schürmann A, Peppler K, et al. TEMPO: A mobile catalyst for rechargeable Li-O_2 batteries. *Journal of the American Chemical Society*, 2014, 136(42): 15054–15064.
132. Lee SH, Park J, Lim H, et al. An advanced separator for Li-O_2 batteries: Maximizing the effect of redox mediators. *Advanced Energy Materials*, 2017, 7(18): 1602417.
133. Lim HD, Song H, Kim J, et al. Superior rechargeability and efficiency of lithium-oxygen batteries: Hierarchical air electrode architecture combined with a soluble catalyst. *Angewandte Chemie International Edition*, 2014, 53(15): 3926–3931.
134. Liu T, Leskes M, Yu W, et al. Cycling Li-O_2 batteries via LiOH formation and decomposition. *Science*, 2015, 350(6260): 530–533.
135. Kwak W, Hirshberg D, Sharon D, et al. Li-O_2 cells with LiBr as an electrolyte and a redox mediator. *Energy and Environmental Science*, 2016, 9(7): 2334–2345.
136. Gao X, Chen Y, Johnson L, et al. Promoting solution phase discharge in Li-O_2 batteries containing weakly solvating electrolyte solutions. *Nature Materials*, 2016, 15(8): 882–888.
137. Zhang Y, Wang L, Zhang X, et al. High-capacity and high-rate discharging of a coenzyme Q10-catalyzed Li-O_2 battery. *Advanced Materials*, 2018, 30(5): 1705571.
138. Zhang X, Li Y, Sun Y, et al. Inverting the triiodide formation reaction by the synergy between strong electrolyte solvation and cathode adsorption for lithium-oxygen batteries. *Angewandte Chemie International Edition*, 2019, 58(51): 18394–18398.
139. Qiao Y, Wu S, Yi J, et al. From O_2– to HO_2–: Reducing by-products and overpotential in Li-O_2 batteries by water addition. *Angewandte Chemie International Edition*, 2017, 56(18): 4960–4964.
140. Zhu YG, Liu Q, Rong Y, et al. Proton enhanced dynamic battery chemistry for aprotic lithium-oxygen batteries. *Nature Communications*, 2017, 8(1): 14308
141. Wu S, Qiao Y, Yang S, et al. Organic hydrogen peroxide-driven low charge potentials for high-performance lithium-oxygen batteries with carbon cathodes. *Nature Communications*, 2017, 8(1): 15607.
142. Gao X, Jovanov ZP, Chen Y, et al. Phenol-catalyzed discharge in the aprotic lithium-oxygen battery. *Angewandte Chemie International Edition*, 2017, 56(23): 6539–6543.
143. Tan P, Jiang HR, Zhu XB, et al. Advanced and challenges in lithium-air batteries. *Applied Energy*, 2017, 204: 780–806.

144. Li X, Zheng J, Xu W, et al. Dendrite-free and performance-enhanced lithium metal batteries through optimizing solvent compositions and adding combinational additives. *Advanced Energy Materials*, 2018, 8(15): 1703022.
145. Ren X, Zhang Y, Xu W, et al. Guided lithium metal deposition and improved lithium coulombic efficiency through synergistic effects of LiAsF6 and cyclic carbonate additives. *ACS Energy Letters*, 2018, 3(1): 14–19.
146. Togasaki N, Momma T, Osaka T. Role of the solid electrolyte interphase on a Li metal anode in a dimethylsulfoxide-based electrolyte for a lithium-oxygen battery. *Journal of Power Sources*, 2015, 294: 588–592.
147. Lee DJ, Lee H, Park JK. Composite protective layer for Li metal anode in high-performance lithium-oxygen batteries. *Electrochemistry Communications*, 2014, 40: 45–48.
148. Zhang X, Zhang Q, Zhou Z. An extremely simple method for protecting lithium anodes in Li-O_2 batteries. *Angewandte Chemie International Edition*, 2018, 57(39): 12814–12818.
149. Sun Y, Liu N, Cui Y. Promises and challenges of nanomaterials for lithium-based rechargeable batteries. *Nature Energy*, 2016, 1(7): 16071.
150. Wang X, Pan Z, Wu Y, et al. Infiltrating lithium into carbon cloth decorated with zinc oxide arrays for dendrite-free lithium metal anode. *Nano Research*, 2018, 12(3): 525–529.
151. Yang XY, Xu JJ, Chang ZW, et al. Blood-capillary-inspired, free-standing, flexible, and low-cost super-hydrophobic N-CNTs@ SS cathodes for high-capacity, high-rate, and stable Li-air batteries. *Advanced Energy Materials*, 2018, 8(12): 1702242.
152. Liao KM, Wu SC, Mu XW, et al. Developing a "water-defendable" and "dendrite-free" lithium-metal anode using a simple and promising GeCl4 pretreatment method. *Advanced Materials*, 2018, 30(36): 1705711.
153. Yan C, Cheng XB, Tian Y, et al. Dual-layered film protected lithium metal anode to enable dendrite-free lithium deposition. *Advanced Materials*, 2018, 30(25): 1707629.
154. Kwak WJ, Rosy HW, Sharon D, et al. Lithium-oxygen batteries and related systems: potential, status, and future. *Chemical Reviews*, 2020, 120(14): 6626–6683.
155. Aurbach D, McCloskey BD, Nazar LF, et al. Advances in understanding mechanisms underpinning lithium-air batteries. *Nature Energy*, 2016, 1(116):140–162.
156. Ma SC, Zhang YL, Cui QH, et al. Understanding oxygen reactions in aprotic Li-O_2 batteries. *Chinese Physics B*, 2016, 25(1): 018204–018214.
157. Guo XX, Zhao N. The role of charge reactions in cyclability of lithium-oxygen batteries. *Advanced Energy Materials*, 2013, 3(11): 1413–1416.
158. Rao Y, Yang JW, Chu SY, et al. Solid-state Li-air batteries: Fundamentals, challenges, and strategies. *Smartmat*, 2023, 4(4): e1205.
159. Suzuki Y, Watanabe K, Sakuma S, et al. Electrochemical performance of an all-solid-state lithium-oxygen battery under humidified oxygen. *Solid State Ionics*, 2016, 289: 72–76.
160. Zhong L, Mitchell RR, Liu Y, et al. In situ transmission electron microscopy observations of electrochemical oxidation of Li_2O_2. *Nano Letters*, 2013, 13(5): 2209–2214.
161. Zheng H, Xiao DD, Li X, et al. New insight in understanding oxygen reduction and evolution in solid-state lithium oxygen batteries using an in situ environmental scanning electron microscope. *Nano Letters*, 2014, 14(8): 4245–4249.
162. Wang H, Zhao N, Bi ZJ, et al. Clear representation of surface pathway reactions at Ag nanowire cathodes in all-solid Li-O_2 batteries. *ACS Applied Materials & Interfaces*, 2021, 13(33): 39157–39164.
163. Luo LL, Liu B, Song SD, et al. Revealing the reaction mechanisms of Li-O_2 batteries using environmental transmission electron microscopy. *Nature Nanotechnology*, 2017, 12(6): 535.

164. Kumar B, Kumar J, Leese R, et al. A solid-state, rechargeable, long cycle life lithium-air battery. *Journal of The Electrochemical Society*, 2010, 157(1): A50–A54.
165. Li C, Liu YJ, Li BJ, et al. Integrated solid electrolyte with porous cathode by facilely one-step sintering for an all-solid-state Li-O_2 battery. *Nanotechnology*, 2019, 30(36): 364003.
166. Kitaura H, Zhou HS. Electrochemical performance of solid-state lithium-air batteries using carbon nanotube catalyst in the air electrode. *Advanced Energy Materials*, 2012, 2(7): 889–894.
167. Liu YJ, Li BJ, Cheng Z, et al. Intensive investigation on all-solid-state li-air batteries with cathode catalysts of single-walled carbon nanotube/RuO_2. *Journal of Power Sources*, 2018, 395: 439–443.
168. Kitaura H, Zhou HS. Electrochemical performance and reaction mechanism of all-solid-state lithium-air batteries composed of lithium, $Li_{1+x}AlyGe_{2-y}(PO_4)_3$ solid electrolyte and carbon nanotube air electrode. *Energy & Environmental Science*, 2012, 5(10): 9077–9084.
169. Wang J, Huang G, Chen K, et al. An adjustable-porosity plastic crystal electrolyte enables high-performance all-solid-state lithium-oxygen batteries. *Angewandte Chemie-International Edition*, 2020, 59(24): 9382–9387.
170. Bonnet-Mercier N, Wong RA, Thomas ML, et al. A structured three-dimensional polymer electrolyte with enlarged active reaction zone for Li-O_2 batteries. *Scientific Reports*, 2014, 4: 7127.
171. Sun X, Song YJ, Liu QS, et al. Tailoring electronic-ionic local environment for solid-state Li-O_2 battery by engineering crystal structure. *Science Advances*, 2022, 8(35): eabq6261.
172. Zhu XB, Zhao TS, Tan P, et al. A high-performance solid-state lithium-oxygen battery with a ceramic-carbon nanostructured electrode. *Nano Energy*, 2016, 26: 565–576.
173. Zhao CT, Zhu YM, Sun Q, et al. Transition of the reaction from three-phase to two-phase by using a hybrid conductor for high-energy-density high-rate solid-state Li-O_2 batteries. *Angewandte Chemie-International Edition*, 2021, 60(11): 5821–5826.
174. Du DF, Zhu Z, Chan KY, et al. Photoelectrochemistry of oxygen in rechargeable Li-O_2 batteries. *Chemical Society Reviews*, 2022, 51(6): 1846–1860.
175. Yang XY, Feng XL, Jin X, et al. An illumination-assisted flexible self-powered energy system based on a Li-O_2 battery. *Angewandte Chemie-International Edition*, 2019, 58(46): 16411–16415.
176. Wang XX, Guan DH, Li F, et al. Magnetic and optical field multi-assisted Li-O_2 batteries with ultrahigh energy efficiency and cycle stability. *Advanced Materials*, 2022, 34(2): 2104792.

4 Lithium-Sulfur Battery

Shaobo Ma and Pengjian Zuo

4.1 SCHEMATIC STRUCTURE AND REACTION MECHANISM OF LI-S BATTERIES

4.1.1 Schematic Structure of Li-S Batteries

Li-S battery is an electrochemical device that stores energy through the conversion reaction between sulfur and lithium.[1–7] Figure 4.1 shows a schematic diagram of the structure and charge–discharge process of a typical Li-S battery.[8,9] Among them, sulfur serves as the positive electrode active material, and metallic lithium acts as the negative electrode. Due to the poor electrical conductivity of sulfur materials, sulfur is usually mixed with conductive agents and binders to form a uniform slurry, and the slurry is further coated on an aluminum current collector to prepare the positive electrode. The negative electrodes generally use metallic lithium foil or lithium strip. The battery separator adopts polyethylene or polypropylene separators, which are commonly used in lithium-ion batteries to prevent short circuits. The commonly used electrolyte for Li-S batteries is an ether-based electrolyte, with a mixture of dimethyl glycol ether (DME) and 1,3-dioxolane (DOL) as the solvent, lithium bis(trifluoromethylsulfonyl)imide (LiTFSI) as the electrolyte salt, and lithium nitrate ($LiNO_3$) as the additive.[6] Sublimed sulfur with the composition of S_8 is the most commonly used positive electrode material in Li-S batteries. Sulfurized polyacrylonitrile (S@pPAN), with ultra-high cycle stability, is a kind of promising sulfur-based

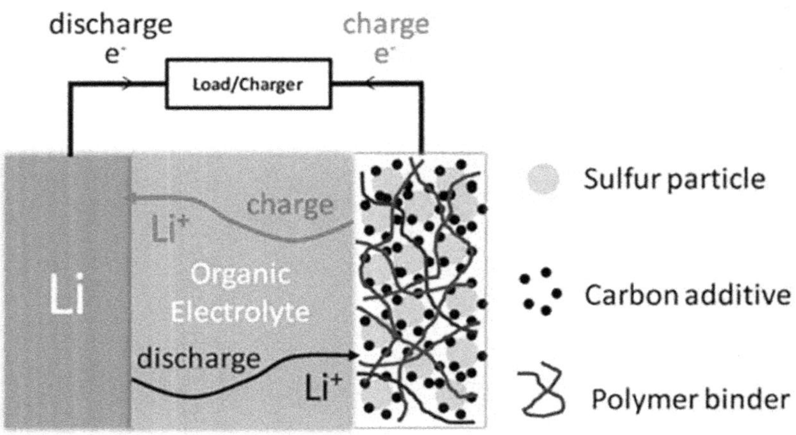

FIGURE 4.1 Schematic of Li-S battery with its discharge/charge process.[8]

DOI: 10.1201/9781003492511-4

cathode material. Other sulfur-based materials, including phosphorus polysulfide, molybdenum polysulfide, and organic sulfur polymers, also exhibit high reversible specific capacity and different reaction mechanisms. Different types of sulfur-based positive electrode materials are applicable to different electrolyte systems and have different reaction mechanisms. The following section will elaborate on sulfur-based materials and their corresponding reaction mechanisms.

4.1.2 Cathode Types and Reaction Mechanism of Li-S Batteries

4.1.2.1 Sublimed Sulfur Cathode

There are more than 30 allotropes of sulfur in nature, among which the most common and stable one is the eight-membered cyclic sulfur molecule (S_8). Usually, S_8 is compounded with carbon-based carriers to be used as the positive electrode of Li-S batteries. When using S_8/C positive electrodes, ester-based electrolytes generally are not applicable, because the lithium polysulfide intermediates and ester solvents are prone to undergo a "nucleophilic addition" reaction, leading to the failure of the positive electrode active material.[10] The ether-based electrolytes are compatible with S_8/C positive electrodes. During the discharge process, metallic lithium loses electrons to generate Li^+, which is solvated in the electrolyte and then transported to the positive electrode side. The S_8 molecules in the positive electrode undergo multistep reduction reactions and combine with Li^+ to finally form solid-phase Li_2S. During the charging process, Li_2S undergoes an oxidation reaction and is reversibly converted into S_8 and Li. Therefore, during the charge–discharge process, the S_8 positive electrode and Li negative electrode undergo an electrochemical redox reaction with a 16-electron transfer process, which can provide a theoretical specific capacity of 1,675 mAh g^{-1}.[11] The overall reaction is shown in Equation (4.1):

$$S_8 + 16Li \rightleftharpoons 8Li_2S \tag{4.1}$$

As shown in Figure 4.2, during the charge–discharge process of Li-S batteries, various intermediate products, such as Li_2S_8, Li_2S_6, Li_2S_4, Li_2S_3 and Li_2S_2 can be generated.[12] Lithium polysulfides (Li_2S_x, $x=3$–8) have high solubility in ether-based electrolytes and mainly exist as liquid-phase products. The solubility of Li_2S_x in the electrolyte increases with the growth of chain length. The reactants S_8, Li_2S_2, and Li_2S have low solubility in the electrolyte and mainly exist in the form of solid-phase substances. Therefore, the charge–discharge process of Li-S batteries is a multi-electron and "solid-liquid-solid" multiphase conversion reaction, and the generated Li_2S_x intermediate products are prone to disproportionation reactions.[13] During the discharge process, the first voltage plateau appears at 2.35 V. The S_8 molecules gain electrons to undergo a ring-opening reaction, and the chain length gradually breaks and shortens, corresponding to the "solid phase-liquid phase" conversion reaction from S_8 to Li_2S_8/Li_2S_6. In this process, each S atom can obtain an average of 0.5 e^-, providing a theoretical specific capacity of 418 mAh g^{-1}. The second voltage plateau appears at 2.1 V, corresponding to the "liquid phase-solid phase" conversion reaction of Li_2S_4 being further reduced to Li_2S_2/Li_2S. In this process, each S atom can obtain 1.5 e^-, which can provide a theoretical specific capacity of 1,254 mAh g^{-1}. During

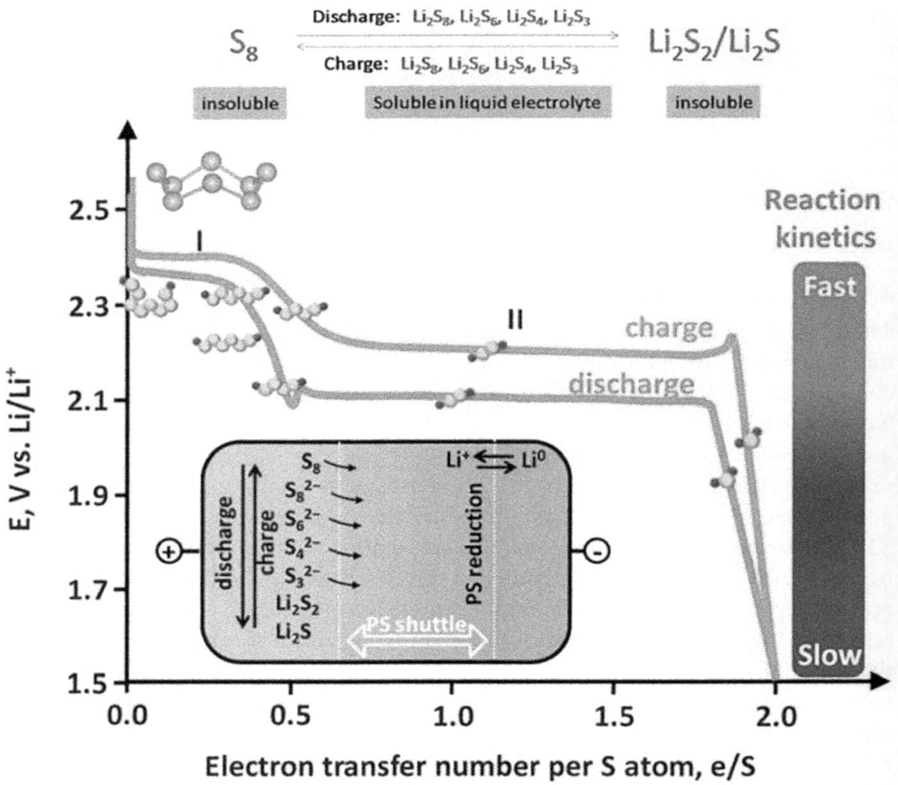

FIGURE 4.2 Reaction paths and (dis-)charge curves of the Li-S battery with S_8 cathode.[12]

the charging process, two voltage plateaus appear at 2.2 and 2.4 V, corresponding to the electrochemical reactions of Li_2S being gradually oxidized to short-chain Li_2S_x, long-chain Li_2S_x, and S^8, which is the reverse process of the discharge reaction.

4.1.2.2 S@pPAN Cathode

S@pPAN is another promising sulfur-based material. It generally undergoes solid-phase conversion reactions in ester electrolytes, thus avoiding the shuttle effect caused by lithium polysulfides. In 2002, Wang first proposed the S@pPAN material and analyzed its structure.[14] Different from conventional S_8/C materials, the short-chain –S@– (x = 2–4) is bonded to the conjugated carbon skeleton through C–S bonds in S@pPAN, and it has excellent cycle stability in ester electrolytes. When S_8 and polyacrylonitrile polymer (PAN) are calcined at 300°C under an inert atmosphere, the C≡N bonds in PAN organic molecules are converted into double bonds forming a polymerized six-membered ring conjugated structure. Meanwhile, the S–S bonds of S_8 break to form short-chain –Sx–, which are connected to the PAN skeleton through C–S bonds. In addition, S_8 can promote the dehydrogenation/cyclization reaction of PAN and generate a large amount of H_2S.[15] As shown in Figure 4.3A,

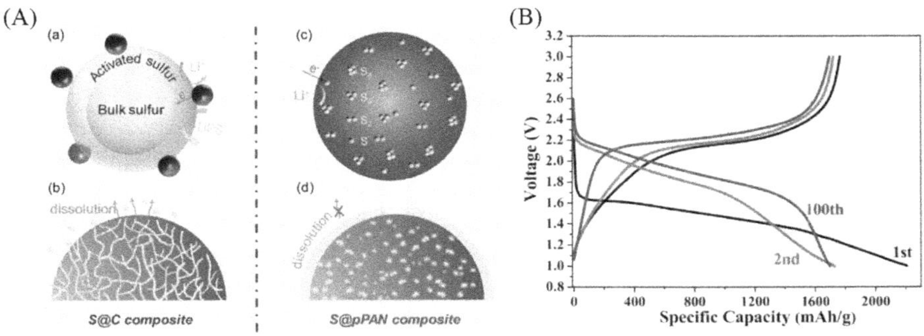

FIGURE 4.3 (A) Schematic presentation of the reaction mechanism of S_8/C and S@pPAN, (B) (dis-)charge curves of S@pPAN at different cycles.[16]

conventional S_8/C materials are suitable for ether electrolytes. During the charge–discharge process, they undergo "solid-liquid-solid" multiphase conversion reactions, and lithium polysulfide intermediates dissolve in ether electrolytes, resulting in a severe "shuttle effect." In contrast, S@pPAN undergoes "solid-solid" conversion reactions in ester electrolytes, without the dissolution of lithium polysulfides and the "shuttle effect."[16] Moreover, compared with ether electrolytes, the solubility of lithium polysulfides in ester electrolytes is lower. In ester electrolytes, a small amount of lithium polysulfides is generated on the electrode surface of S@pPAN at the initial charge-discharge stage, which undergoes a nucleophilic reaction with C=O in ester solvent molecules, forming a cathode–electrolyte interface layer on the cathode surface, which can further improve its cycling stability.[10] Figure 4.3B shows the charge-discharge curves of S@pPAN in ester electrolytes. During the first discharge process, S@pPAN shows a low voltage plateau at ~1.6 V, and the initial discharge capacity exceeds the theoretical specific capacity of S_8. In subsequent cycles, stable discharge/charge voltage plateaus appear at ~2.0 and 2.2 V, indicating high reversibility of the electrochemical reaction.

4.1.2.3 Small-Molecule Sulfur (S_{2-4}) Cathode

Unlike the cyclic S_8 cathode, the small-molecule sulfur cathodes exhibit a linear chain-like structure with a chain length of S_{2-4}. Guo reported a small-molecule sulfur/microporous carbon nanotube cathode material (S/CNT@MPC (micro-porous carbon)), which exhibits high reversible specific capacity and cycle stability in ester electrolytes (Figure 4.4).[17] The CNT@MPC carrier has a uniform microporous structure with a pore size of 0.5 nm and a specific surface area of 936 $m^2 g^{-1}$. Density functional theory (DFT) theoretical calculations indicate that small-molecule sulfur S_{2-4}, with a diameter less than 0.5 nm, can be accommodated in the pore structure of CNT@MPC, while long-chain molecular sulfur S_{5-8}, with diameters larger than 0.5 nm, cannot be loaded into the CNT@MPC carrier. In ester electrolytes, S/CNT@MPC has only one discharge voltage plateau at ~1.9 V, corresponding to the "solid-solid" conversion reaction of short-chain lithium polysulfides being reduced to Li_2S_2/Li_2S, and no long-chain lithium polysulfides can be generated. The first discharge specific

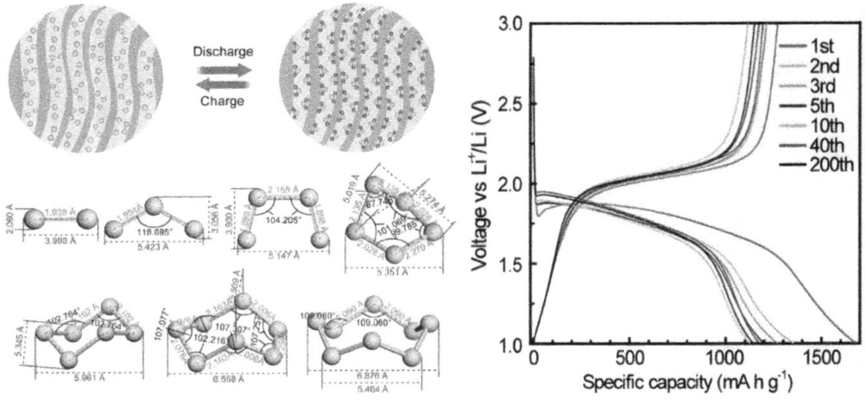

FIGURE 4.4 Reaction mechanism and electrochemical performance of smaller sulfur molecules/microporous carbon nanotubes (CNTs) cathode.[17]

capacity of this material at 0.1°C is 1,670 mAh g^{-1}, and the capacity retention rate is nearly 100% after 200 cycles. Huang reported that the molecular radii of ethylene carbonate (EC) and dimethyl carbonate (DMC) are 5.74 and 7.96 Å, respectively. Solvent molecules, solvated Li^+, and solvated polyanions are difficult to directly enter the interior of microporous materials (0.46 nm), which promotes the occurrence of solid-phase conversion reactions inside microporous materials.[18] However, microporous carbon/sulfur composite cathodes generally have a sulfur content of less than 40 wt%, and their electrochemical performance under high areal loading and high-rate current is poor, making it difficult to achieve commercial application.

4.1.2.4 Other Sulfur-Based Cathode Materials

Phosphorus polysulfide (P_4S_{10+n}) is a new type of sulfur-based cathode material with high reversible specific capacity. Qian reported phosphorus polysulfide cathode materials with different sulfur contents that were prepared by high-temperature calcination of red phosphorus and sublimed sulfur (Figure 4.5).[19] The prepared P_4S_{10+n} cathode has an amorphous structure, and –S*x*– units with different chain lengths can be connected to the interior of the P_2S_{10} molecule. In ether-based electrolytes, the initial capacity of P_4S_{40} material reaches 1,223 mAh g^{-1} at 100 mA g^{-1}, and the capacity remains 720 mAh g^{-1} after 100 cycles. The P_4S_{10+n} molecules exhibit two discharge voltage plateaus, undergoing "solid-liquid-solid" multiphase reactions, which are similar to the electrochemical behavior of S_8. After the first discharge, Li_3PS_4 and Li_2S products are generated. During subsequent charge–discharge processes, Li_3PS_4 and Li_2S undergo reversible Li^+ deintercalation/binding reactions, forming $Li_{3-x}PS_4$ and S_8. Phosphorus polysulfide, as a new type of electrode material, has the advantage of high specific capacity. However, such materials have low active sulfur content, and the electrochemical reaction mechanism of P–S bonds remains unclear.

Molybdenum polysulfides, such as MoS_3,[20] Mo_6S_8,[21] and $MoS_{5.7}$,[22] are common types of sulfur-based cathode materials that can store Li^+ through conversion reactions or intercalation reactions. Lu reported an amorphous MoS_3 cathode material, which generates no soluble lithium polysulfides during charge–discharge processes and exhibits good cycle performance in both ether-based and ester-based electrolytes.[20] Typically, layered MoS_2 materials are used as anodes for Li^+/Na^+ batteries, with a lithium intercalation potential of ~1.1 V, and they do not provide capacity in the high voltage range of 1.2–3.0 V (vs Li/ Li^+).[23] However, the amorphous MoS_3 has a stable voltage plateau at ~1.9 V and a reversible capacity exceeding 600 mAh g^{-1} in both ether and ester electrolytes. Through electrochemical tests and ultraviolet-visible spectroscopy characterization, it is shown that MoS_3 mainly undergoes "solid-solid" conversion reactions during charge–discharge processes, with no generation of soluble lithium polysulfides. In situ X-ray absorption spectroscopy and DFT calculations indicate that both Mo and S combine with Li^+ to provide capacity during the discharge process. The Mo–S bonds remain stable and no Mo–Mo bonds are formed (no elemental substances are generated). Meanwhile, the bridging S positions are more likely to adsorb Li^+, and no long-chain lithium polysulfides or Li_2S are generated. The Chevrel phase Mo_6S_8 with high electronic and ionic conductivity exhibits a theoretical specific capacity of 128 mAh g^{-1}. It stores Li^+ through intercalation–deintercalation reactions, and it mainly involves three-step electrochemical reactions in the voltage range of 1.7–2.8 V, as shown in Equations (4.2)–(4.4).[24] As shown in Figure 4.6, Li reported a composite cathode material of Mo_6S_8 and S_8, which has a dual reaction mechanism of intercalation and conversion. The assembled pouch battery can achieve an energy density of 366 Wh kg^{-1} and 581 Wh L^{-1}, providing a new research idea for promoting the commercial application of Li-S batteries.[21] In the hybrid cathode, Mo_6S_8 has high electronic/Li^+ conductivity, high tap density, and appropriate chemical adsorption for lithium polysulfides, making it an ideal active framework for the Mo_6S_8/S_8 composite cathode. During the charge–discharge process, the hybrid electrode exhibits the dual electrochemical behaviors of Mo_6S_8 and S_8. Through optimization of the electrode construction process, the prepared Mo_6S_8/S_8 composite cathode has a high active material areal loading (>10 mg cm^{-2}), low

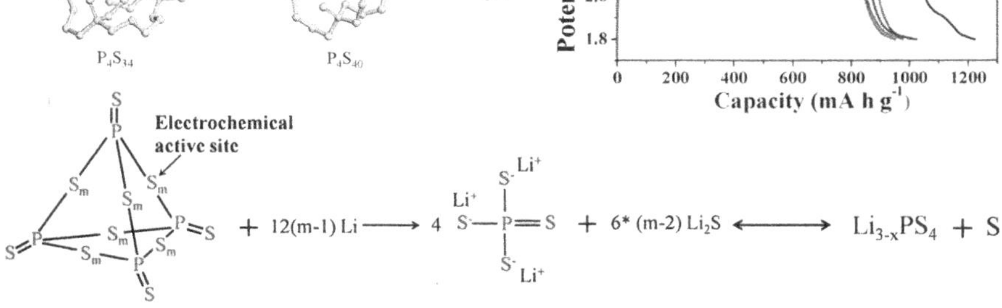

FIGURE 4.5 Sulfur-rich phosphorus sulfide for use in Li-S batteries in an ether electrolyte.[19]

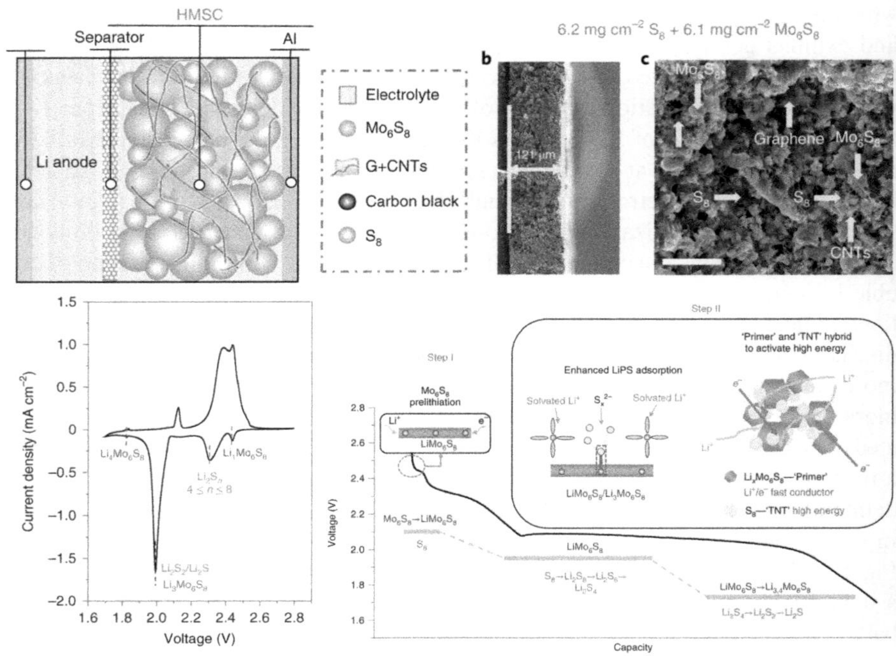

FIGURE 4.6 Chevrel phase Mo_6S_8 and S_8 composite with ultrahigh active area loading for high-energy-density Li-S batteries.[21]

porosity (~55 vol%), low carbon content (~10 wt%), and achieves high discharge specific capacity and cycle stability in ether-based electrolytes under the condition of a low electrolyte/active material ratio (1.2 μL mg^{-1}).

$$Li^+ + e^- + Mo_6S_8 \leftrightarrow Li_1Mo_6S_8 \quad (\sim 2.42\ V) \tag{4.2}$$

$$2Li^+ + 2e^- + Li_1Mo_6S_8 \leftrightarrow Li_3Mo_6S_8 \quad (\sim 2.0\ V) \tag{4.3}$$

$$Li^+ + e^- + Li_3Mo_6S_8 \leftrightarrow Li_4Mo_6S_8 \quad (\sim 1.85\ V) \tag{4.4}$$

Organic sulfur polymers, which cross-link –S*x*– groups with different chain lengths into organic polymer molecules, can be used as sulfur-based cathode materials.[25,26] In the field of rubber production, adding a small amount of sulfur can promote cross-linking reactions of unsaturated chemical bonds in polymer molecules, thereby preparing vulcanized rubber. Drawing on the idea of rubber vulcanization, increasing the addition amount of sulfur to unsaturated organic molecules can prepare sulfur-rich organic sulfur polymer materials, which can be used as cathodes for Li-S batteries.[27] As shown in Figure 4.7, Pyun reported that 1,3-diisopropenylbenzene (DIB) and sublimed sulfur can undergo copolymerization at 185°C. The sulfur content of the prepared S/DIB copolymer material reaches 50 wt%.[28] In ether-based electrolytes, the electrochemical behavior of the S/DIB copolymer is similar to that of S_8, with

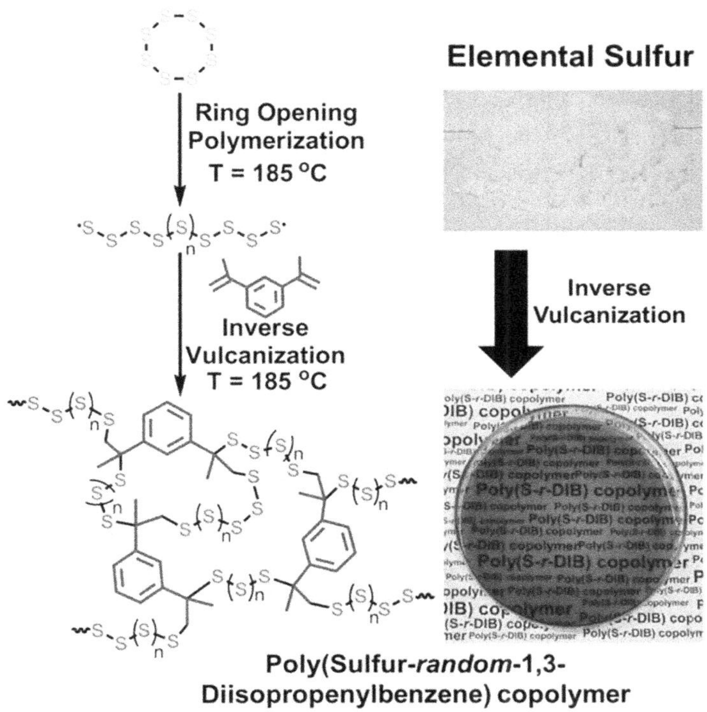

FIGURE 4.7 Synthetic scheme of poly(sulfur-random-1,3-diisopropenylbenzene) copolymers.[28]

two discharge voltage plateaus appearing at ~2.35 and 2.05 V. The discharge specific capacity of this material reaches 1,005 and 635 mAh g^{-1} after 100 and 500 cycles at 0.1°C, respectively. Organic sulfur polymer cathodes usually have low sulfur content and conductivity. Their electrochemical performance can be further improved by adding conductive carbon additives during the in situ polymerization process or constructing a three-dimensional (3D) interconnected conductive network.[29]

4.2 SULFUR CATHODES

4.2.1 Physical Confined Sulfur-Based Cathode Materials

Because of the poor conductivity of sulfur, combining sulfur with other conductive materials is necessary in preparing the sulfur cathodes. They not only provide the efficient physical anchoring of polysulfides for sulfur cathodes, but also prevent the shuttling of dissolved polysulfides. Especially, conductive carbon materials, with high conductivity and porosity, serve as excellent carriers for S_8, which can effectively enhance the reaction activity and reversible specific capacity of sulfur cathodes. The pore size distribution of porous carbon carriers is crucial for sulfur storage capacity and the electrochemical behavior of sulfur cathodes.[30] Based on pore size, porous materials are classified into macropores (> 50 nm), mesopores (2–50 nm), and micropores (< 2 nm). In 2002, Wang reported a sulfur@porous carbon composite with a high sulfur utilization of 90%, which was prepared by co-heating mesoporous activated carbon and sublimed sulfur at 200°C.[31] As shown in Figure 4.8a, microporous carbon materials can only store chain-like small-molecule sulfur (S_{2-4}) with a diameter smaller than their pore size.[32] During charge–discharge processes, they undergo "solid-solid" conversion reactions without the generation of soluble lithium polysulfides. Mesoporous carbon materials, with suitable pore size and large pore volume, are excellent sulfur carriers. Nazar proposed highly ordered mesoporous carbon CMK-3 as a sulfur carrier, and further modified its surface with polyethylene glycol (PEG) to prepare a CMK-3/S-PEG composite cathode (Figure 4.8b).[33] Due to the high electronic conductivity, specific surface area, and pore volume of CMK-3, as well as the restriction effect of the PEG surface modification layer on the dissolution of lithium polysulfides, CMK-3/S-PEG exhibited excellent sulfur utilization and cycle performance. Its initial discharge specific capacity reached 1,320 mAh g^{-1}, significantly improving the electrochemical performance of Li-S batteries. Since then, numerous studies have reported mesoporous carbon sulfur carriers, achieving good electrochemical performance in Li-S batteries. Macroporous carbon materials generally cannot be directly used for sulfur storage, as their large pore size is unfavorable for electron transport and confinement of soluble lithium polysulfides. Hierarchical porous carbon materials, which combine the advantages of mesoporous/macroporous materials (high pore volume, high Li^+ transport properties, low liquid absorption) and microporous structures (high stability), are a promising type of carbon-based

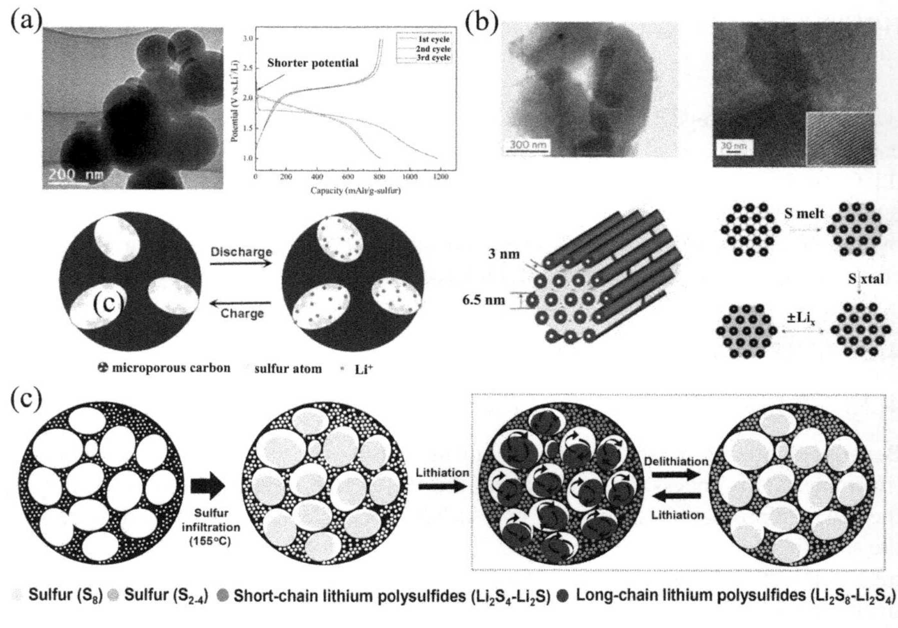

FIGURE 4.8 (a) C/S with microporous pores,[32] (b) mesoporous CMK-3/sulfur composite,[33] (c) C/S composites with hierarchical micro–meso–macro architectures.[36]

material.[34,35] As shown in Figure 4.8c, Chio prepared hierarchical porous carbon materials using ultrasonic spray pyrolysis.[36] The internal structure of the material consists of macropores and mesopores, while the external structure is microporous. S_8 is mainly stored in the mesopores and macropores of the carbon carrier. During charge–discharge processes, S_8 molecules inside undergo "solid-liquid-solid" multiphase conversion reactions, and the external micropores can block the dissolution of long-chain lithium polysulfides from the cathode. Therefore, the prepared cathode material exhibits high cycle performance, with a discharge capacity of 1,412 mAh g^{-1} and a capacity retention rate of 77% after 500 cycles.

Carbon materials with various morphologies can be used as sulfur carriers for the cathode of Li-S batteries, including porous carbon materials, 1-dimensional (D) carbon nanotubes (CNTs), 2D graphene (GO), and 3D composite materials (Figure 4.9). Porous carbon materials, such as Ketjen black and BP-2000, have large pore volume, pore size, and mesoporous distribution, which are commonly used for constructing commercial sulfur cathodes.[34] In Figure 4.9a, through spray drying technology, nanoscale particles can be assembled into micron-scale secondary particles, which further improve the porous structure, electrical transmission, and electrolyte wettability of the material.[37] This is suitable for constructing thick electrodes with high sulfur content and high sulfur areal loading. Zhang prepared cauliflower-shaped IKB-S secondary particles.[38] When the sulfur areal loading is 6–14 mg cm^{-2}, the discharge specific capacity reaches 1,300 mAh g^{-1}, and the energy density of the constructed Li-S pouch battery reaches 504 Wh kg^{-1} and 654 Wh L^{-1}. Two-dimensional

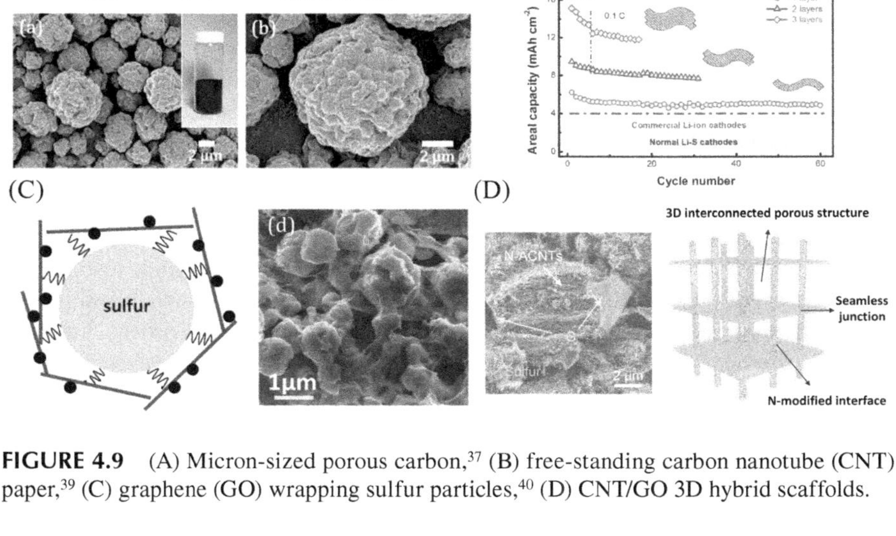

FIGURE 4.9 (A) Micron-sized porous carbon,[37] (B) free-standing carbon nanotube (CNT) paper,[39] (C) graphene (GO) wrapping sulfur particles,[40] (D) CNT/GO 3D hybrid scaffolds.

carbon materials, including CNTs and carbon fibers, are also commonly used as sulfur storage materials and can be used to construct self-supporting electrodes. In Figure 4.9b, Zhang reported CNTs-S self-supporting electrode with sulfur areal loading of 6.3 mg cm^{-2} that exhibits an initial areal capacity of 6.2 mAh cm^{-2}, and the capacity decay rate is 0.2% per cycle after 150 cycles at 0.05°C.[39] Two-dimensional graphene and its modified materials can also be used as carriers for encapsulating elemental sulfur. Cui prepared a GO-coated GO/S composite cathode using a solution reduction method (Figure 4.9c).[40] The carbon black is added to improve conductivity, and a polyethylene glycol surfactant is added to capture lithium polysulfides and limit volume expansion during charge-discharge processes. Multidimensional composite regulation of different types of carbon materials, such as graphene/CNT and CNT/porous carbon, can improve the performance of Li-S batteries.[41,42] In Figure 4.9d, Wei prepared a 3D graphene/carbon nanocomposite cathode material through chemical vapor deposition. By structural regulation, CNTs are uniformly inserted between GO sheets, forming a mesoporous (~8 nm) structure and a 3D network with high electronic conductivity (3,130 S cm^{-1}).[43] Thus, this material exhibits discharge specific capacities of 1,152 and 770 mAh g^{-1} at 1°C and 5°C, respectively.

4.2.2 Chemical Confined Sulfur-Based Cathode Materials

The physical capture of polysulfides as mentioned above suggests their natural deficiency in electrochemical activity, and hence leads to relatively limited improvement in redox cycling stability. Because of their abundant functional groups, metal oxides, nitrides, phosphides, and sulfides exhibit a promising chemical affinity for polysulfides. Strong polar interactions, including polar-metallic and nonmetallic bonding play a key role in suppressing the dissolution and migration of polysulfides by chemical confinement. Although porous carbon materials can significantly enhance the electronic conductivity of active materials, their physical confinement effect on lithium polysulfides can only suppress the shuttle effect to a certain extent. To further strengthen the interaction between the carrier material and lithium polysulfides, researchers have carried out surface modification of carbon materials or polar-material carriers. Through the chemical adsorption of polar materials on lithium polysulfides, their dissolution is restricted, thereby improving the cycling stability of the battery. Meanwhile, the interaction mechanism between polar materials and lithium polysulfides has been clarified by means of theoretical calculations and other methods. Wang calculated the interaction between lithium polysulfides and nitrogen-doped graphene materials according to DFT.[44] It was found that only clustered pyrrolic N can anchor lithium polysulfides through the strong interaction between N and Li.

In addition to elemental doping, some polar metal-based compounds also exhibit strong adsorption capacity for lithium polysulfides. Yuan synthesized mesoporous hollow titanium dioxide materials with high specific surface area, porosity, and conductivity (Figure 4.10).[45] This material can interact with lithium polysulfide anions through surface positive charges, and the adsorption of lithium polysulfides on its surface is reversible. Therefore, the hollow structure not only provides a reversible adsorption–desorption surface and electrochemical reaction sites for lithium polysulfides, but also the internal space can adapt to the volume change during

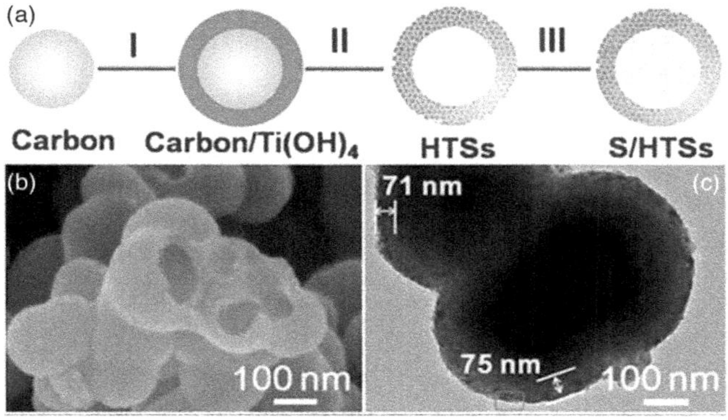

FIGURE 4.10 Schematic illustration of the synthetic process and the microstructure of S/HTSs.[45] (a) Mesoporous hollow titanium dioxide materials interacting with lithium polysulfide. (b) SEM of lithium polysulfides/hollow tinanium dioxide sulfides. (c) TEM of lithium polysulfides/hollow tinanium dioxide sulfides.

charge-discharge processes. Cui successfully coated a titanium disulfide (TiS_2) layer on the surface of lithium sulfide to form a $Li_2S@TiS_2$ eggshell structure.[46] TiS_2 with high electronic conductivity not only provides an electronic pathway for Li_2S thus reducing the decomposition potential during its first charging process, but also adsorbs lithium polysulfides through the polar interaction between sulfur and lithium, thereby inhibiting the shuttle of lithium polysulfide. The cathode material using $Li_2S@TiS_2$ achieves a specific capacity of 503 mAh g^{-1} at 4°C.

Studies have found that some metal oxides exhibit electron transfer characteristics when in contact with lithium polysulfides. Nazar explored the interaction mechanism between a large number of metal oxides and lithium polysulfides, and found that some metal oxides can inhibit the dissolution of lithium polysulfides through surface redox reactions.[47] A representative material is δ-MnO_2. Static adsorption experiments show that δ-MnO_2 can quickly decolorize the lithium polysulfide solution, confirming its strong adsorption capacity for lithium polysulfides.[47] As shown in Figure 4.11a, MnO_2 nanosheets first undergo a redox reaction with lithium polysulfides, generating thiosulfate intermediates on their surface. Then, these intermediates interact with soluble lithium polysulfides to form solid polythionates, and this interaction is reversible without the loss of active materials. This effect can effectively anchor lithium polysulfides, significantly improving the cycle stability of Li-S battery. The S/MnO_2 material with a sulfur content of 75% can provide a specific capacity of 1,300 mAh g^{-1} at 0.05°C. Studies on a large number of metal oxides have revealed that not all metal oxides can interact with lithium polysulfides through this mechanism. Only those with redox potentials within an appropriate range can trigger the reaction.[48] Figure 4.11b shows the relationship between the reaction activity of different metal oxides and lithium polysulfides. Oxides with redox potentials between 2.4 and 3.05 V, such as CuO, MnO_2, and VO_2, can promote the formation of thiosulfates and polythionates. Oxides with too high redox potentials will directly generate irreversible sulfates, while those with too low reduction potentials will not undergo the

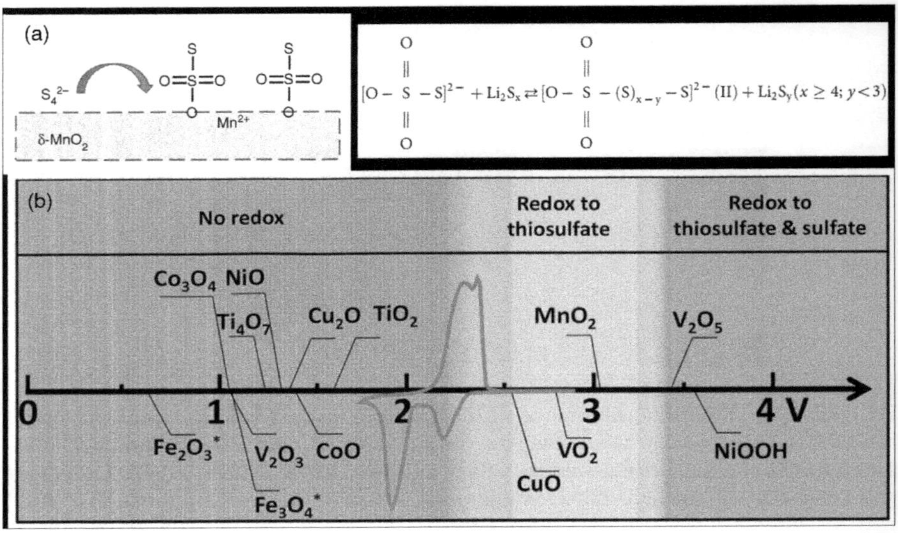

FIGURE 4.11 (a) Reaction mechanism between MnO_2 and S_4^{2-},[47] (b) chemical reactivity of different metal oxides with LiPSs (lithium polysufide).[48]

surface redox reaction. This reaction, regulated by surface redox, is beneficial to fix lithium polysulfides and improve the cycle stability of batteries. In addition, other metal oxides or sulfides, such as Ti_4O_7,[49] MoS_2[50] and MgO,[51] have been gradually explored. However, most polar materials have poor electronic conductivity, which is unfavorable for electron transport and thus reduces the rate performance of the battery. Moreover, the introduction of a large amount of inactive materials will inevitably reduce the energy density of the battery. Therefore, improving the electronic conductivity of the carrier material and reducing the proportion of inactive materials become the key to enhancing battery performance.

4.2.3 Sulfur Composite Cathode with Catalytic Conversion

Facilitating the rapid conversion of polysulfides is a key factor in addressing the shuttle of polysulfides. In recent years, catalytic effects have gradually been introduced into solutions for mitigating the shuttle effect.[52] Catalytic materials can accelerate the rapid and reversible conversion of polysulfides, reducing the contact probability between polysulfides and the electrolyte, thereby enhancing battery efficiency. Lots of materials have been reported to exhibit catalytic properties for polysulfide conversion, such as elemental metals, transition metal compounds, single-atom catalyst (SAC), and nonmetallic catalysts.[53,54]

Studies have shown that elemental metals, including Pt, Ni, Co, have a certain catalytic effect on the kinetic conversion process of lithium polysulfides.[55,56] For instance, Arava proposed the concept of electrocatalysis for the cathode reaction of Li-S batteries. Sulfur carriers modified with elemental metals, such as Pt, Au, and Ni, can accelerate the kinetic process of lithium polysulfide conversion reactions,

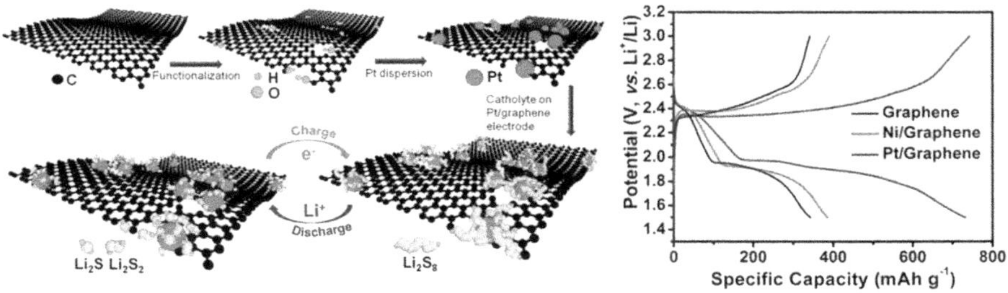

FIGURE 4.12 Illustration of a Pt electrocatalyst anchored on graphene and the corresponding electrochemical performance.[57]

thereby improving the electrochemical performance of the battery.[53,57] Compared with graphene, the initial discharge capacity of the Pt/graphene composite cathode was 1,100 mAh g^{-1}, and the discharge specific capacity after 100 cycles increased by nearly 40% (Figure 4.12). Through electrochemical tests, such as cyclic voltammetry (CV) and Tafel, Pt/graphene has a higher exchange current density and a faster kinetic reaction process. Pt can catalyze the reduction and oxidation processes of lithium polysulfides, thereby reducing the deposition of solid-phase products on the electrode surface.

Transition metal compounds, such as sulfides (MoS_2 and ZnS_2),[58,59] oxides (MnO_2 and Fe_2O_3),[60] and nitrides (VN and TiN),[61] have a certain catalytic effect on the conversion reaction of lithium polysulfides. Arava reported that the edge positions of two-dimensional WS_2 and MoS_2 materials can reduce the polarization voltage of lithium polysulfide conversion and enhance their kinetic conversion process (Figure 4.13a).[62] Zhang reported that CoS_2 can catalyze the redox process of lithium polysulfides. The prepared CoS_2/graphene composite materials show significantly improved electrochemical performance, with the sulfur utilization rate increased by ~60% and the slight capacity decay rate of 0.034% per cycle.[63] Through CV and electrochemical impedance spectropy (EIS) electrochemical tests of Li–Li symmetric batteries, it was found that CoS_2 can significantly increase the current density of the electrode reaction and reduce the charge transfer impedance at the electrode interface. Nazar reported that MnO_2 nanosheets can promote the conversion of soluble lithium polysulfides to solid-phase Li_2S_2/Li_2S (Figure 4.13b). The prepared S/MnO_2 cathode delivers an initial capacity of 1,300 mAh g^{-1} at 0.05°C, and maintains 1,030 mAh g^{-1} after 200 cycles.[47] Recently, Yang reported a TiN/TiO_2 heterostructure cathode material, which accelerates the electrochemical reaction of capture–diffusion–conversion of lithium polysulfides thus realizing high stability of Li-S batteries.[64] For the TiN/TiO_2 heteromaterial, TiO_2 has a strong adsorption effect on lithium polysulfides, and TiN with high conductivity can catalyze the electrochemical reaction of lithium polysulfides to Li_2S_2/Li_2S (Figure 4.13c). With sulfur loadings of 3.1 mg cm^{-2}, the capacity retention within 2,000 cycles can reach 73% at 1°C.

Single-atom materials, including Fe–,[65] Co–,[66,67] and Ni-based[68] SACs, can be used for modifying cathode materials to enhance the electrochemical reaction kinetics of Li-S batteries. Lee reported a spherical SA-Fe@SMCF (sulfur modified

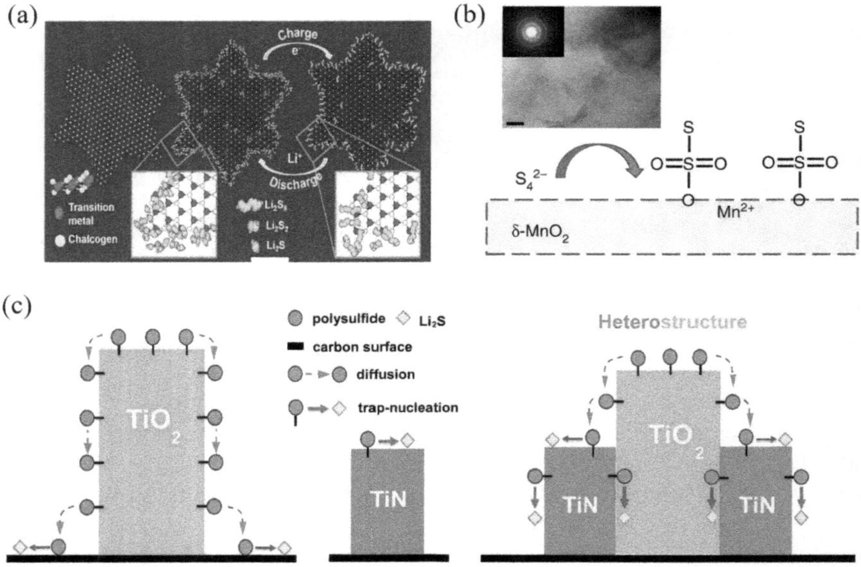

FIGURE 4.13 Different transition metal alloys with catalytic performance for LiPSs conversion: (a) MoS_2 and WS_2 2D materials,[62] (b) ultrathin MnO_2 nanosheet,[47] (c) TiO_2–TiN heterostructure.[64]

carbon fibers) material as a sulfur carrier, where Fe exists in the form of Fe–N–C single-atom active sites. CV and Tafel tests show that the Fe–N–C SAC cathodes can significantly promote the electrochemical reaction kinetics of lithium polysulfides.[69] As shown in Figure 4.14, Amine reported that embedding Co–N–C SAC and polar ZnS nanoparticle dual-binding sites into a highly oriented macroporous conductive framework can effectively fix and catalyze the electrochemical reaction of lithium polysulfides, eliminating the shuttle effect and lithium metal corrosion.[70] A three-phase interface can be formed between the catalytic active sites, the conductive carbon carrier, and the organic electrolyte, enhancing ion transport under high sulfur loading. This structural design improves the cycle stability, rate performance, Coulomb efficiency (CE), and areal capacity of the sulfur cathode. Cui prepared V-NG SAC for high-specific-energy Li-S batteries.[71] DFT calculations show that eight kinds of metal/nitrogen-doped graphene composites (M-NG, M = Fe, Mn, Ru, Zn, Co, Cu, V, and Ag) have good catalytic effects on the Li_2S decomposition reaction, while Cu-NG and Ag-NG have poor stability after adsorbing Li_2S and Li_2S_2. V-NG has the lowest energy barrier for Li_2S decomposition reaction, indicating that it has a good catalytic effect on Li_2S decomposition. Because a single SAC site still shows limitations when offering complementary adsorption/activation sites for diverse polysulfides intermediates. Recently, diatomic catalyst[72] and high-entropy SAC[73] were reported to synergistically regulate d-electron states and the adsorption of intermediate polysulfides, thereby accelerating their conversion kinetics.

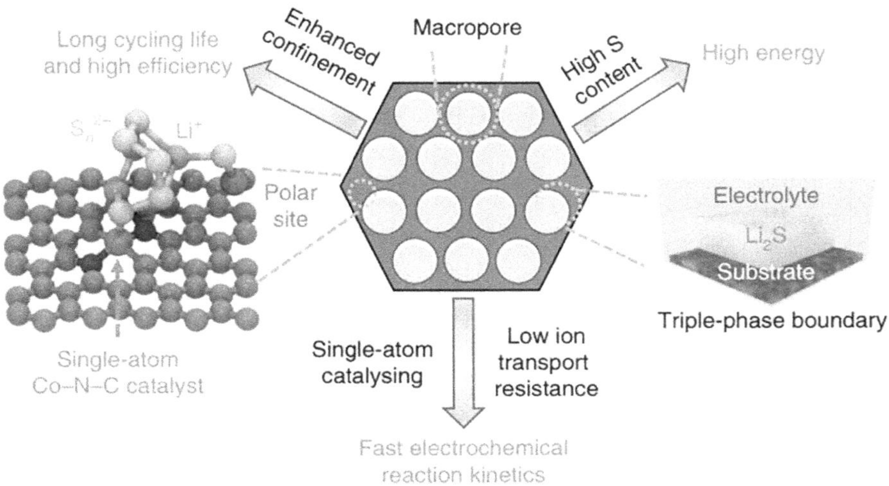

FIGURE 4.14 Macroporous sulfur host with double-end-binding sites.[70]

Nonmetal catalysts, including heteroatom-doped carbon, black phosphorus, and molecularly imprinted polymers, can be used to catalyze the conversion reaction of lithium polysulfides in Li-S batteries. Duan reported a heteroatom-doped porous graphene network model catalyst and detailly explored its electrocatalytic mechanism for sulfur reduction reaction (SRR).[74] By analyzing the activation energy in the SRR, they revealed the difficulty level of various reaction steps. As shown in Figure 4.15, the initial reduction of S_8 to soluble polysulfides has a low activation energy, while the conversion of soluble polysulfides to solid-phase Li_2S_2/Li_2S has a high activation energy. This leads to the accumulation of polysulfides and exacerbates the shuttle effect of polysulfides. N- and S-dual-doped graphene can significantly reduce the activation energy of the SRR process and improve SRR kinetics. DFT theoretical calculations confirm that N and S co-doping can adjust the p-band center of active carbon atoms, enabling optimal adsorption strength for lithium polysulfide intermediates and electrocatalytic activity.

4.2.4 S@pPAN Cathode

Numerous research efforts have been made to enhance the electrochemical performance, interface stability, and safety of S@pPAN through the regulation of material morphology and structure, as well as the development of new binders and electrolytes.[75,76] However, S@pPAN still faces some core issues that remain unsolved,[77,78] mainly including: (1) Low electrical conductivity and a slow kinetic reaction process. This hinders the construction of practical cathodes with high areal loading and high C-rate performance. (2) Unclear molecular structure and reaction mechanisms. There are still significant controversies regarding issues such as the molecular structure of S@pPAN, the occurrence of voltage hysteresis, and the

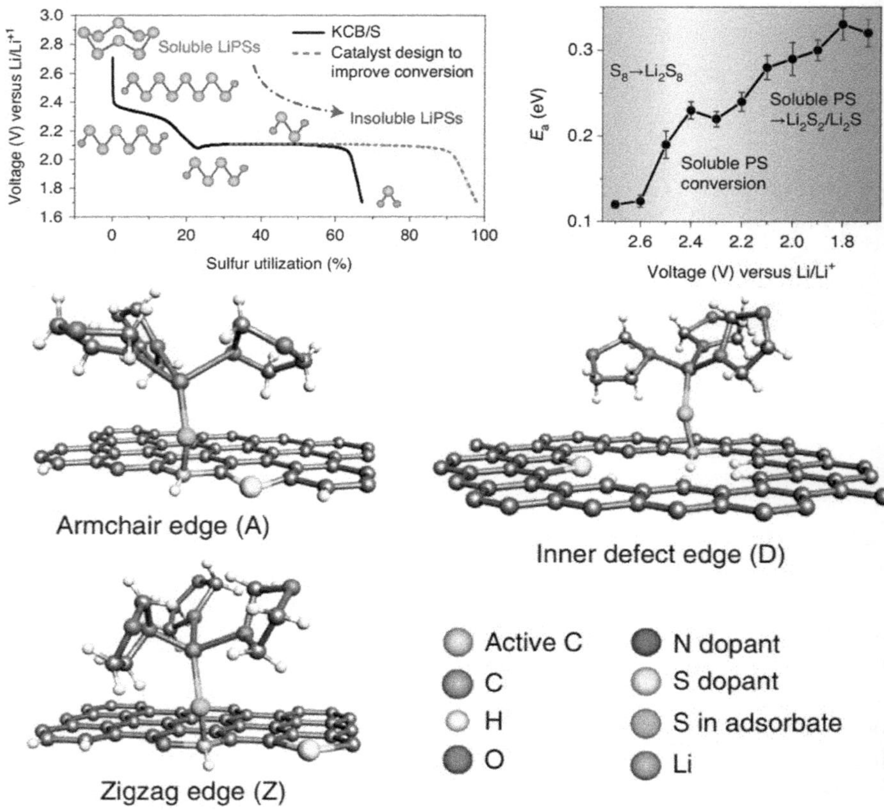

FIGURE 4.15 N, S-graphene materials with a catalytic effect for LiPS conversion.[74]

specific capacity exceeding the theoretical specific capacity of S during the first discharge, and the reaction mechanism during the cycling process. (3) Low sulfur content. The sulfur content in S@pPAN is generally less than 50 wt%, which limits its application in constructing high-energy-density pouch batteries.

Introducing conductive agents (such as carbon black and graphene oxide) into S@pPAN through in situ polymerization/spray drying can effectively improve the conductivity and rate performance of the cathodes. GO, as a 3D conductive network, can enhance the electrochemical performance of GO/S@pPAN (Figure 4.16a).[79] Through electrospinning technology, self-supporting fibrous S@pPAN composites are prepared by adding different conductive additives, and their C-rate electrochemical performance can be significantly optimized. As shown in Figure 4.16b, Liu prepared S@pPAN/CNTs cathodes using electrospinning technology, which achieved a discharge capacity of 1,180 mAh g^{-1} after 1,000 cycles at 800 mA g^{-1}.[80] The porosity and pore size structure of S@pPAN materials also have a certain impact on their electrochemical performance.[81] Xie synthesized a nanoflower-shaped S@pPAN cathode with a high sulfur content of 65%, which delivers a high composite capacity over

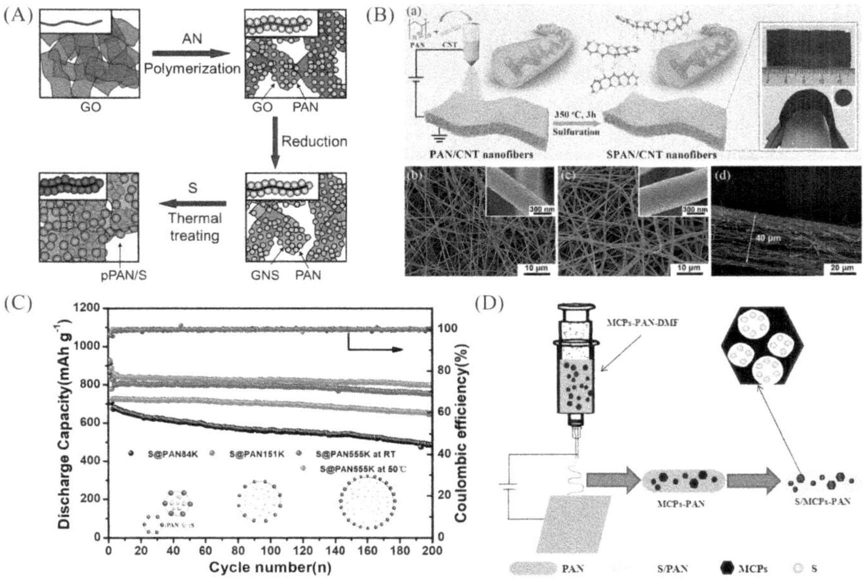

FIGURE 4.16 (A) GO/S@pPAN composite cathode,[79] (B) S@pPAN/CNTs composite cathode prepared by electrospinning,[80] (C) high-molecular-weight PAN precursor,[86] (D) microporous carbon/PAN sulfur host.[87]

1,000 mA h g^{-1} and exceptional stability over 6,000 cycles.[82] Additionally, heteroatom doping modification, such as Se and Te, promotes the generation of S–Se and S–Te eutectics, thereby enhancing the electronic conductivity and electrochemical performance of cathodes.[83,84] For instance, Lou reported a pyrolyzed PAN/selenium disulfide composite with dramatically enhanced active material content (63 wt%) and superior reversible capacity/cycling stability.[85]

It is necessary to develop S@pPAN cathodes with high sulfur content and optimize the electrochemical performance of S@pPAN under high areal loading. Wang studied the relationship between the molecular weights of PAN and sulfur content in S@pPAN and found that PAN with a larger molecular weight is beneficial to increasing the sulfur content of S@pPAN (Figure 4.16c).[86] Using PAN with a molecular weight of 555,000, the prepared S@pPAN achieves a sulfur content of 54.9 wt%. Additionally, Gao prepared a microporous carbon/PAN composite (MCPs-PAN) by encapsulating polyhedral microporous carbon inside PAN via electrospinning technology, achieving a sulfur loading of 52 wt% (Figure 4.16d).[87] The discharge specific capacity reaches 666 mAh g^{-1} after 200 cycles. Chen reported an S@pPAN cathode using guanidine isothiocyanate as the sulfiding agent, and the cathode with a sulfur content of 51.52 wt% delivers a capacity of 897 mAh g^{-1} at 0.1°C.[88] Constructing dense/thick electrodes, reducing electrolyte content and ensuring good electron/ion transport inside the electrodes are essential for the commercialization of Li-S batteries. The development of new multifunctional binders and secondary granulation technologies is conducive to the construction of thick electrodes. Various functional water-based binders, including guar gum/polyacrylic acid cross-linked binders and guar gum/carboxybutyl benzene rubber flexible cross-linked binders, can realize the preparation of high-areal-loading thick electrodes.[89,90] The areal loading of electrode active materials and the compaction density reach 17.5 mg cm^{-2} and 1.15 g cm^{-3}, respectively.[89] Thus, the prepared cathode exhibits a capacity of 6 mAh cm^{-2} after 120 cycles. Using secondary granulation technologies such as spray drying and solution polymerization, micron-sized secondary particles with uniform morphology can be prepared, which can further construct thick electrodes with high active material areal loading. For example, Hu prepared wool ball-shaped SD-C/S materials composed of KB and CNT composites by spray drying. The diameter of the secondary particles is 2–4 µm.[91] When the sulfur content and areal loading are 81 wt% and 2.5 mg cm^{-2}, respectively, the reversible capacity of the cathode still reaches 1,300 mAh g^{-1} at 0.1°C.

4.3 ELECTROLYTES FOR LI-S BATTERIES

The electrolyte is indispensable in all electrochemical devices. The primary role of the electrolyte in battery systems is to serve as a medium for ionic transfer between the pair of electrodes. Meanwhile, when metallic lithium comes into contact with the electrolyte, a solid electrolyte interphase (SEI) layer with specific components is formed on its surface, and the subsequent ion deposition behavior will be continuously affected by the SEI layer. Thus, it is important to maintain a good interfacial contact with the electrodes. In fact, the performance of electrolyte largely determines the applicability of the Li-S batteries. As mentioned above, the dissolution of polysulfides in liquid electrolytes and shuttling between electrodes are a great challenge for Li-S batteries. From the electrolyte perspective, the formation of protective layers at the electrode is also an important step to cope with this problem. In this regard, tremendous efforts have been made including changing electrolyte composition, raising salt concentration, modifying the electrolyte/electrode interfaces, and utilizing novel electrolyte alternatives.

4.3.1 Ether/Ester Electrolyte

Ether/ester liquid electrolytes are conducive to Li^+ transport due to their high dielectric constant and low viscosity. In particular, ether bonds can form supramolecular chelating structures through their oxygen atoms, which have a strong solvating ability for cations, making them the most widely used electrolytes in current Li-S batteries. In electrolyte systems, the formation of dendrites is often inhibited by the selection of electrolyte additives, lithium salts, solvent types, and the regulation of concentrations.

The stable deposition of metallic lithium can be achieved through the selection of electrolyte solvents. Common ether-based electrolytes include DME, tetraethylene glycol dimethyl ether (TEGDME), polyethylene glycol dimethyl ether (PEGDME), and fluoroethers. Carbone found that the increase in ether chain length is beneficial to improving the thermal stability of the battery, and TEGDME has lower resistance, excellent stability, and high capacity. However, a single solvent is difficult to meet the basic requirements of liquid electrolytes for Li-S batteries, and the optimization of electrolytes based on mixed solvents has important practical significance. Wang explored the influence of the ratio of DME to DOL on the electrochemical performance of Li-S batteries.[92] It was found that when the optimal volume ratio of DME/DOL is 2:1, the Li-S battery delivers the highest reversible capacity and cycling stability. Fluorinated ether electrolytes have become a research hotspot due to their advantages, such as low viscosity and high oxidation potential. They are usually used as cosolvents and can participate in the construction of a LiF-rich SEI layer on the surface of metallic lithium, such as HFE (1,1,2,2-tetrafluoroethyl 2,2,3,3-tetrafluoropropyl ether) and TTE (1,1,2,2-tetrafluoro-3-(1,1,2,2-tetrafluoroethoxy)propane).[93] Wang proposed a new electrolyte designed by mixing OFE (1H,1H,5H-octafluoropentyl-1,1,2,2-tetrafluoroethyl ether) with DME for Li-S batteries, which not only effectively reduces flammability but also achieves an average CE of up to 99.3% during long-term cycling of metallic lithium.[94] Archer also proposed an electrolyte mixed with silicon tetrachloride ($SiCl_4$) and propylene carbonate (PC), which constructs an SEI layer composed of silicon-bonded ring-opening organic carbonate oligomers and binary lithium salts on the surface of metallic lithium. The Li-Li symmetric battery can achieve stable cycling under the condition of 3.0 mA cm^{-2}.[95]

Electrolyte additives with small amounts can significantly inhibit dendrite growth and enhance the interfacial stability of metallic lithium. Lithium nitrate ($LiNO_3$) is the most widely used electrolyte additive in Li-S batteries, which can react with metallic lithium to form insoluble Li_xNO_y products on its surface, thus preventing direct contact with polysulfides and inhibiting side reactions.[96,97] Other functional electrolyte additives are widely investigated in Li-S batteries, including lanthanum nitrate ($La(NO_3)_3$),[98] lithium bis(oxalato)borate (LiBOB), thionyl chloride ($SOCl_2$),[99] lithium difluoro(oxalate)borate (LiDFOB),[100] lithium difluorophosphate (LiDFP),[101] and boric acid (H_3BO_3).[102] In addition, some electrolyte additives not only participate in the construction of the SEI layer, but also can stabilize the dendrite-free formation of metallic lithium through electrostatic shielding/adsorption effects. For instance, Lai reported that cetyltrimethylammonium chloride as an electrolyte additive. During lithium deposition, the cationic surfactant can accumulate at the protrusions of the lithium surface through electrostatic attraction, while the long-chain end of the molecule forms a nonpolar, lithium-repellent outer protective layer.[103] This repels Li^+ from depositing on the protrusions and forces them to deposit in adjacent areas, thus avoiding excessive deposition at the tips to form dendrites (Figure 4.17). Thus, the Li-Li symmetric battery can stably cycle for over 300 hours at 1.0 mA cm^{-2}. Zhang proposed using the principle of electrostatic shielding to inhibit lithium dendrites.[104] By adding alkali metal ions (such as cesium ions Cs^+ and rubidium ions Rb^+) to the electrolyte, these alkali metal ions form a positively charged electrostatic shielding layer at the protrusions on the lithium surface, which promotes the deposition of lithium ions in adjacent areas and prevents excessive dendrite growth. Furthermore, certain additives can form a stable SEI layer on the surface of metallic lithium through adsorption, such as octaphenyl polyoxyethylene[105] and 3,5-bis(trifluoromethyl)benzenethiol.[106] Besides, dual-salt systems can also effectively stabilize lithium metal anodes. Yu reported a mixed dual-salt electrolyte consisting of 0.1 mol L^{-1} LiDFP

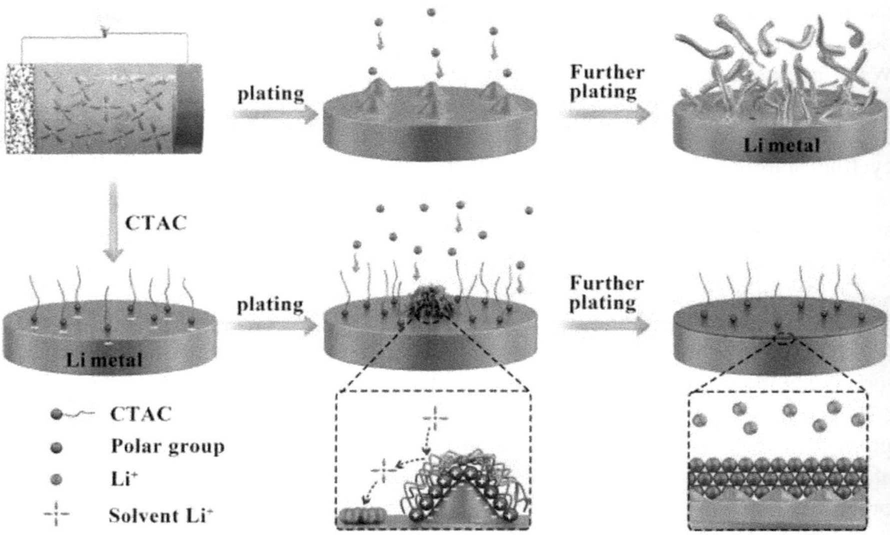

FIGURE 4.17 Schematic illustration of the effect of cetyltrimethylammonium chloride addition on the Li decomposition process.[103]

and 0.4 mol L^{-1} of one of the three lithium salts (LiBOB, LiFSI, or LiTFSI). By regulating the composition of the SEI layer on the metallic lithium, stable cycling is achieved.[107]

High-concentration electrolytes (HCE) refer to electrolytes with a relatively high content of lithium salts, usually greater than 3 mol L^{-1}, and have been widely used in the protection of metallic lithium. Their high lithium salt concentration and high Li^+ transference number can not only effectively inhibit the shuttle effect but also stabilize the lithium anode. Zhang proposed a new type of electrolyte composed of 4 mol L^{-1} LiFSI dissolved in DME, which has enabled Li–Li symmetric batteries to stably cycle for more than 6,000 times under 10 mA cm^{-2}.[108] Meanwhile, Xu designed a LiFSI–acetonitrile (AN)–VC (molar ratio 0.52: 1.0: 0.09) electrolyte, which not only minimizes the corrosion of metallic lithium but also achieves CE up to 99.2% in symmetric batteries.[109] However, HCE suffer from issues such as high cost, excessive viscosity, poor separator wettability, and low ionic conductivity. Therefore, localized high-concentration electrolytes (LHCE) have gradually come into view. LHCE are achieved by diluting the electrolyte with an inert cosolvent in which lithium salts are not soluble. Thus, LHCE not only maintain the high-concentration Li^+ coordination environment but also exhibit low electrolyte viscosity and improved separator wettability. Choi proposed 1,2-difluorobenzene as a diluent, and designed a 2 mol L^{-1} LiFSI solution in a mixture of DMC and 1,2-difluorobenzene (with a volume ratio of 3:7).[110] As shown in Figure 4.18, this electrolyte participates in the construction of a LiF-rich SEI layer on the surface of the lithium metal anode. It achieves a dendrite-free surface after multiple cycles, which is smoother than that under the cycling of HCE (5.5 mol L^{-1} LiFSI-DMC).

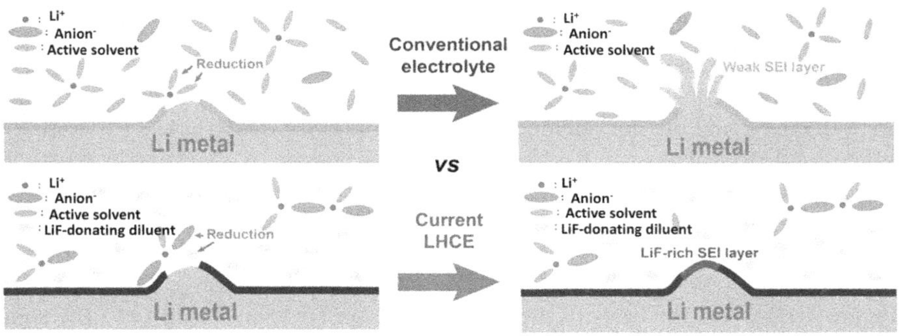

FIGURE 4.18 Schematic illustration of the mechanism of SEI layer formation in the conventional electrolyte and the current localized high-concentration electrolytes.[110]

4.3.2 Solid-State Electrolyte

Solid-state electrolyte (SSE) is expected to solve a series of key scientific problems in liquid Li-S batteries and may promote the industrialization of Li-S batteries.[111] Firstly, the use of SSE can prevent the dissolution and diffusion of lithium polysulfide intermediates in liquid electrolyte, thus inhibiting the shuttling of lithium polysulfide. Secondly, SSE has high mechanical strength, which can block lithium dendrites and improve the cycling stability of batteries. Thirdly, SSE has no risk of liquid leakage and is not easy to burn, which can significantly improve the safety performance of batteries. However, solid-state lithium-sulfur batteries (SSLSBs) also face many challenges, mainly including low ionic conductivity of SSE, poor chemical/electrochemical stability of SSE and poor compatibility at the electrolyte/electrode interface.[112] SSE used in SSLSBs are mainly divided into sulfide SSE, oxide SSE, and polymer SSE.

Polymer electrolytes exhibit the unique advantages of high flexibility, ease of processing, and good interfacial contact, and they have been extensively studied in SSLSBs.[113] Currently, the polymer electrolytes mainly include polyethylene oxide (PEO), PAN, polymethyl methacrylate and polyvinylidene fluoride (PVDF). However, some key issues, such as insufficient ionic conductivity at room temperature, low mechanical strength, and the shuttle effect of polysulfides, limit their applications. Adding inorganic salts or introducing strong polar groups can reduce the crystallinity of polymer electrolytes, thereby improving their ionic conductivity and mechanical properties. For instance, Min reported that nano-In_2O_3 fillers can significantly improve the ionic conductivity of the PEO electrolyte; meanwhile, the in situ formed Li–In alloy interlayer at the anode interface enables rapid Li^+ diffusion and smooth lithium deposition, thus achieving high-performance SSLSBs.[114] Santiago proposed that adding $Li[N(SO_2CF_2H)(SO_2CF_3)]$ (LiDFTFSI) inorganic salt additive to the polymer SSE can enhance the ionic conductivity of SSE, and the in situ forms of a LiF layer improves the compatibility/stability of the electrolyte/electrode interface.[115] Fan deposited an Al_2O_3 interlayer on the surface of the PEO electrolyte (PEO–LiTFSI) using atomic layer deposition (ALD) technology.[116] The

Al_2O_3 coating layer increases the Li^+ transference number of the SSE and effectively inhibits the shuttle effect of polysulfides.

Sulfide solid electrolytes have attracted extensive attention due to their excellent ionic conductivity (10^{-3} to 10^{-2} S cm^{-1}) and relatively low activation energy (0.15–0.35 eV), and are considered highly promising SSE materials.[117–118] However, sulfide solid electrolytes still face many challenges in terms of ionic conductivity, air stability, and chemical/electrochemical stability at the electrolyte/electrode interface. Doping and interface modification are effective strategies to improve the thermodynamic and kinetic stability of sulfide SSE. Recently, Pang reported a lithium thioborophosphate iodide (LBPSI) glass-phase solid electrolytes using LiI as a dopant (Figure 4.19a).[5] On the basis of the reversible redox between I^- and I_2/I_3^-, the solid electrolyte—as well as serving as a superionic conductor—functions as a surficial redox mediator that facilitates the sluggish reactions at the solid–solid two-phase boundaries, thereby substantially increasing the density of active sites. Zhao proposed a Sb_2O_3/LiI dual-doping strategy to prepare multifunctional sulfide SSE ($Li_7P_{2.9}Sb_{0.1}S_{10.75}O_{0.25}I_{0.2}$).[119] Through bridging oxygen atoms and the distorted lattice configuration of Sb-S tetrahedrons, the sulfur–oxygen structural units of POS_3^{3-} and POS_3^{3-} are enhanced, which significantly improves the stability of S atoms in SSE and reduces the generation of H_2S. Additionally, introducing a suitable interlayer between the anode and the SSE can enhance the interfacial stability and electrochemical performance of SSLSBs. For instance, Wang constructed a $Mg_{16}Bi_{84}$ interlayer at the Li/LPSCl interface, which significantly improved the interfacial stability between the lithium anode and the sulfide SSE.[120]

Oxide solid electrolytes, such as NASICON-type ($Li_{1.3}Al_{0.3}Ti_{1.7}O_{12}$, LATP), garnet-type ($Li_7La_3Zr_2O_{12}$, LLZO) and perovskite-type electrolytes ($Li_{0.33}La_{0.56}TiO_3$, LLTO),

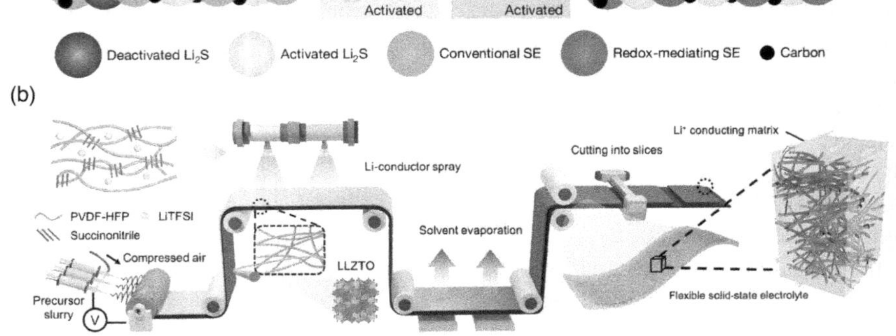

FIGURE 4.19 (a) Schematic showing the design principle for the fast-charging mechanism of ASSLSBs,[5] (b) the preparation process of the inorganic/organic hybrid SSE.[127]

exhibit advantages of a wide voltage window, high ionic conductivity, and high safety.[121] However, oxide solid electrolytes still face many problems, such as poor electrode–electrolyte interface contact, space charge layers, lithium dendrites, and difficulties in scaling-up preparation.[122] To address these challenges, approaches such as structural design, physical polishing/re-sintering, elemental doping (such as Ta/Al co-doping), and interfacial engineering (such as Al_2O_3 via ALD) are used to improve the performance of oxide SSE.[122,123] For instance, Huang reported a Bi-doped LATP electrolyte to mitigate Ti^{4+} reduction of LATP by lithium metal, thereby enhancing the stability of Li/SSE interfaces.[124] Hu reported a 3D bilayer garnet SSE framework for advanced SSLSBs.[125] The dense layer is reduced in thickness to a few microns and still retains good mechanical stability, thereby enabling the safe use of Li metal anodes. The thick porous layer serves as a host for a high loading of the sulfur cathode and provides pathways for continuous ion transport.

Hybrid solid electrolytes with various components can combine the advantages of different electrolytes.[126] Recently, Zhou developed a hybrid solid electrolyte by impregnating a mixed solution of PEO, succinonitrile, and LiTFSI into the 3D porous skeleton of PVDF–HFP and LLZTO.[127] This structural design is conducive to achieving uniform distribution of Li^+ flux and stable local current density, thereby promoting uniform lithium deposition. Meanwhile, this electrolyte has good flexibility/mechanical strength and an ultra-thin thickness of 18 μm. Judez reported a new battery design structure, by adding SSE thin layer containing Al_2O_3 on the lithium anode, while using a PEO electrolyte without fillers on the side of the cathode.[128] This ingenious design allows polysulfides to diffuse freely near the cathode while forming a stable solid electrolyte interphase layer on the anode side to protect the lithium anode. In the future, more research can be focused on hybrid SSE to enhance its ionic conductivity, electrode–electrolyte interface compatibility, and the stability of lithium anode.

4.3.3 Ionic Liquid Electrolyte

Ionic liquids (ILs) are another type of liquid electrolyte, which can be used in Li-S batteries. They are entirely composed of large-volume ions with delocalized charges, featuring large radii, relatively loose structures, and weak interactions between anions/cations, which generally allow them to exist in a liquid state at room temperature. In addition, IL electrolytes possess high thermal/chemical stability, excellent ionic conductivity, and outstanding performance in inhibiting the shuttling of lithium polysulfide. Yan systematically analyzed the influence of cations in TFSI-based ILs on the electrochemical performance of Li-S batteries.[129] The study showed that although the N-methyl-N-propylpyrrolidinium cation (P13) with low ionic conductivity is beneficial for inhibiting the dissolution of polysulfides during the charge-discharge process of Li-S batteries, the excessively high overpotential caused by the slow diffusion of Li^+ still deteriorates the stability of the battery. With the decreasing concentration of lithium salt in ILs, the Li-S batteries exhibit a high discharge capacity of 1,580 mAh g^{-1} and good cycle stability.

In addition, based on the solvating ability of organic solvents, studies have found that incorporating organic solvents into ILs helps alter the interaction between

Li^+ and anions, generating weaker and more dynamic clusters and improving Li^+ transport. Xiong found that adding 10 wt% DOL to N-methylbutylpyrrolidinium diimide salt not only promoted the increased capacity and extended cycle life for Li-S batteries, but also induced the decreased flash point, glass transition temperature, and viscosity of the electrolyte.[130] Pal proposed that the addition of DME in ILs is not only beneficial for creating more coordination options for Li^+, but also can improve the solubility of lithium salts.[131] In addition, studies have shown that mixed electrolytes are effective in eliminating polysulfide dissolution, so ILs/organic liquid mixed electrolyte systems are promising candidates for further exploration in Li-S batteries.

4.4 LITHIUM METAL ANODES

Lithium metal is a promising anode, which has the advantages of low density, high theoretical specific capacity, low electrochemical potential, and strong reducibility. These advantages endow lithium metal with active chemical/electrochemical properties, a high discharge voltage platform, and high energy density, but at the same time, they also bring problems of instability and uncontrollable high reactivity in practical applications.[132] At present, there are two main issues hindering the interfacial stability of lithium metal anodes, including the growth of dendrites and the instability of the SEI layer. As shown in Figure 4.20, the uneven deposition of Li^+ triggers continuous dendrite growth, leading to the accumulation of dead lithium on the surface of metallic lithium.[133] This will thicken the SEI layer and increase the ion/electron transport paths, which further causes increased voltage polarization. Meanwhile, the continuously growing dendrites expand the specific surface area of metallic lithium, enhancing contact with the electrolyte and triggering more side reactions. Furthermore, lithium dendrites may pierce the separator and result in short circuits. The short circuit can easily cause safety accidents, such as thermal runaway and explosions, and this insecurity is the biggest obstacle to the application of lithium anodes. Due to the hostless nature of lithium metal anodes, huge volume expansion occurs during cycling. The generated mechanical stress continuously damages the SEI layer, and the exposed fresh lithium will react with the electrolyte to form a new

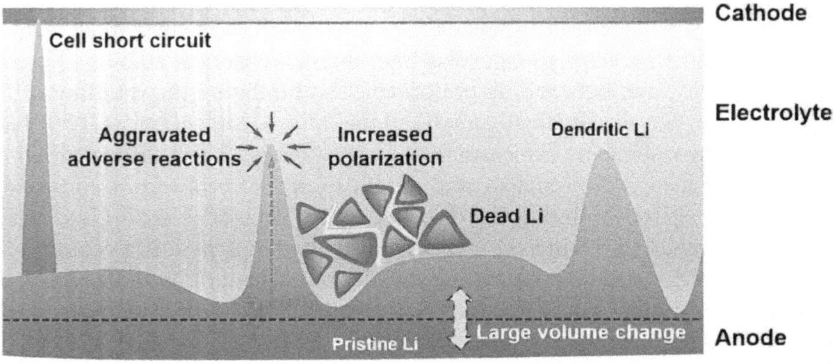

FIGURE 4.20 Scheme of dilemma for Li metal anode in rechargeable batteries.[133]

layer repeatedly. Therefore, to achieve highly stable lithium metal anodes, it is essential to reasonably inhibit dendrite growth and maintain the stability of the electrode/electrolyte interface.[134]

4.4.1 Three-Dimensional (3D) Design of Lithium Anode

Metallic lithium undergoes unlimited expansion during cycling, which not only damages the stability of the surface SEI layer but also continuously consumes metallic lithium and the electrolyte. Additionally, high current density is detrimental to the uniformity of ion deposition and leads to the rapid growth of dendrites.[135] Considering the above two aspects, the development of 3D structured anode current collectors is highly necessary. The 3D structure has a large specific surface area, which can reduce the effective current density, and the unique micro/mesoporous structure of 3D current collectors can effectively alleviate the volume effect caused by repeated cycling and inhibit dendrite growth. 3D carbon skeletons have been widely studied as matrix materials due to their good conductivity, high specific surface area, and unique microporous structure. Lou embedded silver (Ag) nanoparticles in a nitrogen-doped carbon fiber structure as the lithium anode skeleton, achieving stable cycling for 500 times under different current densities.[136] Fan proposed a one-step carbonization method for a cobalt (Co)-based metal-organic framework to modify the metallic lithium anode (Figure 4.21).[137] Utilizing the lithiophilicity of Co, the affinity between lithium and carbon materials is effectively improved, thereby regulating Li^+ deposition behavior and achieving stable cycling. Jiang innovatively prepared a 3D porous copper current collector using controllable laser etching technology and coated its surface with a layer of lithiophilic polydopamine.[138] This not only reduces the current density and nucleation overpotential but also provides uniform nucleation

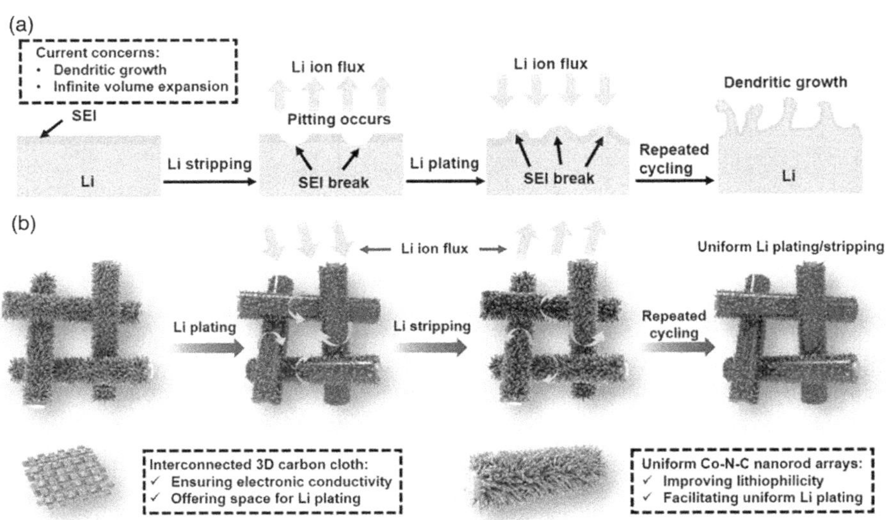

FIGURE 4.21 Description of the synergistic effect of interconnected 3D carbon cloth and Co–N–C NRAs on Li stripping/plating behavior.[137]

sites. The modified Li-Li symmetric battery maintains a high CE of 96.4% within 150 cycles under 2.0 mA cm^{-2}. Furthermore, there are other innovative 3D structures used to stabilize metallic lithium anodes, such as MXene skeletons,[139] CNF grown on nickel foam surfaces,[140] 3D hydrophobic graphene skeletons,[141] CNTs/Si micro-cages,[142] magnesium-doped lithium–boron alloy skeletons,[143] and porphyrin-based skeletons.[144]

4.4.2 Lithium Anode with Artificial SEI Layer

The high reactivity of metallic lithium leads to spontaneous reactions with the electrolyte, resulting in the formation of SEI layer on its surface. Effectively regulating the SEI layer on the surface of metallic lithium is crucial for the interfacial stability of metallic lithium. A well-formed SEI layer can effectively regulate the uniform deposition of Li^+ and isolate further reactions between the lithium anode and the electrolyte, thereby stabilizing the interface and improving the cycle life.

4.4.2.1 Physical Coating Modification of SEI Layer

Physical coating is a very simple method to construct a stable SEI layer on the lithium anode. Usually, a protective layer is coated on the surface of metallic lithium before assembling the battery. The composition and thickness of this modified layer can be adjusted and controlled, and it does not undergo side reactions with lithium when coated on the surface of metallic lithium. It is a stable and scalable surface modification method, mainly including ordinary blade coating, molecular layer deposition (MLD)/ALD, vacuum evaporation, spin coating, and other methods.

Organic-polymer-modified coating layers generally have good flexibility, chemical stability, and excellent film-forming properties. To improve Li^+ conductivity, components such as lithium salts are often added to the polymer layer. Polyvinyl alcohol (PVA), with low cost, is commonly used as a coating layer for metallic

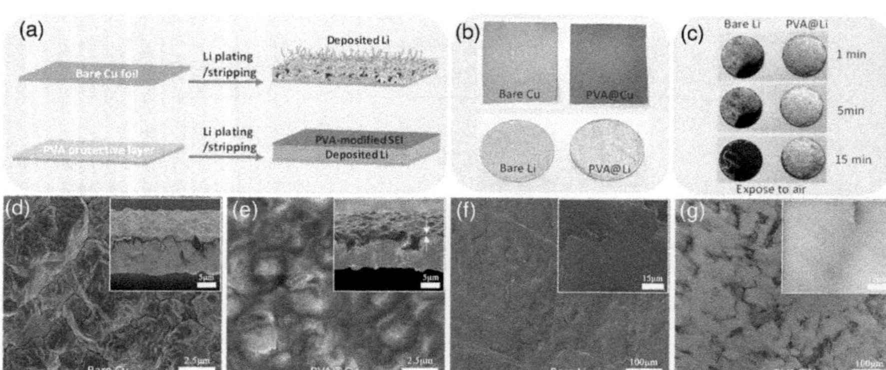

FIGURE 4.22 Schematic illustration of Li plating/stripping behavior on Cu foils with and without polyvinyl alcohol protection layer.[145] (a) Li plating on bare Cu foil and on PVA protective layer. (b) Surface images of bare Cu, Li and Li plated on Cu and PVA. (c) Electrode surface images exposed to air. (d) SEM of Cu surface and cross section. (e) SEM of PVA coated on Cu. (f) SEM of bare Li surface. (g) SEM of Li plated on PVA.

lithium (Figure 4.22).[145] For instance, Wang reported that metallic lithium coated with PVA can effectively suppress the decomposition of the electrolyte on the lithium anode surface and exhibit stable cycling performance in Li-S batteries.[145,146] Meanwhile, Li discovered that the hydroxyl groups (–OH) in PVA molecules have a strong adsorption effect on Li^+, which can effectively regulate the Li^+ deposition process and avoid the excessive growth of lithium dendrites.[147] Polysaccharide-based natural polymers are also commonly used as coating layers for lithium metal. Sun reported that both agarose and tamarind seed polysaccharide coatings can effectively inhibit lithium dendrite growth, and their high elastic modulus can also effectively alleviate the volume expansion of the lithium anode.[148,149] In addition, other polymer layers, such as Nafion films,[150] polyvinylidene fluoride (PVDF),[151] polydimethylsiloxane (PDMS),[152] and polyimide (PI),[153] can also effectively stabilize metallic lithium anodes. Reasonable structural design of polymer coating layers can further promote the practical application of lithium anodes. Sun prepared an inorganic/organic coating layer with controllable gradient inorganic zinc distribution on the lithium anode according to MLD technique.[154] The outer layer consists of an electron-insulating polymer, while inner lithiumphilic zinc sites can promote uniform nucleation and further uniform deposition of lithium. Gao adopted a vacuum evaporation method to construct a homogeneous LiF protective layer on the surface of metallic lithium, which can regulate the Li^+ flux and thus improve the cycling stability of the lithium anode.[155] Cui obtained a polycrystalline nanodiamond film with weak adhesion to the substrate by spraying colloidal diamond grains onto a copper substrate pretreated with a graphene oxide release layer.[156] Through nanoindentation tests, it was measured that the nanodiamond interfacial layer has an extremely high modulus exceeding 200 GPa. Due to its excellent mechanical strength, the nanodiamond interfacial layer can effectively inhibit dendrite growth and form dense lithium deposition.

4.4.2.2 Chemical Reaction Modification of SEI Layer

Functional artificial interfacial layers composed of a single component or multiple components are prepared on the surface of metallic lithium through chemical reactions. These chemical reactions are not limited to solid–liquid phase reactions between metallic lithium and reactant solutions, but also include solid–gas phase reactions with gases such as nitrogen (N_2), oxygen (O_2), selenium disulfide (SeS_2), and elemental sulfur (S) at room temperature or high temperature. For instance, Guo reported a liquid-phase treatment method to pretreat metallic lithium with a polyphosphoric acid solution, constructing a stable SEI layer composed of lithium phosphate (Li_3PO_4) on the lithium surface (Figure 4.23).[157] This layer has a Young's modulus higher than 6.0 GPa, which can effectively inhibit dendrite growth. The organic *n*-decylphosphonic acid can form a scale-like SEI layer on the lithium anode through the reaction between intramolecular phosphoric acid groups and metallic lithium. In Li-S batteries, this layer can not only stabilize metallic lithium but also inhibit the shuttle of lithium polysulfide.[158] In addition, Guo also proposed that the chemical reaction between ammonia borane (NH_3BH_3) and metallic lithium generates a layer of lithium ammonia borane, which spontaneously forms a polymer interfacial layer on the surface of metallic lithium through self-polymerization.[159] The modified symmetric battery can stably cycle for more than 700 hours at 3.0 mA cm^{-2}. Covalent organic frameworks (COFs) can also be used for modifying the interfacial

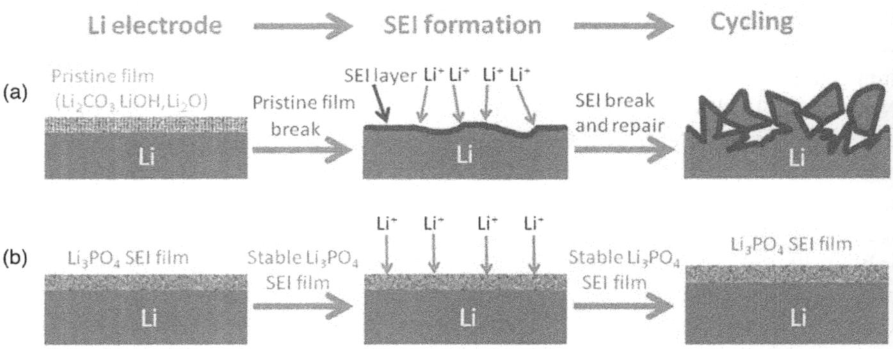

FIGURE 4.23 Schematics of the different Li anode structures: (a) The general Li metal anode resolution during SEI formation and cycling.[157] (b) Li_3PO_4-modified Li metal anode resolution during SEI formation and cycling.

layers of metallic lithium. Chen designed to in situ constructs a COF layer on the surface of metallic lithium through a Schiff base reaction, which can effectively inhibit interfacial side reactions.[160] This layer with a Young's modulus of 6.8 GPa can effectively suppress dendrite growth, and its microporous structure is also conducive to the uniform deposition of Li^+. Silane coupling agents can also be used to form stable and uniform SEI layers on metallic lithium by reacting with lithium to form Li–O–Si bonds.[161] In terms of solid–gas phase modification, Zhang proposed to construct a lithium nitride (Li_3N) based SEI layer through the chemical reaction between highly reactive metallic lithium and N_2. Cui successfully built a uniform single-component Li_2S interfacial layer on the lithium surface through the chemical reaction between metallic lithium and gaseous sulfur.[157]

4.4.3 Lithium-Rich Alloy

Modification of metallic lithium through alloying is found to be an effective way to construct a stable lithium anode. The metallic lithium alloying exhibits the advantages of facilitating ionic transport, relieving the volumetric changes of lithium metal, and decreasing the local current density.[162] Moreover, alloying can improve the chemical stability of the anode due to the lattice rearrangement, and change the electrochemical potential of the anode. Lithium-rich alloy (Li_xM) and lithium halide interfacial layers on the lithium surface have been extensively studied. The general preparation process involves preparing a metal halide into a solution with specific concentration, and after reacting with metallic lithium for a certain period of time, a biphasic mixed interfacial modified layer of lithium-rich alloy and lithium halide can be obtained on the lithium surface. The halogens are generally fluorine (F), chlorine (Cl), and iodine (I), while M includes not only metal elements such as aluminum (Al), indium (In), Zn, bismuth (Bi), magnesium (Mg) and tin (Sn), but also nonmetal elements such as arsenic (As).[162] M chloride as the main reactant and summarized the chemical reaction occurring therein and the subsequent alloying reaction into the following equations[163]:

$$xLi + MCl_x \rightarrow M + xLiCl \quad (4.5)$$

$$yLi + zM \rightarrow Li_yM_z \quad (4.6)$$

Additionally, the lithium-rich alloy phase provides a Li^+ transport channel for smooth deposition under the modified layer, while lithium chloride (LiCl), acting as an electron-insulating layer, can effectively prevent the reduction of Li^+ on the surface of the modified layer.[164,165] In terms of interfacial layer modification, Xie innovatively proposed a simple method to construct polymer/alloy layer on the surface of metallic lithium.[166] It only requires immersing metallic lithium in a 0.1 mol L^{-1} tin tetrachloride ($SnCl_4$) solution dissolved in tetrahydrofuran (THF). The schematic mechanism is shown in Figure 4.24. Firstly, metallic lithium undergoes a displacement reaction with $SnCl_4$, followed by an alloying reaction between metallic lithium and Sn,

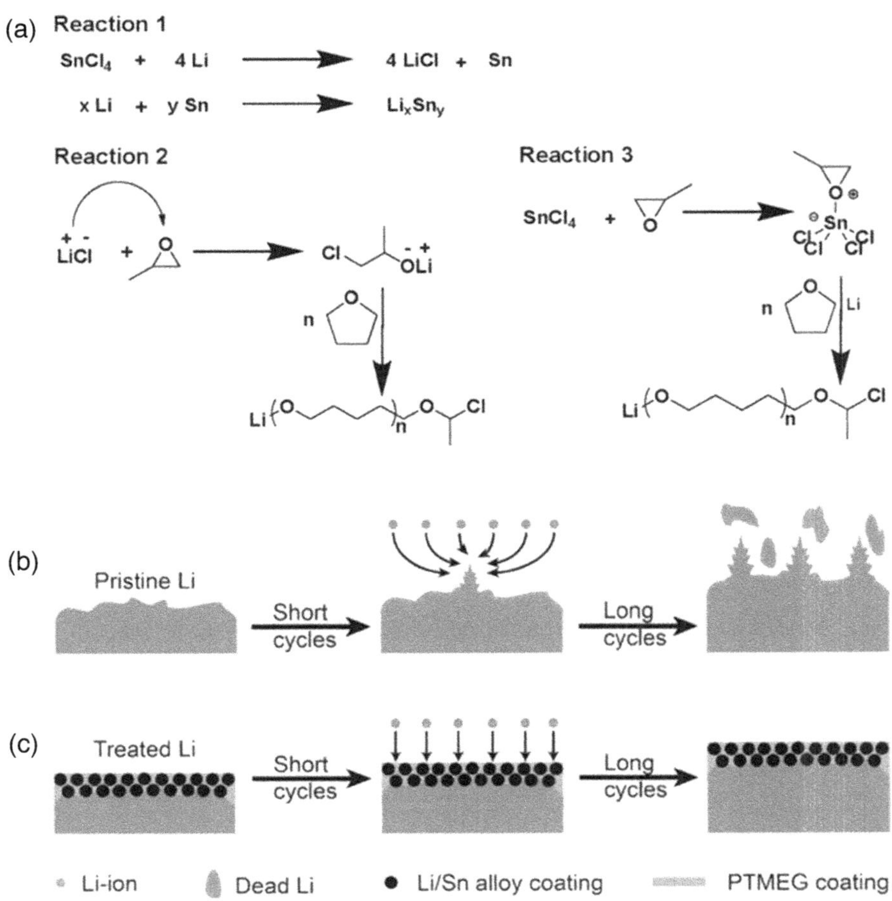

FIGURE 4.24 (a) Possible reaction mechanism for the formation of the polytetramethylene ether glycol–Li/Sn alloy hybrid layer, (b and c) Deposition of pristine Li and treated Li.[166]

generating lithium–tin alloy and LiCl. LiCl as a nucleophile, together with the Lewis acid $SnCl_4$, induces the ring-opening polymerization of THF to form polytetramethylene ether glycol (PTMEG). Eventually, a biphasic interfacial layer composed of a mixture of polymer and inorganic components (PTMEG–Li/Sn alloy) is formed on the surface of metallic lithium, which can effectively improve the cycling performance of the lithium anode.[166] Similarly, Archer prepared a multi-component modified layer composed of lithium–aluminum alloy, lithium iodide, and poly(1,3-dioxolane) (poly-DOL) on the lithium surface, by using the Lewis acid-aluminum iodide (AlI_3) to induce the ring-opening polymerization of DOL and the displacement-alloying reaction between AlI_3 and metallic lithium.[167]

4.5 PERSPECTIVES FOR LI-S BATTERIES

The high-specific-energy characteristic of Li-S batteries is highly attractive. Despite significant improvements made in aspects such as cathode configuration, the lithium anode, and the electrolyte, the commercialization process of Li-S batteries still faces numerous challenges. Specifically, the issues of polysulfide dissolution and migration between the cathode and anode sides are the focus of attention in the improvement of Li-S batteries. Meanwhile, the safety and dendrite of the lithium anode still pose significant challenges preventing the commercialization of Li-S batteries. To realize the high energy density and stable cycling performance of Li-S batteries, many critical issues of the key materials for Li-S batteries remain to be carefully considered in the future research.

4.5.1 For Sulfur Cathodes

For the enhancement of the electrical conductivity of high load sulfur cathodes and the minimization of the polysulfide species diffusion, a variety of sulfur–carbon composite materials and metal-based nanomaterials including metal oxides, sulfides, carbides, nitrides, and phosphides have displayed the desired performance. The structural design and surface modification of the cathode can effectively improve the cycle stability and rate performance of Li-S batteries. In the future, the following aspects should be considered to develop advanced sulfur cathodes.

1. To evaluate the electrochemical performance of batteries, the high sulfur loading, proper coupling of the cathode with electrolyte, electrolyte-to-sulfur ratio, and lithium anode mass are considered as key parameters. Although various reports are available on the optimization of these parameters from an individual perspective, their standardization for a practical battery system is still missing.
2. The combination of physical confinement and chemical adsorption is an effective strategy to suppress the shuttling of lithium polysulfide. For the design of host materials, their electronic conductivity, specific surface area, and pore size distribution should be comprehensively considered to balance cathode conductivity, shuttle effect, sulfur loading, volume change, manufacturing cost, and other parameters. Electrocatalysts should be further

explored to effectively accelerate the conversion kinetics of lithium polysulfides, and the catalytic mechanism needs to be further clarified.
3. The synthesis of the cathode materials at the laboratory scale within a controlled parametric environment may not meet the requirements at the industrial level. A scalable production methodology is highly desirable with cost-effectiveness and eco-friendliness to build advanced sulfur cathodes with the advantages of excellent electrical conductivity, well-established interaction with polysulfides, outstanding electrocatalysis, confinement of polysulfides, and porosity of cathode materials to buffer volume expansion.
4. S@pPAN cathode with improved electrical conductivity exhibits extremely high cycling stability in an ester electrolyte. To achieve S@pPAN cathode with high sulfur content, high sulfur areal loading, and high C-rate performance will be an interesting and rewarding research.

4.5.2 For Electrolytes

The choice of suitable electrolytes for Li-S batteries is apparently very important, as it plays additional roles in passivating highly reactive Li-metal anodes, solvating intermediate polysulfide moieties, stabilizing radical species, and channeling reaction mechanisms. To achieve the target of 500 Wh kg^{-1}, the primary requirement is to use less electrolyte or SSE, further development of advanced liquid and SSE electrolytes is needed to make the Li-S batteries successful. Some suggestions are given as follows for future development of electrolytes.

1. Liquid electrolytes with relatively high solubility of polysulfides provide faster redox reaction kinetics. To solve the polysulfide shuttle and its resultant issues, solutions including electrolyte component modification, new functional solvent development, and seeking new additives should receive more attention.
2. To improve moderately solvating electrolytes, further efforts, including the addition of redox mediators, appropriate cosolvents, and the incorporation of efficient electrocatalysts, might offer effective strategies to improve battery performance in lean electrolyte conditions.
3. Highly solvating electrolyte provides an alternative strategy to reinforce polysulfide dissolution and decrease the electrolyte/sulfur ratio. However, the incompatibility between metallic lithium and high-DN solvents leads to fast capacity fading and battery failure. Creative design related to solvents, salts, and additives is required to maintain lithium stability and polysulfide solubility.
4. SSE can effectively decrease the polysulfide shuttle and lithium dendrite formation. Breakthroughs of advanced SSE with high ionic conductivity/stability and an ideal electrode/electrolyte interface are expected to further boost Li-S battery development. The bulk/surface modification of SSE and hybrid electrolyte design are useful strategies. Meanwhile, theoretical work using machine learning and big data could be tremendously useful to rapidly screen novel SSE.

4.5.3 For Lithium Anode

Lithium metal as a high-capacity anode material has regained research prominence and has also promoted persistent efforts to completely realize the commercialization of rechargeable lithium metal batteries. The essential strategies including the anode modification, anodic surface treatment, and the structural design for the preservation of lithium anodes from dendrites, have been well studied. Various technologies including the addition of interlayer, artificial film, sandwich method, and compound techniques, have been employed to improve the anode performance. The keys toward stabilizing anode structures are given as follows.

1. A better understanding of the SEI is a critical research need for producing high-efficiency deposition of metallic Li. More information regarding the morphology, formation mechanism, and properties of the SEI layer should be investigated by advanced characterization techniques, especially in in situ mode.
2. More attention should be paid to suppressing the volume change of lithium metal by using 3D structured scaffolds/matrices. However, scaffold materials should have a relatively low electrical conductivity to prevent direct lithium deposition. Additionally, it is highly desirable to develop some helpful techniques and conduct appropriate experiments to quantify the dead lithium and lithium dendrites.
3. Developing the interlayer with high mechanical strength is equally important for enhancing the blockage of dendrite formation. An effective strategy to make the hybrid interlayer is urgently required with faster Li^+ transportation and polysulfide encapsulation.
4. The artificial thin film is a more effective approach in comparison with the SEI film technology. However, the remarkable lithium volume expansion is a major drawback of this strategy. It is suggested that the hybrid artificial layers or alloy lithium anode can be an effective way to fabricate anodes for better performance of Li-S batteries.

REFERENCES

1. Dunn, B.; Kamath, H.; Tarascon, J.-M., Electrical energy storage for the grid: A battery of choices. *Science* **2011**, *334* (6058), 928–935.
2. Tarascon, J. M.; Armand, M., Issues and challenges facing rechargeable lithium batteries. *Nature* **2001**, *414* (6861), 359–367.
3. Lin, D.; Liu, Y.; Cui, Y., Reviving the lithium metal anode for high-energy batteries. *Nature Nanotechnology* **2017**, *12* (3), 194–206.
4. Zhou, J.; Holekevi Chandrappa, M. L.; Tan, S.; Wang, S.; Wu, C.; Nguyen, H.; Wang, C.; Liu, H.; Yu, S.; Miller, Q. R. S.; Hyun, G.; Holoubek, J.; Hong, J.; Xiao, Y.; Soulen, C.; Fan, Z.; Fullerton, E. E.; Brooks, C. J.; Wang, C.; Clément, R. J.; Yao, Y.; Hu, E.; Ong, S. P.; Liu, P., Healable and conductive sulfur iodide for solid-state Li–S batteries. *Nature* **2024**, *627* (8003), 301–305.
5. Song, H.; Münch, K.; Liu, X.; Shen, K.; Zhang, R.; Weintraut, T.; Yusim, Y.; Jiang, D.; Hong, X.; Meng, J.; Liu, Y.; He, M.; Li, Y.; Henkel, P.; Brezesinski, T.; Janek, J.; Pang, Q., All-solid-state Li–S batteries with fast solid–solid sulfur reaction. *Nature* **2025**, *637* (8047), 846–853.

6. Pang, Q.; Liang, X.; Kwok, C. Y.; Nazar, L. F., Advances in lithium–sulfur batteries based on multifunctional cathodes and electrolytes. *Nature Energy* **2016**, *1* (9), 16132.
7. Yao, W.; Liao, K.; Lai, T.; Sul, H.; Manthiram, A., Rechargeable metal-sulfur batteries: Key materials to mechanisms. *Chemical Reviews* **2024**, *124* (8), 4935–5118.
8. Manthiram, A.; Fu, Y.; Chung, S.-H.; Zu, C.; Su, Y.-S., Rechargeable lithium–sulfur batteries. *Chemical Reviews* **2014**, *114* (23), 11751–11787.
9. Zhou, S.; Shi, J.; Liu, S.; Li, G.; Pei, F.; Chen, Y.; Deng, J.; Zheng, Q.; Li, J.; Zhao, C.; Hwang, I.; Sun, C.-J.; Liu, Y.; Deng, Y.; Huang, L.; Qiao, Y.; Xu, G.-L.; Chen, J.-F.; Amine, K.; Sun, S.-G.; Liao, H.-G., Visualizing interfacial collective reaction behaviour of Li–S batteries. *Nature* **2023**, *621* (7977), 75–81.
10. Yim, T.; Park, M.-S.; Yu, J.-S.; Kim, K. J.; Im, K. Y.; Kim, J.-H.; Jeong, G.; Jo, Y. N.; Woo, S.-G.; Kang, K. S.; Lee, I.; Kim, Y.-J., Effect of chemical reactivity of polysulfide toward carbonate-based electrolyte on the electrochemical performance of Li–S batteries. *Electrochimica Acta* **2013**, *107*, 454–460.
11. Han, Z.; Chen, A.; Li, Z.; Zhang, M.; Wang, Z.; Yang, L.; Gao, R.; Jia, Y.; Ji, G.; Lao, Z.; Xiao, X.; Tao, K.; Gao, J.; Lv, W.; Wang, T.; Li, J.; Zhou, G., Machine learning-based design of electrocatalytic materials towards high-energy lithium–sulfur batteries development. *Nature Communications* **2024**, *15*, 8433.
12. Wang, D.-W.; Zeng, Q.; Zhou, G.; Yin, L.; Li, F.; Cheng, H.-M.; Gentle, I. R.; Lu, G. Q. M., Carbon–sulfur composites for Li–S batteries: Status and prospects. *Journal of Materials Chemistry A* **2013**, *1* (33), 9382–9394.
13. Liu, R.; Wei, Z.; Peng, L.; Zhang, L.; Zohar, A.; Schoeppner, R.; Wang, P.; Wan, C.; Zhu, D.; Liu, H.; Wang, Z.; Tolbert, S. H.; Dunn, B.; Huang, Y.; Sautet, P.; Duan, X., Establishing reaction networks in the 16-electron sulfur reduction reaction. *Nature* **2024**, *626* (7997), 98–104.
14. Wang, J.; Yang, J.; Xie, J.; Xu, N., A novel conductive polymer–sulfur composite cathode material for rechargeable lithium batteries. *Advanced Materials* **2002**, *14* (13–14), 963–965.
15. Wang, J.; Yang, J.; Wan, C.; Du, K.; Xie, J.; Xu, N., Sulfur composite cathode materials for rechargeable lithium batteries. *Advanced Functional Materials* **2003**, *13* (6), 487–492.
16. Yang, H.; Chen, J.; Yang, J.; Wang, J., Prospect of sulfurized pyrolyzed poly(acrylonitrile) (S@pPAN) cathode materials for rechargeable lithium batteries. *Angewandte Chemie International Edition* **2020**, *59* (19), 7306–7318.
17. Xin, S.; Gu, L.; Zhao, N.-H.; Yin, Y.-X.; Zhou, L.-J.; Guo, Y.-G.; Wan, L.-J., Smaller sulfur molecules promise better lithium–sulfur batteries. *Journal of the American Chemical Society* **2012**, *134* (45), 18510–18513.
18. Li, Z.; Yuan, L.; Yi, Z.; Sun, Y.; Liu, Y.; Jiang, Y.; Shen, Y.; Xin, Y.; Zhang, Z.; Huang, Y., Insight into the electrode mechanism in lithium-sulfur batteries with ordered microporous carbon confined sulfur as the cathode. *Advanced Energy Materials* **2014**, *4* (7), 1301473.
19. Li, X.; Liang, J.; Lu, Y.; Hou, Z.; Cheng, Q.; Zhu, Y.; Qian, Y., Sulfur-rich phosphorus sulfide molecules for use in rechargeable lithium batteries. *Angewandte Chemie International Edition* **2017**, *56* (11), 2937–2941.
20. Ye, H.; Ma, L.; Zhou, Y.; Wang, L.; Han, N.; Zhao, F.; Deng, J.; Wu, T.; Li, Y.; Lu, J., Amorphous MoS_3 as the sulfur-equivalent cathode material for room-temperature Li–S and Na–S batteries. *Proceedings of the National Academy of Sciences* **2017**, *114* (50), 13091–13096.
21. Xue, W.; Shi, Z.; Suo, L.; Wang, C.; Wang, Z.; Wang, H.; So, K. P.; Maurano, A.; Yu, D.; Chen, Y.; Qie, L.; Zhu, Z.; Xu, G.; Kong, J.; Li, J., Intercalation-conversion hybrid cathodes enabling Li–S full-cell architectures with jointly superior gravimetric and volumetric energy densities. *Nature Energy* **2019**, *4* (5), 374–382.

22. Wang, X.; Du, K.; Wang, C.; Ma, L.; Zhao, B.; Yang, J.; Li, M.; Zhang, X.-X.; Xue, M.; Chen, J., Unique reversible conversion-type mechanism enhanced cathode performance in amorphous molybdenum polysulfide. *ACS Applied Materials & Interfaces* **2017**, *9* (44), 38606–38611.
23. Stephenson, T.; Li, Z.; Olsen, B.; Mitlin, D., Lithium ion battery applications of molybdenum disulfide (MoS_2) nanocomposites. *Energy & Environmental Science* **2014**, *7* (1), 209–231.
24. Mei, L.; Xu, J.; Wei, Z.; Liu, H.; Li, Y.; Ma, J.; Dou, S., Chevrel phase Mo_6T_8 (T = S, Se) as electrodes for advanced energy storage. *Small* **2017**, *13* (34), 1701441.
25. Ma, T.; Yue, H.; Xiao, Y.; Huang, Y.; Li, X.; Gao, X.; He, N.; Zhan, C.; Nan, D., A review of organic sulfur applications in lithium-sulfur batteries. *Journal of Power Sources* **2025**, *625*, 235717.
26. Deng, Y.; Huang, Z.; Feringa, B. L.; Tian, H.; Zhang, Q.; Qu, D.-H., Converting inorganic sulfur into degradable thermoplastics and adhesives by copolymerization with cyclic disulfides. *Nature Communications* **2024**, *15* (1), 3855.
27. Chung, W. J.; Griebel, J. J.; Kim, E. T.; Yoon, H.; Simmonds, A. G.; Ji, H. J.; Dirlam, P. T.; Glass, R. S.; Wie, J. J.; Nguyen, N. A.; Guralnick, B. W.; Park, J.; Somogyi, Á.; Theato, P.; Mackay, M. E.; Sung, Y.-E.; Char, K.; Pyun, J., The use of elemental sulfur as an alternative feedstock for polymeric materials. *Nature Chemistry* **2013**, *5* (6), 518–524.
28. Simmonds, A. G.; Griebel, J. J.; Park, J.; Kim, K. R.; Chung, W. J.; Oleshko, V. P.; Kim, J.; Kim, E. T.; Glass, R. S.; Soles, C. L.; Sung, Y.-E.; Char, K.; Pyun, J., Inverse vulcanization of elemental sulfur to prepare polymeric electrode materials for Li–S batteries. *ACS Macro Letters* **2014**, *3* (3), 229–232.
29. Kim, H.; Lee, J.; Ahn, H.; Kim, O.; Park, M. J., Synthesis of three-dimensionally interconnected sulfur-rich polymers for cathode materials of high-rate lithium–sulfur batteries. *Nature Communications* **2015**, *6* (1), 7278.
30. Raza, H.; Bai, S.; Cheng, J.; Majumder, S.; Zhu, H.; Liu, Q.; Zheng, G.; Li, X.; Chen, G., Li-S batteries: Challenges, achievements and opportunities. *Electrochemical Energy Reviews* **2023**, *6* (1), 29.
31. Wang, J. L.; Yang, J.; Xie, J. Y.; Xu, N. X.; Li, Y., Sulfur–carbon nano-composite as cathode for rechargeable lithium battery based on gel electrolyte. *Electrochemistry Communications* **2002**, *4* (6), 499–502.
32. Zhang, B.; Qin, X.; Li, G. R.; Gao, X. P., Enhancement of long stability of sulfur cathode by encapsulating sulfur into micropores of carbon spheres. *Energy & Environmental Science* **2010**, *3* (10), 1531–1537.
33. Ji, X.; Lee, K. T.; Nazar, L. F., A highly ordered nanostructured carbon–sulphur cathode for lithium–sulphur batteries. *Nature Materials* **2009**, *8* (6), 500–506.
34. Tomer, V. K.; Malik, R.; Tjong, J.; Sain, M., State and future implementation perspectives of porous carbon-based hybridized matrices for lithium sulfur battery. *Coordination Chemistry Reviews* **2023**, *481*, 215055.
35. Nong, S.; Huang, D.; Li, Y.; Yang, R.; Xie, J.; Li, J.; Huang, H.; Liang, X.; Li, G.; Lan, Z.; Liu, H.; Huang, D.; Xu, S.; Guo, J.; Zhou, W., Multifunction-balanced porous carbon and its application in sulfur-loading host and separator modification for lithium–sulfur batteries. *Journal of Energy Storage* **2024**, *81*, 110296.
36. Jung, D. S.; Hwang, T. H.; Lee, J. H.; Koo, H. Y.; Shakoor, R. A.; Kahraman, R.; Jo, Y. N.; Park, M.-S.; Choi, J. W., Hierarchical porous carbon by ultrasonic spray pyrolysis yields stable cycling in lithium–sulfur battery. *Nano Letters* **2014**, *14* (8), 4418–4425.
37. Gueon, D.; Hwang, J. T.; Yang, S. B.; Cho, E.; Sohn, K.; Yang, D.-K.; Moon, J. H., Spherical macroporous carbon nanotube particles with ultrahigh sulfur loading for lithium–sulfur battery cathodes. *ACS Nano* **2018**, *12* (1), 226–233.

38. Ma, Y.; Zhang, H.; Wu, B.; Wang, M.; Li, X.; Zhang, H., Lithium sulfur primary battery with super high energy density: Based on the cauliflower-like structured C/S cathode. *Scientific Reports* **2015**, *5* (1), 14949.
39. Yuan, Z.; Peng, H.-J.; Huang, J.-Q.; Liu, X.-Y.; Wang, D.-W.; Cheng, X.-B.; Zhang, Q., Hierarchical free-standing carbon-nanotube paper electrodes with ultrahigh sulfur-loading for lithium–sulfur batteries. *Advanced Functional Materials* **2014**, *24* (39), 6105–6112.
40. Wang, H.; Yang, Y.; Liang, Y.; Robinson, J. T.; Li, Y.; Jackson, A.; Cui, Y.; Dai, H., Graphene-wrapped sulfur particles as a rechargeable lithium–sulfur battery cathode material with high capacity and cycling stability. *Nano Letters* **2011**, *11* (7), 2644–2647.
41. Wu, C.-C.; Chung, S.-H., Graphene/CNT foam as a high-loading sulfur cathode for lean-electrolyte lithium–sulfur cells. *Energy & Fuels* **2025**, *39* (24), 11970–11980.
42. Fan, Y.; Su, L.; Li, C.; Zhang, P.; Song, M.; Gao, L.; Li, F., Graphene-like porous carbon sheet/carbon nanotube composite as sulfur host for lithium-sulfur batteries. *Ionics* **2024**, *30* (12), 7997–8005.
43. Tang, C.; Zhang, Q.; Zhao, M.-Q.; Huang, J.-Q.; Cheng, X.-B.; Tian, G.-L.; Peng, H.-J.; Wei, F., Nitrogen-doped aligned carbon nanotube/graphene sandwiches: Facile catalytic growth on bifunctional natural catalysts and their applications as scaffolds for high-rate lithium-sulfur batteries. *Advanced Materials* **2014**, *26* (35), 6100–6105.
44. Gong, Y.; Ma, W.; Xu, Z.; Wang, Y., Understanding the interaction of N-doped graphene and sulfur compounds in a lithium–sulfur battery: A density functional theory investigation. *New Journal of Chemistry* **2022**, *46* (25), 12300–12310.
45. Li, J.; Ding, B.; Xu, G.; Hou, L.; Zhang, X.; Yuan, C., Enhanced cycling performance and electrochemical reversibility of a novel sulfur-impregnated mesoporous hollow TiO_2 sphere cathode for advanced Li–S batteries. *Nanoscale* **2013**, *5* (13), 5743–5746.
46. Seh, Z. W.; Yu, J. H.; Li, W.; Hsu, P.-C.; Wang, H.; Sun, Y.; Yao, H.; Zhang, Q.; Cui, Y., Two-dimensional layered transition metal disulphides for effective encapsulation of high-capacity lithium sulphide cathodes. *Nature Communications* **2014**, *5* (1), 5017.
47. Liang, X.; Hart, C.; Pang, Q.; Garsuch, A.; Weiss, T.; Nazar, L. F., A highly efficient polysulfide mediator for lithium–sulfur batteries. *Nature Communications* **2015**, *6* (1), 5682.
48. Liang, X.; Kwok, C. Y.; Lodi-Marzano, F.; Pang, Q.; Cuisinier, M.; Huang, H.; Hart, C. J.; Houtarde, D.; Kaup, K.; Sommer, H.; Brezesinski, T.; Janek, J.; Nazar, L. F., Tuning transition metal oxide–sulfur interactions for long life lithium sulfur batteries: The "Goldilocks" principle. *Advanced Energy Materials* **2016**, *6* (6), 1501636.
49. Pang, Q.; Kundu, D.; Cuisinier, M.; Nazar, L. F., Surface-enhanced redox chemistry of polysulphides on a metallic and polar host for lithium-sulphur batteries. *Nature Communications* **2014**, *5* (1), 4759.
50. Li, Z.; Sami, I.; Yang, J.; Li, J.; Kumar, R. V.; Chhowalla, M., Lithiated metallic molybdenum disulfide nanosheets for high-performance lithium–sulfur batteries. *Nature Energy* **2023**, *8* (1), 84–93.
51. Ponraj, R.; Kannan, A. G.; Ahn, J. H.; Kim, D.-W., Improvement of cycling performance of lithium–sulfur batteries by using magnesium oxide as a functional additive for trapping lithium polysulfide. *ACS Applied Materials & Interfaces* **2016**, *8* (6), 4000–4006.
52. Zeng, P.; Li, X.; Zhao, B.; Gao, J.; Feng, W.; Wang, Q.; Song, Y., Electronic structure engineering in electrocatalysts: Enabling regulated redox mediation for advanced lithium-sulfur chemistry. *Advanced Energy Materials* **2025**, *15* (32), 2501603.
53. Wang, P.; Xi, B.; Xiong, S., Insights into the optimization of catalytic active sites in lithium–sulfur batteries. *Accounts of Chemical Research* **2024**, *57* (15), 2093–2104.
54. Li, H.; Meng, R.; Ye, C.; Tadich, A.; Hua, W.; Gu, Q.; Johannessen, B.; Chen, X.; Davey, K.; Qiao, S.-Z., Developing high-power Li–S batteries via transition metal/carbon nanocomposite electrocatalyst engineering. *Nature Nanotechnology* **2024**, *19* (6), 792–799.

55. Yang, H.; Xu, Y.; Zhao, Y.; Geng, C.; Ge, H.; Wang, L.; Lv, W.; Yang, C.; Chen, W.; Yang, Q.-H., Catalysis-driven sulfur conversion: From electrolyte-flooded to solid-state batteries. *Advanced Functional Materials* **2025**, *35* (27), 2426089.
56. Kim, S.; Lim, W.-G.; Jung, H.; Jeong, Y. C.; Park, C.-Y.; Yang, S. B.; Lee, C. H.; Wang, D.; Sohn, K.; Han, J. W.; Lee, J., Protective catalytic layer powering activity and stability of electrocatalyst for high-energy lithium-sulfur pouch cell. *Nature Communications* **2025**, *16* (1), 1649.
57. Al Salem, H.; Babu, G.; V. Rao, C.; Arava, L. M. R., Electrocatalytic polysulfide traps for controlling redox shuttle process of Li–S batteries. *Journal of the American Chemical Society* **2015**, *137* (36), 11542–11545.
58. Shen, Z.; Jin, X.; Tian, J.; Li, M.; Yuan, Y.; Zhang, S.; Fang, S.; Fan, X.; Xu, W.; Lu, H.; Lu, J.; Zhang, H., Cation-doped ZnS catalysts for polysulfide conversion in lithium–sulfur batteries. *Nature Catalysis* **2022**, *5* (6), 555–563.
59. He, Q.; Chen, W.; Fan, B.; Wei, Q.; Zou, Y., Abundant adsorption and catalytic sites of the CoS_2/MoS_2 heterostructure for enhanced reversible kinetics in polysulfide conversion. *Journal of Energy Chemistry* **2025**, *107*, 570–581.
60. Zhu, H.; Chen, S.; Yao, X.; Yang, K.; Zhao, W.; Chen, T.; Yang, L.; Pan, F., Upcycling spent cathode materials to bifunctional catalysts for high-stability lithium–sulfur batteries. *Advanced Functional Materials* **2024**, *34* (29), 2401470.
61. Liu, Q.; Wu, Y.; Li, D.; Peng, Y.-Q.; Liu, X.; Li, B.-Q.; Huang, J.-Q.; Peng, H.-J., Dilute alloying to implant activation centers in nitride electrocatalysts for lithium–sulfur batteries. *Advanced Materials* **2023**, *35* (7), 2209233.
62. Babu, G.; Masurkar, N.; Al Salem, H.; Arava, L. M. R., Transition metal dichalcogenide atomic layers for lithium polysulfides electrocatalysis. *Journal of the American Chemical Society* **2017**, *139* (1), 171–178.
63. Yuan, Z.; Peng, H.-J.; Hou, T.-Z.; Huang, J.-Q.; Chen, C.-M.; Wang, D.-W.; Cheng, X.-B.; Wei, F.; Zhang, Q., Powering lithium–sulfur battery performance by propelling polysulfide redox at sulfiphilic hosts. *Nano Letters* **2016**, *16* (1), 519–527.
64. Zhou, T.; Lv, W.; Li, J.; Zhou, G.; Zhao, Y.; Fan, S.; Liu, B.; Li, B.; Kang, F.; Yang, Q.-H., Twinborn TiO_2–TiN heterostructures enabling smooth trapping–diffusion–conversion of polysulfides towards ultralong life lithium–sulfur batteries. *Energy & Environmental Science* **2017**, *10* (7), 1694–1703.
65. Zhang, J.; Yang, D.; Li, C.; Gong, Q.; Bi, W.; Lai, W.-H.; Li, S.; Lei, Y.; Zhou, G.; Cabot, A.; Wang, G., Promoting sulfur redox kinetics of atomically dispersed Fe–NC electrocatalysts by carbon vacancies toward robust lithium–sulfur batteries. *Energy & Environmental Science* **2025**, *18*, 7905–7915.
66. You, T.; Sun, H.; Hua, W.; Geng, S.; Hu, Z.; Shang, Y.; Huang, Q.; Dai, S.; Chen, K., Insights into co-catalytic single-atom-support interactions for boosting sulfur reduction electrocatalysis. *Angewandte Chemie International Edition* **2025**, *64* (16), e202425144.
67. Xu, W.; Lang, S.; Wang, K.; Zeng, R.; Li, H.; Feng, X.; Krumov, M. R.; Bak, S.-M.; Pollock, C. J.; Yeo, J.; Du, Y.; Abruña, H. D., Fundamental mechanistic insights into the catalytic reactions of Li–S redox by Co single-atom electrocatalysts via operando methods. *Science Advances* **2023**, *9* (33), eadi5108.
68. Kong, X.; Li, Y.; Cai, G.; Liu, W.; Xu, J.; Liu, C.; Zhang, G.; Wang, Y.; Lu, Z.; Zhang, J.; Wu, X.; Zhang, D.; Luo, H.; Jin, S.; Ji, H., Asymmetric coordinated single-atom catalysts offering zero-order sulfur redox kinetics for high performance Li–S batteries. *Angewandte Chemie International Edition* **2025**, *64* (37), e202510212.
69. Lim, W.-G.; Park, C.-Y.; Jung, H.; Kim, S.; Kang, S. H.; Lee, Y.-G.; Jeong, Y. C.; Yang, S. B.; Sohn, K.; Han, J. W.; Lee, J., Cooperative electronic structure modulator of Fe single-atom electrocatalyst for high energy and long cycle Li–S pouch cell. *Advanced Materials* **2023**, *35* (10), 2208999.

70. Zhao, C.; Xu, G.-L.; Yu, Z.; Zhang, L.; Hwang, I.; Mo, Y.-X.; Ren, Y.; Cheng, L.; Sun, C.-J.; Ren, Y.; Zuo, X.; Li, J.-T.; Sun, S.-G.; Amine, K.; Zhao, T., A high-energy and long-cycling lithium–sulfur pouch cell via a macroporous catalytic cathode with double-end binding sites. *Nature Nanotechnology* **2021**, *16* (2), 166–173.
71. Zhou, G.; Zhao, S.; Wang, T.; Yang, S.-Z.; Johannessen, B.; Chen, H.; Liu, C.; Ye, Y.; Wu, Y.; Peng, Y.; Liu, C.; Jiang, S. P.; Zhang, Q.; Cui, Y., Theoretical calculation guided design of single-atom catalysts toward fast kinetic and long-life Li–S batteries. *Nano Letters* **2020**, *20* (2), 1252–1261.
72. Zhang, H.; Wan, F.; Li, X.; Zhang, M.; Zhang, N.; Wang, P.; Xiong, S.; Feng, J.; Xi, B., Atomically dispersed Co–Ru dimer catalyst boosts conversion of polysulfides toward high-performance lithium–sulfur batteries. *Advanced Materials* **2025**, *37* (28), 2500950.
73. Zhang, Y.; Yu, T.; Xiao, R.; Tang, P.; Fang, R.; Li, Z.; Cheng, H.-M.; Sun, Z.; Li, F., The role of long-range interactions between high-entropy single-atoms in catalyzing sulfur conversion reactions. *Advanced Materials* **2025**, *37* (10), 2413653.
74. Peng, L.; Wei, Z.; Wan, C.; Li, J.; Chen, Z.; Zhu, D.; Baumann, D.; Liu, H.; Allen, C. S.; Xu, X.; Kirkland, A. I.; Shakir, I.; Almutairi, Z.; Tolbert, S.; Dunn, B.; Huang, Y.; Sautet, P.; Duan, X., A fundamental look at electrocatalytic sulfur reduction reaction. *Nature Catalysis* **2020**, *3* (9), 762–770.
75. Wang, S.; Lu, B.; Cheng, D.; Wu, Z.; Feng, S.; Zhang, M.; Li, W.; Miao, Q.; Patel, M.; Feng, J.; Hopkins, E.; Zhou, J.; Parab, S.; Bhamwala, B.; Liaw, B.; Meng, Y. S.; Liu, P., Structural transformation in a sulfurized polymer cathode to enable long-life rechargeable lithium–sulfur batteries. *Journal of the American Chemical Society* **2023**, *145* (17), 9624–9633.
76. Liu, J.; Lu, H.; Kong, X.; Guan, Y.; Wang, Q.; Chong, B.; Yang, J.; NuLi, Y.; Duan, H.; Wang, J., Formation mechanism and molecular structure of sulfurized polyacrylonitrile. *Advanced Materials* **2025**, *37* (29), 2503534.
77. Liu, J.; Lu, H.; Wang, Q.; Kong, X.; Hao, C.; Yang, J.; NuLi, Y.; Duan, H.; Wang, J., Roles of the polymer backbone for sulfurized polyacrylonitrile cathodes in rechargeable lithium batteries. *Journal of the American Chemical Society* **2025**, *147* (1), 426–435.
78. Phan, A. L.; Le, P. M. L.; Wang, C., Realizing high-energy and long-life Li/SPAN batteries. *Joule* **2024**, *8* (6), 1601–1618.
79. Yin, L.; Wang, J.; Lin, F.; Yang, J.; Nuli, Y., Polyacrylonitrile/graphene composite as a precursor to a sulfur-based cathode material for high-rate rechargeable Li–S batteries. *Energy & Environmental Science* **2012**, *5* (5), 6966–6972.
80. Wang, X.; Qian, Y.; Wang, L.; Yang, H.; Li, H.; Zhao, Y.; Liu, T., Sulfurized polyacrylonitrile cathodes with high compatibility in both ether and carbonate electrolytes for ultrastable lithium–sulfur batteries. *Advanced Functional Materials* **2019**, *29* (39), 1902929.
81. Moschner, R.; Gerle, M.; Danner, T.; Simanjuntak, E. K.; Michalowski, P.; Latz, A.; Nojabaee, M.; Kwade, A.; Friedrich, K. A., Impact of the sulfurized polyacrylonitrile cathode microstructure on the electrochemical performance of lithium–sulfur batteries. *Advanced Science* **2025**, *12* (15), 2415436.
82. Wu, Q.; Wu, Y.; Yan, H.; Zhong, W.; Qin, M.; Zhu, H.; Cheng, S.; Xie, J., A cross-linked nanoflower network and Se-doping enabling sulfur-rich SPAN towards lithium–sulfur batteries beyond 600 W h kg^{-1}. *Energy & Environmental Science* **2025**, *18* (10), 4905–4915.
83. Li, S.; Zeng, Z.; Yang, J.; Han, Z.; Hu, W.; Wang, L.; Ma, J.; Shan, B.; Xie, J., High performance room temperature sodium–sulfur battery by eutectic acceleration in tellurium-doped sulfurized polyacrylonitrile. *ACS Applied Energy Materials* **2019**, *2* (4), 2956–2964.

84. Chen, X.; Peng, L.; Wang, L.; Yang, J.; Hao, Z.; Xiang, J.; Yuan, K.; Huang, Y.; Shan, B.; Yuan, L.; Xie, J., Ether-compatible sulfurized polyacrylonitrile cathode with excellent performance enabled by fast kinetics via selenium doping. *Nature Communications* **2019**, *10* (1), 1021.
85. Li, Z.; Zhang, J.; Lu, Y.; Lou, X. W., A pyrolyzed polyacrylonitrile/selenium disulfide composite cathode with remarkable lithium and sodium storage performances. *Science Advances* **2018**, *4* (6), eaat1687.
86. Lei, J.; Chen, J.; Zhang, H.; Naveed, A.; Yang, J.; Nuli, Y.; Wang, J., High molecular weight polyacrylonitrile precursor for S@pPAN composite cathode materials with high specific capacity for rechargeable lithium batteries. *ACS Applied Materials & Interfaces* **2020**, *12* (30), 33702–33709.
87. Zhang, Y.-Z.; Wu, Z.-Z.; Pan, G.-L.; Liu, S.; Gao, X.-P., Microporous carbon polyhedrons encapsulated polyacrylonitrile nanofibers as sulfur immobilizer for lithium–sulfur battery. *ACS Applied Materials & Interfaces* **2017**, *9* (14), 12436–12444.
88. Wang, Y.; Shuai, Y.; Chen, K., Diphenyl guanidine as vulcanization accelerators in sulfurized polyacrylonitrile for high performance lithium-sulfur battery. *Chemical Engineering Journal* **2020**, *388*, 124378.
89. Yang, H.; Chen, J.; Yang, J.; Nuli, Y.; Wang, J., Dense and high loading sulfurized pyrolyzed poly (acrylonitrile)(S@pPAN) cathode for rechargeable lithium batteries. *Energy Storage Materials* **2020**, *31*, 187–194.
90. Chen, J.; Zhang, H.; Yang, H.; Lei, J.; Naveed, A.; Yang, J.; Nuli, Y.; Wang, J., Towards practical Li–S battery with dense and flexible electrode containing lean electrolyte. *Energy Storage Materials* **2020**, *27*, 307–315.
91. Ma, J.; Fang, Z.; Yan, Y.; Yang, Z.; Gu, L.; Hu, Y.-S.; Li, H.; Wang, Z.; Huang, X., Novel large-scale synthesis of a C/S nanocomposite with mixed conducting networks through a spray drying approach for Li–S batteries. *Advanced Energy Materials* **2015**, *5* (16), 1500046.
92. Wang, W.; Wang, Y.; Huang, Y.; Huang, C.; Yu, Z.; Zhang, H.; Wang, A.; Yuan, K., The electrochemical performance of lithium–sulfur batteries with $LiClO_4$ DOL/DME electrolyte. *Journal of Applied Electrochemistry* **2010**, *40* (2), 321–325.
93. Su, C.-C.; He, M.; Amine, R.; Amine, K., A selection rule for hydrofluoroether electrolyte cosolvent: Establishing a linear free-energy relationship in lithium–sulfur batteries. *Angewandte Chemie International Edition* **2019**, *58* (31), 10591–10595.
94. Zheng, J.; Ji, G.; Fan, X.; Chen, J.; Li, Q.; Wang, H.; Yang, Y.; DeMella, K. C.; Raghavan, S. R.; Wang, C., High-fluorinated electrolytes for Li–S batteries. *Advanced Energy Materials* **2019**, *9* (16), 1803774.
95. Zhao, Q.; Tu, Z.; Wei, S.; Zhang, K.; Choudhury, S.; Liu, X.; Archer, L. A., Building Organic/inorganic hybrid interphases for fast interfacial transport in rechargeable metal batteries. *Angewandte Chemie International Edition* **2018**, *57* (4), 992–996.
96. Zhang, S. S., A new finding on the role of $LiNO_3$ in lithium-sulfur battery. *Journal of Power Sources* **2016**, *322*, 99–105.
97. Li, W.; Yao, H.; Yan, K.; Zheng, G.; Liang, Z.; Chiang, Y.-M.; Cui, Y., The synergetic effect of lithium polysulfide and lithium nitrate to prevent lithium dendrite growth. *Nature Communications* **2015**, *6* (1), 7436.
98. Liu, S.; Li, G.-R.; Gao, X.-P., Lanthanum nitrate as electrolyte additive to stabilize the surface morphology of lithium anode for lithium–sulfur battery. *ACS Applied Materials & Interfaces* **2016**, *8* (12), 7783–7789.
99. Li, S.; Dai, H.; Li, Y.; Lai, C.; Wang, J.; Huo, F.; Wang, C., Designing Li-protective layer via $SOCl_2$ additive for stabilizing lithium-sulfur battery. *Energy Storage Materials* **2019**, *18*, 222–228.
100. Yu, L.; Chen, S.; Lee, H.; Zhang, L.; Engelhard, M. H.; Li, Q.; Jiao, S.; Liu, J.; Xu, W.; Zhang, J.-G., A localized high-concentration electrolyte with optimized solvents and

lithium difluoro(oxalate)borate additive for stable lithium metal batteries. *ACS Energy Letters* **2018**, *3* (9), 2059–2067.
101. Shi, P.; Zhang, L.; Xiang, H.; Liang, X.; Sun, Y.; Xu, W., Lithium difluorophosphate as a dendrite-suppressing additive for lithium metal batteries. *ACS Applied Materials & Interfaces* **2018**, *10* (26), 22201–22209.
102. Huang, Z.; Ren, J.; Zhang, W.; Xie, M.; Li, Y.; Sun, D.; Shen, Y.; Huang, Y., Protecting the Li-metal anode in a Li-O_2 battery by using boric acid as an SEI-forming additive. *Advanced Materials* **2018**, *30* (39), 1803270.
103. Dai, H.; Xi, K.; Liu, X.; Lai, C.; Zhang, S., Cationic surfactant-based electrolyte additives for uniform lithium deposition via lithiophobic repulsion mechanisms. *Journal of the American Chemical Society* **2018**, *140* (50), 17515–17521.
104. Ding, F.; Xu, W.; Graff, G. L.; Zhang, J.; Sushko, M. L.; Chen, X.; Shao, Y.; Engelhard, M. H.; Nie, Z.; Xiao, J.; Liu, X.; Sushko, P. V.; Liu, J.; Zhang, J.-G., Dendrite-free lithium deposition via self-healing electrostatic shield mechanism. *Journal of the American Chemical Society* **2013**, *135* (11), 4450–4456.
105. Dai, H.; Gu, X.; Dong, J.; Wang, C.; Lai, C.; Sun, S., Stabilizing lithium metal anode by octaphenyl polyoxyethylene-lithium complexation. *Nature Communications* **2020**, *11* (1), 643.
106. Wei, J.-Y.; Zhang, X.-Q.; Hou, L.-P.; Shi, P.; Li, B.-Q.; Xiao, Y.; Yan, C.; Yuan, H.; Huang, J.-Q., Shielding polysulfide intermediates by an organosulfur-containing solid electrolyte interphase on the lithium anode in lithium–sulfur batteries. *Advanced Materials* **2020**, *32* (37), 2003012.
107. Zheng, H.; Xiang, H.; Jiang, F.; Liu, Y.; Sun, Y.; Liang, X.; Feng, Y.; Yu, Y., Lithium difluorophosphate-based dual-salt low concentration electrolytes for lithium metal batteries. *Advanced Energy Materials* **2020**, *10* (30), 2001440.
108. Qian, J.; Henderson, W. A.; Xu, W.; Bhattacharya, P.; Engelhard, M.; Borodin, O.; Zhang, J.-G., High rate and stable cycling of lithium metal anode. *Nature Communications* **2015**, *6* (1), 6362.
109. Peng, Z.; Cao, X.; Gao, P.; Jia, H.; Ren, X.; Roy, S.; Li, Z.; Zhu, Y.; Xie, W.; Liu, D.; Li, Q.; Wang, D.; Xu, W.; Zhang, J.-G., High-power lithium metal batteries enabled by high-concentration acetonitrile-based electrolytes with vinylene carbonate additive. *Advanced Functional Materials* **2020**, *30* (24), 2001285.
110. Yoo, D.-J.; Yang, S.; Kim, K. J.; Choi, J. W., Fluorinated aromatic diluent for high-performance lithium metal batteries. *Angewandte Chemie International Edition* **2020**, *59* (35), 14869–14876.
111. Kim, J. T.; Su, H.; Zhong, Y.; Wang, C.; Wu, H.; Zhao, D.; Wang, C.; Sun, X.; Li, Y., All-solid-state lithium–sulfur batteries through a reaction engineering lens. *Nature Chemical Engineering* **2024**, *1* (6), 400–410.
112. Lee, J.; Zhao, C.; Wang, C.; Chen, A.; Sun, X.; Amine, K.; Xu, G.-L., Bridging the gap between academic research and industrial development in advanced all-solid-state lithium–sulfur batteries. *Chemical Society Reviews* **2024**, *53* (10), 5264–5290.
113. Shao, J.; Li, Y.; Guo, J.; Duan, J.; Liang, H.; Dou, Y.; Lu, S.; Liu, J., Solid polymer electrolytes for all-solid-state lithium–sulfur batteries: Different designs dependent on their interaction with sulfur cathodes. *Advanced Materials* **2025**, *37* (24), 2415864.
114. Zhang, X.; Zhang, H.; Geng, Y.; Shi, Z.; Zhu, S.; Xu, Q.; Min, Y., A multifunctional nano filler for solid polymer electrolyte toward stable cycling for lithium-metal anodes in lithium–sulfur batteries. *Chemical Engineering Journal* **2022**, *444*, 136328.
115. Santiago, A.; Castillo, J.; Garbayo, I.; Saenz de Buruaga, A.; Coca Clemente, J. A.; Qiao, L.; Cid Barreno, R.; Martinez-Ibañez, M.; Armand, M.; Zhang, H.; Li, C., Salt additives for improving cyclability of polymer-based all-solid-state lithium–sulfur batteries. *ACS Applied Energy Materials* **2021**, *4* (5), 4459–4464.
116. Fan, Z.; Ding, B.; Zhang, T.; Lin, Q.; Malgras, V.; Wang, J.; Dou, H.; Zhang, X.; Yamauchi, Y., Solid/solid interfacial architecturing of solid polymer electrolyte–based

all-solid-state lithium–sulfur batteries by atomic layer deposition. *Small* **2019**, *15* (46), 1903952.

117. Kato, Y.; Hori, S.; Saito, T.; Suzuki, K.; Hirayama, M.; Mitsui, A.; Yonemura, M.; Iba, H.; Kanno, R., High-power all-solid-state batteries using sulfide superionic conductors. *Nature Energy* **2016**, *1* (4), 16030.
118. Yu, Z.; Singh, B.; Yu, Y.; Nazar, L. F., Suppressing argyrodite oxidation by tuning the host structure for high-areal-capacity all-solid-state lithium–sulfur batteries. *Nature Materials* **2025**, *24* (7), 1082–1090.
119. Zhao, B.-S.; Chen, P.; Gao, X.-P., Bulk and interface-strengthened $Li_7P_{2.9}Sb_{0.1}S_{10.65}O_{0.15}I_{0.2}$ electrolyte via dual-source doping for all-solid-state lithium-sulfur batteries. *Science China Materials* **2023**, *66* (2), 513–521.
120. Wan, H.; Wang, Z.; Zhang, W.; He, X.; Wang, C., Interface design for all-solid-state lithium batteries. *Nature* **2023**, *623* (7988), 739–744.
121. Jun, K.; Chen, Y.; Wei, G.; Yang, X.; Ceder, G., Diffusion mechanisms of fast lithium-ion conductors. *Nature Reviews Materials* **2024**, *9* (12), 887–905.
122. Chen, Y.; Qian, J.; Wang, K.; Li, L.; Wu, F.; Chen, R., Cutting-edge developments at the interface of inorganic solid-state electrolytes. *Advanced Materials* **2025**, *37* (39), 2502653.
123. Xu, Y.; Wang, K.; Zhang, X.; Ma, Y.; Peng, Q.; Gong, Y.; Yi, S.; Guo, H.; Zhang, X.; Sun, X.; Gao, H.; Xin, S.; Guo, Y.-G.; Ma, Y., Improved Li-ion conduction and (electro)chemical stability at garnet-polymer interface through metal-nitrogen bonding. *Advanced Energy Materials* **2023**, *13* (14), 2204377.
124. Zhang, X.; Wang, J.; Hu, D.; Du, W.; Hou, C.; Jiang, H.; Wei, Y.; Liu, X.; Jiang, F.; Sun, J.; Yuan, H.; Huang, X., High-performance lithium metal batteries based on composite solid-state electrolytes with high ceramic content. *Energy Storage Materials* **2024**, *65*, 103089.
125. Fu, K.; Gong, Y.; Hitz, G. T.; McOwen, D. W.; Li, Y.; Xu, S.; Wen, Y.; Zhang, L.; Wang, C.; Pastel, G.; Dai, J.; Liu, B.; Xie, H.; Yao, Y.; Wachsman, E. D.; Hu, L., Three-dimensional bilayer garnet solid electrolyte based high energy density lithium metal–sulfur batteries. *Energy & Environmental Science* **2017**, *10* (7), 1568–1575.
126. Zheng, F.; Li, C.; Li, Z.; Cao, X.; Luo, H.; Liang, J.; Zhao, X.; Kong, J., Advanced composite solid electrolytes for lithium batteries: Filler dimensional design and ion path optimization. *Small* **2023**, *19* (21), 2206355.
127. Nie, L.; Zhu, J.; Wu, X.; Zhang, M.; Xiao, X.; Gao, R.; Wu, X.; Zhu, Y.; Chen, S.; Han, Z.; Yu, Y.; Wang, S.; Ling, S.; Zhou, G., A large-scale fabrication of flexible, ultrathin, and robust solid electrolyte for solid-state lithium-sulfur batteries. *Advanced Materials* **2024**, *36* (29), 2400115.
128. Judez, X.; Eshetu, G. G.; Gracia, I.; López-Aranguren, P.; González-Marcos, J. A.; Armand, M.; Rodriguez-Martinez, L. M.; Zhang, H.; Li, C., Understanding the role of nano-aluminum oxide in all-solid-state lithium-sulfur batteries. *ChemElectroChem* **2019**, *6* (2), 326–330.
129. Yan, Y.; Yin, Y.; Guo, Y.; Wan, L.-J., Effect of cations in ionic liquids on the electrochemical performance of lithium-sulfur batteries. *Science China Chemistry* **2014**, *57* (11), 1564–1569.
130. Xiong, S.; Scheers, J.; Aguilera, L.; Lim, D.-H.; Xie, K.; Jacobsson, P.; Matic, A., Role of organic solvent addition to ionic liquid electrolytes for lithium–sulphur batteries. *RSC Advances* **2015**, *5* (3), 2122–2128.
131. Pal, U.; Girard, G. M. A.; O'Dell, L. A.; Roy, B.; Wang, X.; Armand, M.; MacFarlane, D. R.; Howlett, P. C.; Forsyth, M., Improved Li-ion transport by DME chelation in a novel ionic liquid-based hybrid electrolyte for Li–S battery application. *The Journal of Physical Chemistry C* **2018**, *122* (26), 14373–14382.

132. Xu, W.; Wang, J.; Ding, F.; Chen, X.; Nasybulin, E.; Zhang, Y.; Zhang, J.-G., Lithium metal anodes for rechargeable batteries. *Energy & Environmental Science* **2014**, *7* (2), 513–537.
133. Cheng, X.-B.; Zhang, R.; Zhao, C.-Z.; Zhang, Q., Toward safe lithium metal anode in rechargeable batteries: A review. *Chemical Reviews* **2017**, *117* (15), 10403–10473.
134. Zhao, Y.; Ye, Y.; Wu, F.; Li, Y.; Li, L.; Chen, R., Anode interface engineering and architecture design for high-performance lithium–sulfur batteries. *Advanced Materials* **2019**, *31* (12), 1806532.
135. Chen, L.; Fan, X.; Ji, X.; Chen, J.; Hou, S.; Wang, C., High-energy Li metal battery with lithiated host. *Joule* **2019**, *3* (3), 732–744.
136. Fang, Y.; Zhang, S. L.; Wu, Z.-P.; Luan, D.; Lou, X. W., A highly stable lithium metal anode enabled by Ag nanoparticle–embedded nitrogen-doped carbon macroporous fibers. *Science Advances* **2021**, *7* (21), eabg3626.
137. Wang, T.-S.; Liu, X.; Wang, Y.; Fan, L.-Z., High areal capacity dendrite-free Li anode enabled by metal–organic framework-derived nanorod array modified carbon cloth for solid state Li metal batteries. *Advanced Functional Materials* **2021**, *31* (2), 2001973.
138. Jiang, J.; Pan, Z.; Kou, Z.; Nie, P.; Chen, C.; Li, Z.; Li, S.; Zhu, Q.; Dou, H.; Zhang, X.; Wang, J., Lithiophilic polymer interphase anchored on laser-punched 3D holey Cu matrix enables uniform lithium nucleation leading to super-stable lithium metal anodes. *Energy Storage Materials* **2020**, *29*, 84–91.
139. Chen, X.; Shang, M.; Niu, J., Inter-layer-calated thin Li metal electrode with improved battery capacity retention and dendrite suppression. *Nano Letters* **2020**, *20* (4), 2639–2646.
140. Zhang, Z.; Wang, J.; Yan, X.; Zhang, S.; Yang, W.; Zhuang, Z.; Han, W.-Q., In-situ growth of hierarchical N-doped CNTs/Ni Foam scaffold for dendrite-free lithium metal anode. *Energy Storage Materials* **2020**, *29*, 332–340.
141. Dong, L.; Nie, L.; Liu, W., Water-stable lithium metal anodes with ultrahigh-rate capability enabled by a hydrophobic graphene architecture. *Advanced Materials* **2020**, *32* (14), 1908494.
142. Zuo, T.-T.; Yin, Y.-X.; Wang, S.-H.; Wang, P.-F.; Yang, X.; Liu, J.; Yang, C.-P.; Guo, Y.-G., Trapping lithium into hollow silica microspheres with a carbon nanotube core for dendrite-free lithium metal anodes. *Nano Letters* **2018**, *18* (1), 297–301.
143. Wu, C.; Huang, H.; Lu, W.; Wei, Z.; Ni, X.; Sun, F.; Qing, P.; Liu, Z.; Ma, J.; Wei, W.; Chen, L.; Yan, C.; Mai, L., Mg doped Li–Li B alloy with in situ formed lithiophilic LiB skeleton for lithium metal batteries. *Advanced Science* **2020**, *7* (6), 1902643.
144. Li, B. Q.; Chen, X. R.; Chen, X.; Zhao, C. X.; Zhang, R.; Cheng, X. B.; Zhang, Q., Favorable lithium nucleation on lithiophilic framework porphyrin for dendrite-free lithium metal anodes. *Research* **2019**, *2019*, 4608940.
145. Zhao, Y.; Wang, D.; Gao, Y.; Chen, T.; Huang, Q.; Wang, D., Stable Li metal anode by a polyvinyl alcohol protection layer via modifying solid-electrolyte interphase layer. *Nano Energy* **2019**, *64*, 103893.
146. Fan, L.; Guo, Z.; Zhang, Y.; Wu, X.; Zhao, C.; Sun, X.; Yang, G.; Feng, Y.; Zhang, N., Stable artificial solid electrolyte interphase films for lithium metal anode via metal–organic frameworks cemented by polyvinyl alcohol. *Journal of Materials Chemistry A* **2020**, *8* (1), 251–258.
147. Wu, H.; Yao, Z.; Wu, Q.; Fan, S.; Yin, C.; Li, C., Confinement effect and air tolerance of Li plating by lithiophilic poly(vinyl alcohol) coating for dendrite-free Li metal batteries. *Journal of Materials Chemistry A* **2019**, *7* (39), 22257–22264.
148. You, J.-H.; Zhang, S.-J.; Deng, L.; Li, M.-Z.; Zheng, X.-M.; Li, J.-T.; Zhou, Y.; Huang, L.; Sun, S.-G., Suppressing Li dendrite by a protective biopolymeric film from tamarind seed polysaccharide for high-performance Li metal anode. *Electrochimica Acta* **2019**, *299*, 636–644.

149. Zhang, S.-J.; Gao, Z.-G.; Wang, W.-W.; Lu, Y.-Q.; Deng, Y.-P.; You, J.-H.; Li, J.-T.; Zhou, Y.; Huang, L.; Zhou, X.-D.; Sun, S.-G., A natural biopolymer film as a robust protective layer to effectively stabilize lithium-metal anodes. *Small* **2018**, *14* (31), 1801054.
150. Song, J.; Lee, H.; Choo, M.-J.; Park, J.-K.; Kim, H.-T., Ionomer-liquid electrolyte hybrid ionic conductor for high cycling stability of lithium metal electrodes. *Scientific Reports* **2015**, *5* (1), 14458.
151. Sun, F.; Zhou, D.; He, X.; Osenberg, M.; Dong, K.; Chen, L.; Mei, S.; Hilger, A.; Markötter, H.; Lu, Y.; Dong, S.; Marathe, S.; Rau, C.; Hou, X.; Li, J.; Stan, M. C.; Winter, M.; Dominko, R.; Manke, I., Morphological reversibility of modified Li-based anodes for next-generation batteries. *ACS Energy Letters* **2020**, *5* (1), 152–161.
152. Zhu, B.; Jin, Y.; Hu, X.; Zheng, Q.; Zhang, S.; Wang, Q.; Zhu, J., Poly(dimethylsiloxane) Thin film as a stable interfacial layer for high-performance lithium-metal battery anodes. *Advanced Materials* **2017**, *29* (2), 1603755.
153. Shen, F.; Wang, K.; Yin, Y.; Shi, L.; Zeng, D.; Han, X., PAN/PI functional double-layer coating for dendrite-free lithium metal anodes. *Journal of Materials Chemistry A* **2020**, *8* (13), 6183–6189.
154. Sun, Y.; Zhao, C.; Adair, K. R.; Zhao, Y.; Goncharova, L. V.; Liang, J.; Wang, C.; Li, J.; Li, R.; Cai, M.; Sham, T.-K.; Sun, X., Regulated lithium plating and stripping by a nano-scale gradient inorganic–organic coating for stable lithium metal anodes. *Energy & Environmental Science* **2021**, *14* (7), 4085–4094.
155. Liu, S.; Ma, Y.; Wang, J.; Zuo, P.; Du, C.; Yin, G.; Gao, Y., Regulating Li deposition by constructing homogeneous LiF protective layer for high-performance Li metal anode. *Chemical Engineering Journal* **2022**, *427*, 131625.
156. Liu, Y.; Tzeng, Y.-K.; Lin, D.; Pei, A.; Lu, H.; Melosh, N. A.; Shen, Z.-X.; Chu, S.; Cui, Y., An ultrastrong double-layer nanodiamond interface for stable lithium metal anodes. *Joule* **2018**, *2* (8), 1595–1609.
157. Li, N.-W.; Yin, Y.-X.; Yang, C.-P.; Guo, Y.-G., An artificial solid electrolyte interphase layer for stable lithium metal anodes. *Advanced Materials* **2016**, *28* (9), 1853–1858.
158. Shen, X.; Qian, T.; Chen, P.; Liu, J.; Wang, M.; Yan, C., Bioinspired polysulfiphobic artificial interphase layer on lithium metal anodes for lithium sulfur batteries. *ACS Applied Materials & Interfaces* **2018**, *10* (36), 30058–30064.
159. Wang, Z.; Wang, Y.; Zhang, Z.; Chen, X.; Lie, W.; He, Y.-B.; Zhou, Z.; Xia, G.; Guo, Z., Building artificial solid-electrolyte interphase with uniform intermolecular ionic bonds toward dendrite-free lithium metal anodes. *Advanced Functional Materials* **2020**, *30* (30), 2002414.
160. Chen, D.; Huang, S.; Zhong, L.; Wang, S.; Xiao, M.; Han, D.; Meng, Y., In situ preparation of thin and rigid COF film on Li anode as artificial solid electrolyte interphase layer resisting Li dendrite puncture. *Advanced Functional Materials* **2020**, *30* (7), 1907717.
161. Wang, Y.; Wang, Z.; Zhao, L.; Fan, Q.; Zeng, X.; Liu, S.; Pang, W. K.; He, Y.-B.; Guo, Z., Lithium metal electrode with increased air stability and robust solid electrolyte interphase realized by silane coupling agent modification. *Advanced Materials* **2021**, *33* (14), 2008133.
162. Liu, H.; Cheng, X.-B.; Huang, J.-Q.; Kaskel, S.; Chou, S.; Park, H. S.; Zhang, Q., Alloy anodes for rechargeable alkali-metal batteries: Progress and challenge. *ACS Materials Letters* **2019**, *1* (2), 217–229.
163. Guo, W.; Han, Q.; Jiao, J.; Wu, W.; Zhu, X.; Chen, Z.; Zhao, Y., In situ construction of robust biphasic surface layers on lithium metal for lithium–sulfide batteries with long cycle life. *Angewandte Chemie International Edition* **2021**, *60* (13), 7267–7274.
164. Ye, H.; Zheng, Z.-J.; Yao, H.-R.; Liu, S.-C.; Zuo, T.-T.; Wu, X.-W.; Yin, Y.-X.; Li, N.-W.; Gu, J.-J.; Cao, F.-F.; Guo, Y.-G., Guiding uniform Li plating/stripping through lithium–aluminum alloying medium for long-life Li metal batteries. *Angewandte Chemie International Edition* **2019**, *58* (4), 1094–1099.

165. Pathak, R.; Chen, K.; Gurung, A.; Reza, K. M.; Bahrami, B.; Pokharel, J.; Baniya, A.; He, W.; Wu, F.; Zhou, Y.; Xu, K.; Qiao, Q., Fluorinated hybrid solid-electrolyte-interphase for dendrite-free lithium deposition. *Nature Communications* **2020**, *11* (1), 93.
166. Jiang, Z.; Jin, L.; Han, Z.; Hu, W.; Zeng, Z.; Sun, Y.; Xie, J., Facile generation of polymer–alloy hybrid layers for dendrite-free lithium-metal anodes with improved moisture stability. *Angewandte Chemie International Edition* **2019**, *58* (33), 11374–11378.
167. Ma, L.; Kim, M. S.; Archer, L. A., Stable artificial solid electrolyte interphases for lithium batteries. *Chemistry of Materials* **2017**, *29* (10), 4181–4189.

5 The Advanced Position for Metal-Air Batteries in the Future Grid

Taiyuan Wang, Zhan Yan, Jianxin Gao, Qianfeng Liu, Min Liu, Xueliang Wang, Shanshan Gao, and Erdong Wang

5.1 INTRODUCTION

In the global transition toward renewable energy systems, the growing imperative for long-duration energy storage to buffer the intermittent nature of solar and wind generation has intensified the focus on battery chemistries offering both high specific energy and economic viability. In this context, metal-air batteries (MABs) are emerging as a particularly promising technology [1]: they generate electricity via the oxidation of a metallic anode coupled with the reduction of atmospheric oxygen at a porous air cathode, enabling a lightweight, open-air architecture in which ambient oxygen substitutes for stored cathode material, thereby reducing cell mass and cost. Compared to conventional lithium-ion batteries, MABs deliver energy densities of approximately 300 Wh kg^{-1} in zinc-air and theoretically up to several thousand Wh kg^{-1} in lithium-air or aluminum-air chemistries.

Further, the use of abundant and inexpensive metals such as zinc (Zn), iron (Fe), aluminum (Al), and magnesium (Mg) confers substantial cost advantages [2], while minimizing environmental impact relative to lithium-based alternatives. Thus, by combining elevated theoretical energy density, reliance on earth-abundant materials, and scalable, low-cost architecture, MABs represent a compelling and distinctive class of electrochemical energy storage capable of addressing the long-duration demands of renewable energy integration more effectively than conventional battery systems.

The working principle of rechargeable MABs typically involves the oxidation and reduction of the metallic anode and the oxygen cathode. As MABs incorporate design features from both fuel cells and conventional batteries, such as metal-ion cells, they require a liquid electrolyte between electrodes that are separated by a membrane. However, the electrochemical reactions occurring at the electrodes and electrolytes differ depending on the type of metallic anode, as has been reported. For example, metals such as Zn and Fe exhibit kinetic stability in aqueous electrolytes when passivated by corresponding hydroxides or oxides, explaining their prevalence

 DOI: 10.1201/9781003492511-5

in aqueous MABs research. The electrochemical reactions for aqueous MABs are governed by the following equations:

Metal anode:

$$M \rightleftharpoons M^{n+} + ne^-$$

Air cathode:

$$O_2 + 2H_2O + 4e^- \rightleftharpoons 4OH^-$$

Overall reaction:

$$4M + nO_2 + 2nH_2O \rightleftharpoons 4M^{n+} + 4nOH^-$$

Here, M stands for Zn, Fe, Al, and Mg metals, and n denotes the number of metal ion charge transfers (Figure 5.1). It is important to note that these electrochemical reactions represent the discharge process from left to right, which can be reversed upon recharging the MABs. To put it another way, during recharging, the metal is deposited back onto the anode and oxygen is released at the cathode.

A fundamental distinction between aqueous and nonaqueous MABs resides in their oxygen reduction reaction (ORR) and oxygen evolution reaction (OER) mechanisms. In nonaqueous systems, ORR entails the reduction of molecular oxygen at the cathode electrocatalyst surface to form superoxide (O_2^-) or peroxide (O_2^{2-}) species. These subsequently combine with metal cations to yield corresponding metal superoxides (MO_2), peroxides (M_2O_2), or mixed-phase compounds as discharge products. Beyond these mechanistic differences in redox chemistry, nonaqueous systems exhibit a heightened sensitivity of lithium, sodium, and potassium anodes to atmospheric H_2O and CO_2. Consequently, such batteries typically require operation in purified oxygen environments, leading to their conventional designation as Li-O_2, Na-O_2, and K-O_2 batteries.

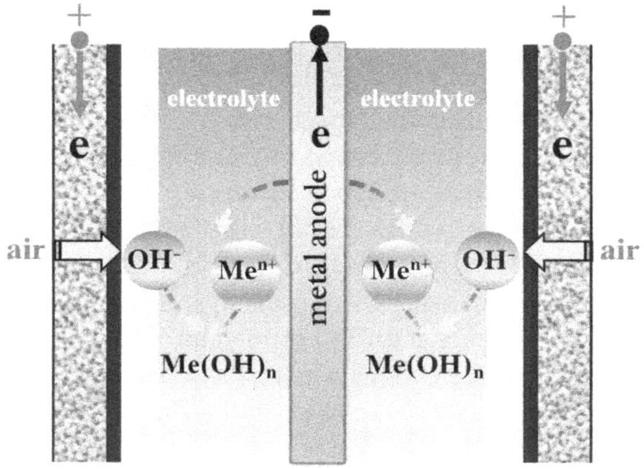

FIGURE 5.1 Working principle and structure of typical metal-air batteries.

By contrast, aqueous MABs demonstrate significant advantages in technological maturity, practical implementation viability, and scalability for grid-scale energy storage (GSES) applications relative to their nonaqueous counterparts. Aqueous MABs combine very low material costs, high theoretical energy density, simple cell architecture, and inherent safety, making them a compelling option for GSES. However, compared with Li-ion batteries, current MAB prototypes exhibit lower round-trip energy efficiency, higher power-capacity costs, and shorter cycle life. These advantageous attributes make MABs particularly well-suited for long-duration energy storage applications, which require high stored-energy-to-power ratios. Moreover, MABs negative electrodes such as Zn, Fe, Al, and Mg use commodity metals produced at scale for other industries, implying that, once commercialized, MABs production could ramp more rapidly than Li-ion manufacturing.

5.2 ZINC–AIR BATTERIES

Zn-air, Fe-air, Al-air, and Mg-air batteries are well-established aqueous MABs. Among these, Zn-air batteries possess the most extensive research history [3], with the first prototype developed in 1878. Despite the commercialization of primary Zn-air batteries in 1932, substantive scientific advancement remained limited until the industrialization of secondary systems circa 2012, when the development of rechargeable Zn-air devices constituted the paramount breakthrough [4]. The advancement of the field during the 20th century was impeded by significant challenges, namely the formation of zinc dendrites and the low energy efficiencies (EE) of only about 60%, compounded by the absence of substantial innovation in the field of bifunctional catalysts [5]. In comparison with Al-air and Mg-air systems, Zn-air batteries exhibit superior electrochemical stability in aqueous electrolytes, primarily attributable to their established rechargeability. The theoretical energy density of 1,353 Wh kg^{-1} has positioned Zn-air batteries as a prominent research focus [6]. Consequently, they are projected to offer significant commercial potential. The prevailing scientific consensus suggests that advanced Zn-air configurations have the potential to exceed the performance of LIBs in large-scale energy storage (LSES) applications [7].

The basic structural schematic of a typical zinc-air battery (Figure 5.2) comprises a zinc anode, a porous air cathode containing suitable catalysts, and an aqueous electrolyte. The Zn anode can be designed in various forms, including foam, fiber, sponge, coating, and alloying, while the air cathode offers flexibility in catalyst design, including metal-free carbon, M–N–C, perovskite, and spinel structures. The gas diffusion electrode (GDE) consists of a catalyst nanoparticle layer on a nanoporous carbon substrate with a gas diffusion polymer as the current collector [9]. It plays an important role in the electrocatalytic reactions of the air cathode in OER/ORR. The electrode reactions that take place in the Zn-air cell can be expressed by the following equations:

Zn anode:

$$Zn + 4OH^- \leftrightarrow Zn(OH)_4^{2-} + 2e^-$$

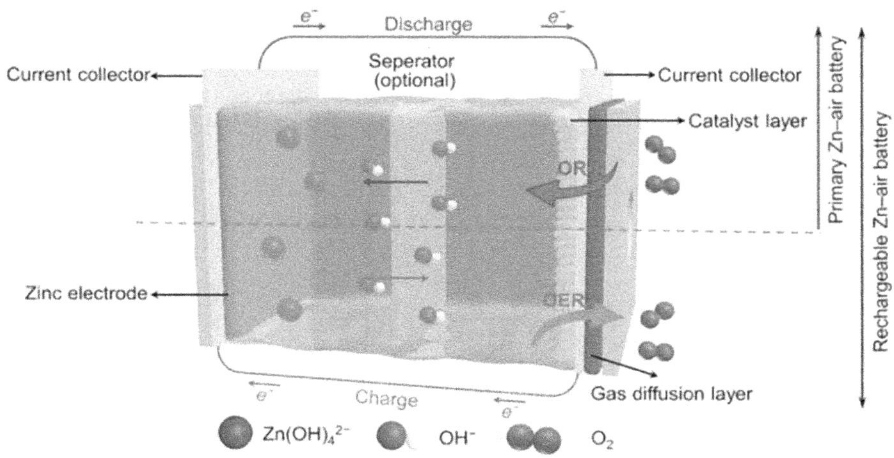

FIGURE 5.2 Schematic configuration of Zn-air battery [8].

$$Zn(OH)_4^{2-} \leftrightarrow ZnO + H_2O + 2OH^-$$

Air cathode:

$$O_2 + 4e^- + 2H_2O \leftrightarrow 4OH^-$$

Overall anode:

$$Zn + 2OH^- \leftrightarrow ZnO + H_2O + 2e^-$$

Overall:

$$2Zn + O_2 \leftrightarrow 2ZnO$$

During discharge, atmospheric oxygen diffuses into the cathode, where it undergoes reduction via the ORR to generate hydroxide ions (OH^-). These hydroxide ions then migrate toward the zinc anode, where they react to form soluble zincate ions ($Zn(OH)_4^{2-}$). The zincate species then decompose to yield insoluble zinc oxide (ZnO) as the primary discharge product, leading to supersaturation of the liquid electrolyte. At the same time, the anodic oxidation of metallic zinc releases electrons that flow through the external circuit toward the air electrode. Notably, the migration of $Zn(OH)_4^{2-}$ from the zinc anode to the air cathode increases polarization and reduces cycling efficiency [10]. To mitigate this issue, a separator is introduced between the electrodes. This selectively permits the transport of OH^- ions while effectively inhibiting the crossover of $Zn(OH)_4^{2-}$, thereby enhancing electrochemical performance. During charging, the aforementioned reactions proceed in reverse: zinc redeposits at the anode and oxygen evolves at the interface between the electrode and electrolyte via the OER.

5.2.1 Air Cathode and Electrocatalysts of Zn—Air Batteries

The air cathode is critical to MABs performance and longevity. Oxygen from the air diffuses to the electrode's three-phase interface and is reduced by an ORR catalyst. Thus, both the cathode's architecture and the ORR catalyst design strongly influence cell efficiency and durability.

A porous electrode structure maximizes ORR activity by balancing gas and electrolyte access. Early zinc-air cells (circa 1878) already employed platinized-carbon cathodes; by 1932, Heise and Schumacher introduced a three-layer porous-carbon electrode treated with wax to prevent flooding, which is a design still in use today [11] (Figure 5.3).

Cathode humidity must be carefully controlled: excessive moisture floods the electrode, whereas dryness causes electrolyte evaporation. Optimizing the hydrophobic/hydrophilic balance via polytetrafluoroethylene (PTFE) loading, the choice of carbon support, and processing conditions enables the cathode to resist flooding and minimize drying. For example, Toray carbon paper uses graphitized-carbon binders for conductivity and PTFE treatment for tunable wettability [12]. Gas-diffusion layers are made as thin as possible, with minimal catalyst–carbon loading entrapped in metal meshes (e.g., QuantumSphere's MetAir® cathode) to facilitate O_2 transport.

Numerous studies have probed how carbon type and loading affect electrode performance. Maja et al. compared carbon blacks (Vulcan XC-72R, Black Pearls 3700, and Shawinigan Black) in a bilayer GDE; Shawinigan-based electrodes achieved the most uniform PTFE distribution and hydrophobic particle networks, sustaining gas-electrolyte equilibrium over prolonged operation [13]. Eom et al. evaluated several activated carbons (Darco G-60N/A, Norit SX Ultra, YP-17, and BP-20) and demonstrated that the ratio of macro- and mesopore area to micropore area, rather than the total BET surface area, governs ORR performance. Darco G-60N exhibited superior ORR activity despite its lower BET surface area, a finding attributed to its favorable pore-size distribution [14]. Nevertheless, the fabrication of multilayer GDEs using PTFE binders is complex and susceptible to microcracking [15].

Sluggish ORR kinetics (O=O bond energy of 498 kJ mol^{-1}) necessitate catalysts to lower overpotentials and boost power density. Catalysts fall into three classes: precious metals, metal oxides, and carbonaceous materials. Pt was the first ORR catalyst in zinc-air cells, and Pt/C remains the benchmark. Pd [16], Ag [17], Au [18],

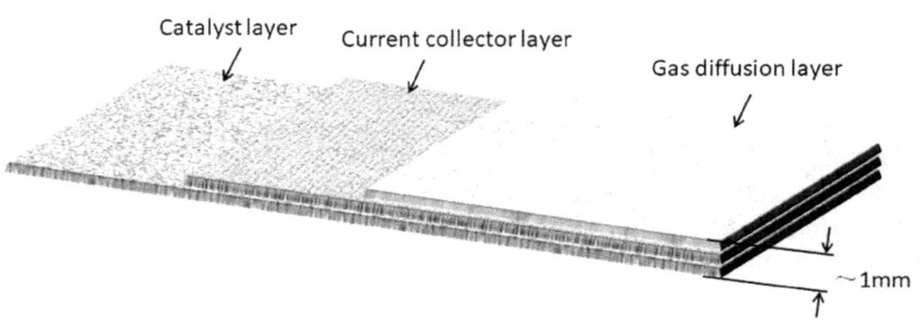

FIGURE 5.3 The typical three-layer structure of the air cathode.

and their alloys have also been explored. Ag is particularly attractive because its cost is approximately 1% that of platinum; nevertheless, achieving competitive catalytic performance requires nanoscale structuring, alloying, or dispersion on carbon supports to enhance the adsorption of reaction intermediates [19]. At high KOH concentrations (e.g., 30 wt%), 10 wt% Ag/C can even outperform Pt/C [20], driving its adoption in commercial cells (e.g., 5% $AgMnO_4 + MnO_2$ catalysts delivering 51 mA cm^{-2} at 1.1 V) [21].

Binary and ternary oxides (spinel, perovskite, etc.) offer low cost and reasonable stability. MnO_x (e.g., Mn_3O_4, MnO_2, and MnOOH) is the most prevalent in hearing-aid Zn-air batteries, where γ-MnO_2 affords >400 Wh kg^{-1}. ORR activity correlates with Mn valence and phase ($\alpha > \beta > \gamma$-MnO_2) and improves for nanostructures and preferential facets. Composite catalysts such as Ag–Mn_3O_4/C [22] and MnO_x–Co_3O_4/C [23] further enhance kinetics. Perovskites ($LaNiO_3$ and $LaMnO_3$) achieve high activity, although $LaMnO_3$ suffers hydroxide-induced degradation; substitution (e.g., $La_{0.6}Sr_{0.4}Fe_{0.6}Mn_{0.4}O_3$) boosts durability (−0.26 V vs. Hg/HgO at 500 mA cm^{-2} over 70 hours) [24]. Spinels such as Co_3O_4 require carbon support; hybrids like Co_3O_4–rGO [64] or Co/NCNT [25] can outperform Pt/C in 10 M NaOH at 80°C.

Pristine carbon is ORR inactive, but heteroatom doping (N, P) introduces active sites. M–N–C materials, especially Fe–N_x and Co–N_x, have emerged as leading non-precious catalysts [26]. Although polypyrrole–Co catalysts showed 0.15–0.20 V overpotentials [27], the optimization of polymeric precursors and pyrolysis has improved activity. While the exact atomic active site remains debated, mounting evidence implicates metal–N_x moieties as the primary ORR centers [28].

5.2.2 Anode Electrochemistry of Zn-Air Batteries

Heavy metals or heavy-metal oxides/hydroxides/nitrides are commonly added to the zinc anode for the inhibition of hydrogen evolution and passivation. The enlarged overpotential for the hydrogen evolution reaction (HER) on the additives, such as In [29], Sn [30], Bi [31], and La [32], decreases the rate of hydrogen production. The additives possessing more positive reduction potentials compared to zinc, such as Bi, Pb, and In, stay in metallic form and thus maintain the conductivity of the zinc anode during the discharge of zinc-air fuel cells, which lowers the local current density and avoids the formation of a compact ZnO film. Therefore, both hydrogen evolution and zinc passivation can be reduced with appropriate heavy-metal additions.

According to the type of zinc-air fuel cells (mechanically or hydraulically refuelable) and the requirements of current density and specific capacity, zinc is presented in various forms, such as plate, powder, pellet, dendrite, or fiber. A plate-shaped zinc anode with high mechanical strength is suitable for the mechanically refuelable zinc-air fuel cell, in which the zinc anode can be easily replaced physically with a fresh one. Solid plate of zinc anode is frequently inconvenient for zinc-air fuel cells due to its limited surface area. To increase the surface of zinc anode, particular strategies can be adopted, resulting in an enhancement of fuel cell performance. The most frequently used method is to fabricate a porous anode using small zinc particles. A porosity of 60%–80% is commonly found in the zinc anodes of zinc-air fuel cells, which achieves a well balance between reactivity and conductivity. It is worth

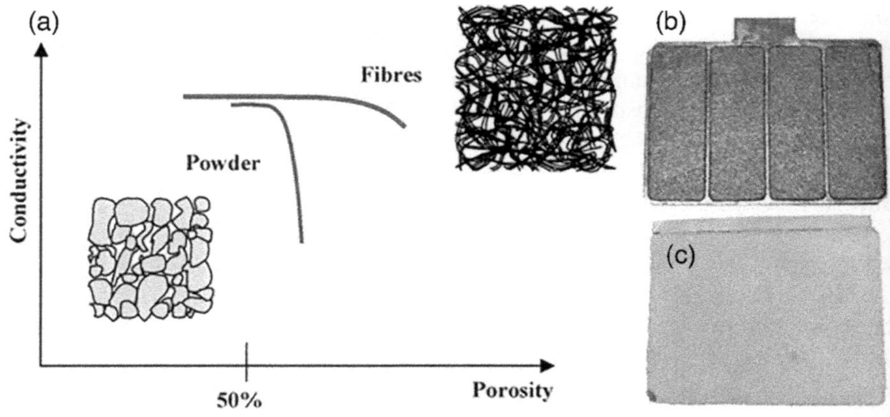

FIGURE 5.4 Qualitative conductivity–porosity relationship (a) and photographs (b–c) of fibrous (b) and powder (c) zinc electrodes.

noticing that the morphology of zinc materials brings different behaviors between conductivity and porosity, as reported in Figure 5.4a [33]. Porosity of anodes prepared by mold pressing zinc fibers (Figure 5.4b) can vary over a much greater range without decreasing the conductivity. As for zinc powders, a gelling agent is necessary to increase the porosity and mechanical strength, as shown in Figure 5.4c. In some of these porous zinc anodes, additives of carbon-based materials, such as super P [34], acetylene black, and carbon nanotube [35], are used to compensate for the decrease in conductivity due to insulating additives. Depositing zinc materials onto a three-dimensional template is another method to prepare porous zinc anode. Conductive polyurethane [36], copper [37], and nickel foam [38] are commonly applied, which lead to a much greater porosity of over 90%.

In hydraulically rechargeable zinc-air fuel cells, the anodes are continually supplied with fresh metal in a flowing electrolyte. Both the size distribution and the shape of zinc particles are important for the anode kinetics. Zinc particles, made by mechanically cutting zinc wires, electrodeposition, or direct casting, are preferred to aiming at better fluid dynamic properties.

5.2.3 Applications of Zn-Air Batteries

Primary zinc-air battery systems have long replaced lead–acid, alkaline, and mercury batteries in small-scale applications, including hearing aids, military field electronics, and portable telephones, thanks to their superior energy density and safety profile. Attention has now shifted toward exploiting the high energy density and extended cycle life of secondary zinc-air chemistries for electric vehicles (EVs). In the context of EVs, Zn-air systems currently offer the most favorable energy-to-cost and energy-to-weight ratios reported to date. Another advantage of Zn-air technology is its low inherent safety risk, arising from the use of aqueous or solid-state electrolytes. Over the past few decades, several commercial entities have improved Zn-air battery performance. For instance, Electric Fuel, an industry leader in the field of portable power solutions, has demonstrated Zn-air cells with specific energies approaching 200 Wh kg^{-1}, sufficient to propel a passenger bus for 8 hours on a single charge without emissions [8]. More recently, this technology has been adapted for defense applications, achieving energy densities of around 400 Wh kg^{-1} and reliable operation under extreme temperature conditions [39]. Another German company, Dreisbach Electromotive Inc., has integrated 60 kWh Zn-air modules with 10 kWh Ni–Cd cells to power Chrysler minivans, yielding a range of 322 km at a cruising speed of 64 km h^{-1}.

Significant progress has been made by energy technology firms in developing cost-effective, reliable zinc-air solutions for LSES in both industrial and ancillary markets. NantEnergy, for instance, provides solid-state Zn-air microgrid systems, both grid-tied and off-grid, which are integrated with renewable energy sources, such as photovoltaic arrays. These systems have an operational lifespan of over ten years, are fully recyclable, and have competitive energy storage costs of less than \$100 kWh^{-1}. They also eliminate the need for fossil fuels, avoid the risk of thermal runaway or gas evolution, and support remote monitoring and data analytics. Notably, the raw material cost of Zn-air electrodes is around 85% lower than that of Li-ion counterparts, reflecting the natural abundance of zinc, and this contributes to a reduction in overall manufacturing costs of almost 95% relative to Li-ion systems (Figure 5.5).

Similarly, the Canadian company Zinc8 has patented a modular Zn-air architecture that avoids traditional cell stacking: capacity is increased simply by enlarging the electrolyte reservoir that stores particulate zinc fuel. This design reportedly reduces energy-capacity costs to as low as \$45 kWh^{-1} and achieves negligible capacity fade over an operational lifetime of roughly 2,000 hours with 8-hour cycling. These attributes highlight the suitability of Zn-air cells for distributed energy storage in the residential, commercial, and industrial sectors. In 2020, the New York Power Authority, in partnership with Zinc8, began commercialization efforts to decarbonize the state's power infrastructure. Meanwhile, Eos Energy Storage Systems has collaborated with Con Edison and National Grid to pursue Zn-air technologies for grid-scale deployment. EOS Energy Storage's Aurora product, a 1 MW/4 MWh zinc-air battery designed for utility-scale grid storage, achieves a cost as low as \$160 kWh^{-1}. Collaborating with Caterpillar and Indonesian utilities, EOS Energy Storage and Fluid Energy have collectively deployed over 250 MWh of zinc-air batteries, enabling solar PV energy storage in 500 remote communities.

The \$160 kWh^{-1} benchmark is notable given the nascent stage of rechargeable zinc-air battery manufacturing. By comparison, lithium-ion batteries are not projected to reach this cost threshold until 2025 at the earliest. A recent economic analysis identified zinc-air batteries as the most economically viable option for smart grid energy storage applications. Their high energy density, low cost, operational safety, and demonstrated market viability underscore the growing interest in this technology, particularly when contrasted with alternative battery systems.

Due to their low mass production cost, Zn-air batteries are set to enter broader energy markets and could potentially be the "holy grail" of energy storage. While Li-ion remains the dominant technology, concerns over dwindling lithium supplies and rising material costs have increased interest in Zn-air systems as a viable,

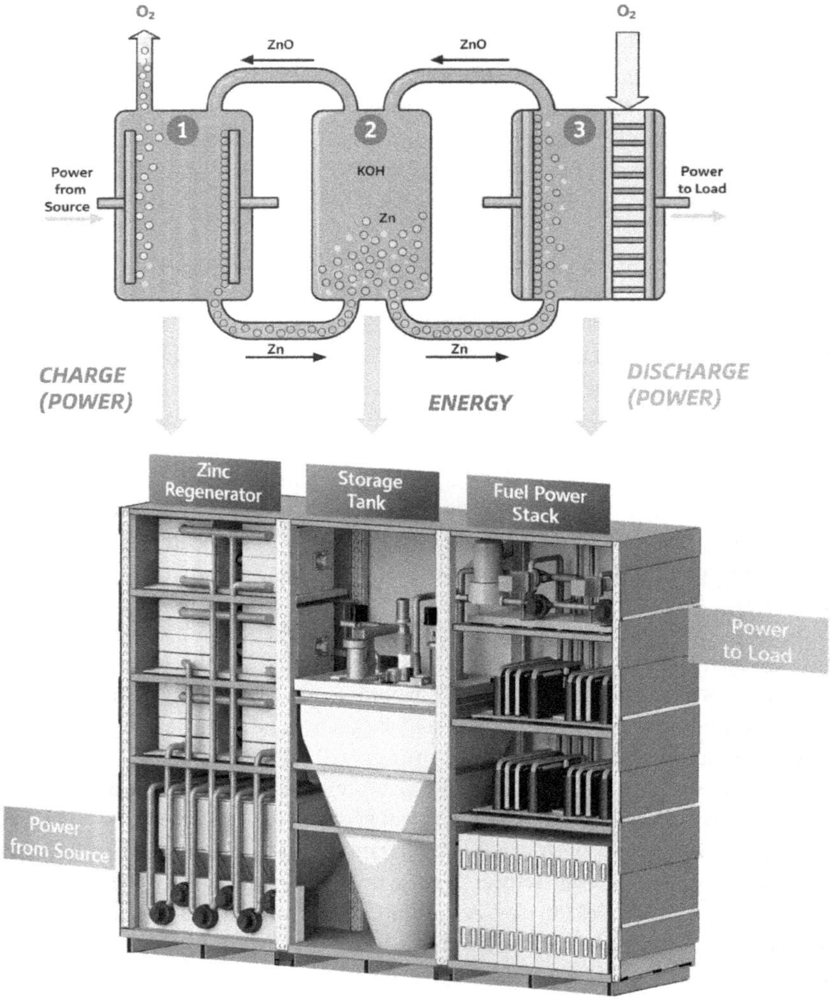

FIGURE 5.5 Zinc8's fuel-tank-based Zn-air battery design.

economical, and environmentally sustainable solution for GSES. These systems are capable of balancing supply and demand, as well as mitigating peak-pricing challenges.

5.3 IRON-AIR BATTERIES

It is evident that magnesium-air and aluminum-air batteries demonstrate higher theoretical energy densities in comparison to the majority of alternative metal-air systems. However, their practical energy output is significantly limited by parasitic hydrogen evolution resulting from spontaneous metal–water reactions. Moreover, considerable obstacles are faced in the pursuit of efficient rechargeability within aqueous electrolytes. Zinc-air batteries are renowned for their exceptional stability and electrochemical reversibility, characteristics that have led to their extensive utilization in low-power applications, including hearing aids and digital watches. However, issues such as zinc dendrite formation and morphological changes in the electrode during cycling contribute to poor capacity retention and reduced operational lifespan [40]. Lithium-air batteries have attracted the most attention due to their remarkably high theoretical energy density (3,458 Wh kg^{-1}), which significantly exceeds that of conventional lithium-ion batteries. However, the high cost and scarcity of lithium, in conjunction with its extreme reactivity and associated safety risks, have prompted a surge of interest in sodium–air and potassium–air alternatives. These alkali-metal systems exhibit analogous electrochemical behavior; however, they are encumbered by challenges pertaining to flammability, moisture sensitivity, and elevated fabrication costs, which are attributable to the necessity for specialized materials and environments.

Among these systems, iron-air batteries stand out as a particularly promising candidate for LSES. Although they offer lower theoretical energy density, their practical advantages are compelling: long cycle life, excellent rechargeability, intrinsic safety, and the use of abundant, low-cost iron resources. These attributes make iron-air batteries especially attractive for sustainable and scalable deployment in grid-level and long-duration energy storage applications. In addition, iron exhibits excellent charge/discharge characteristics and effectively suppresses the formation of undesirable dendritic precipitates during repeated cycling [41]. This enhances the structural stability of the electrode and contributes to an extended battery lifespan. Compared to other metal-air systems, Fe-air batteries also offer easier rechargeability, longer negative electrode durability, and reduced susceptibility to corrosion, further strengthening their potential as a reliable and cost-effective energy storage solution [42] (Figure 5.6).

The redox reaction at the surface of the negative electrode during charge/discharge is mainly dependent on the electrochemical reactions of Fe in an alkaline aqueous solution. Two steps of chemical reactions are involved, where the first step is the redox of $Fe(OH)_2$ and Fe as depicted in the following equation:

$$Fe + 2OH^- \rightarrow Fe(OH)_2 + 2e^-$$

The second step involves the redox reaction of $Fe(OH)_2$ and FeOOH or Fe_3O_4 as shown in the following equations.

$$3Fe(OH)_2 + 2OH^- \rightarrow Fe_3O_4 + H_2O + 2e^-$$

$$Fe(OH)_2 + OH^- \rightarrow FeOOH + H_2O + e^-$$

Air cathode:

$$O_2 + 2H_2O + 4e^- \leftrightarrow 4OH^-$$

In iron-air secondary batteries (FeABs), the iron electrode serves as the anode. These cells typically feature two air-breathing cathodes arranged in-line on opposite faces

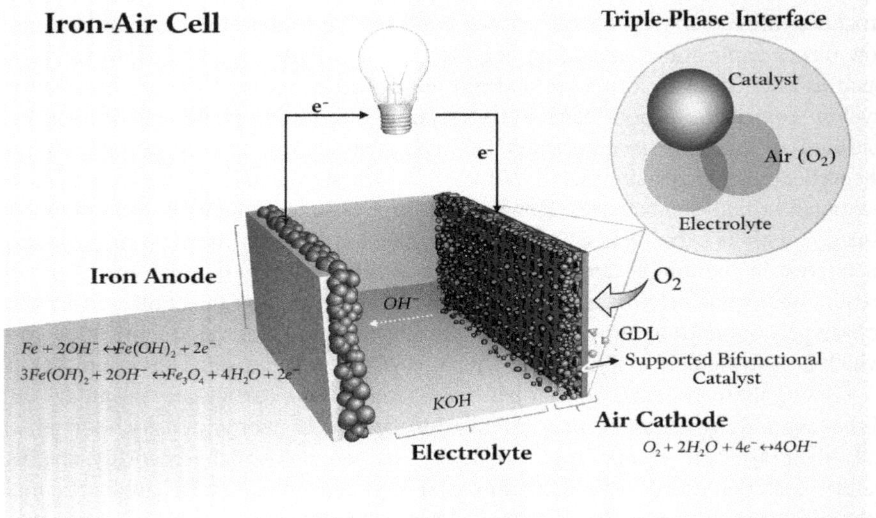

FIGURE 5.6 A schematic illustration of an iron-air cell, including magnification of the triple-phase interface in the air cathode [2].

of the cell. During cycling, FeAB cathodes are engineered to facilitate oxygen reduction and subsequent oxygen deposition. However, the electrodes' susceptibility to oxidative degradation at high potentials poses a major barrier to practical metal-air systems. To minimize overpotential during oxygen redox, electrodes must exhibit robust electrochemical stability.

The iron anode in iron-air battery systems has traditionally been composed primarily of finely powdered iron oxides, notably Fe_3O_4, or metallic iron. Adding conductive carbon additives to these powders increases both surface area and electronic conductivity. However, several challenges hinder efficient iron electrodes: suboptimal iron utilization, parasitic hydrogen evolution requiring overcharge to reach full capacity, corrosion of iron, and surface photocatalytic reactions during charging. Moreover, the formation of electrically insulating iron hydroxide ($Fe(OH)_2$) on the electrode surface diminishes discharge rates by lowering cell voltage under high current. To overcome underutilization, electrode nanostructuring—such as using iron nanoparticles—has emerged to boost accessible electrochemical surface area [43]. Indeed, nanosized Fe_3O_4 electrodes on carbon nanofibers achieved 786 mAh g^{-1} specific capacity and 76% cycle efficiency by cycle 30, owing to their high surface area (>300 $m^2 g^{-1}$) and conductive carbon matrix [44]. Likewise, carbonyl iron electrodes with ultra-high purity have shown improved discharge capacity in FeABs [45].

The air cathode in secondary iron-air batteries must catalyze both the on ORR and OER. High catalytic activity demands materials with a large surface area and durable stability in strongly alkaline electrolytes (5–8 M KOH) [46]. While Pt and Pd remain benchmarks for ORR, and IrO_2 and RuO_2 for OER, recent efforts target cost-effective alternatives such as bimetallic oxides, specifically spinels, pyrochlores, and perovskites, due to their bifunctional four-electron oxygen redox activity [47]. To remedy their low intrinsic conductivity, these catalysts are dispersed on porous conductive substrates. Carbon supports (e.g., Ketjen black, acetylene black, and Vulcan carbon) are favored for high surface area and conductivity, whereas corrosion-resistant supports like nickel powders or Ebonex ensure stability under high oxidative stress despite their lower surface area. For instance, LSFO perovskite on porous carbon achieved notable bifunctional performance, and LCCO and $LaMnO_3$ perovskites surpassed commercial Pt/C in ORR metrics. Conversely, pyrochlores such as $Pb_2Ru_2O_{6.5}$ display excellent activity but degrade under extended negative-potential cycling [48].

The potential of iron-air batteries for low-cost, environmentally benign, grid-scale storage has revived interest in a technology first explored decades ago. During discharge, iron-air batteries convert iron to iron oxide by consuming atmospheric oxygen. Conversely, during charging, these batteries electrochemically reduce iron oxide back to metallic iron. The merits of iron-air systems include inherent safety, modular scalability, long cycle life, and low cost for grid-scale applications. According to Form Energy, a prominent manufacturer in the field, has developed substantial potential for iron-air to revolutionize grid-scale storage. The company has targeted commercial production by mid-2024. Form Energy's iron-air system, which is a rechargeable cell that comprises an iron anode and dual air cathodes, can deliver over 100 hours of discharge using only iron, water, and air. Its nonflammable and earth-abundant components, along with the elimination of critical minerals, provide a sustainable, multiday, and cost-effective alternative to lithium-ion batteries for GSES (Figure 5.7).

FIGURE 5.7 Fe-air batteries produced by Form Energy.

5.4 ALUMINUM–LUMINUM–UMINU

Compared to Zn-air batteries, which have a theoretical energy density of 1.3 $kWh{\cdot}kg^{-1}$, Al-air systems offer a higher theoretical energy density of up to 8.1 $kWh{\cdot}kg^{-1}$, although this remains lower than that of Li-air batteries (11.4 $kWh{\cdot}kg^{-1}$) [49]. Al-air batteries are valued for their cost-effectiveness, the abundance, light weight, non-toxicity, and recyclability of aluminum, as well as their relatively low electrochemical equivalence (2.98 $Ah{\cdot}g^{-1}$) compared to other metals such as lithium (3.86 $Ah{\cdot}g^{-1}$). Furthermore, aluminum exhibits a high open-circuit potential (–1.87 V vs. SHE at pH 14.6) that remains stable in the absence of parasitic reactions [50].

However, unlike Zn- and Fe-air batteries, Al-air batteries are not electrochemically rechargeable in aqueous electrolytes, as metallic aluminum cannot be efficiently recovered from its oxides. Instead, they are typically operated as primary systems and "recharged" mechanically by replacing the consumed Al anode, which makes them more suitable for niche applications such as long-endurance military power units and emergency backup systems rather than conventional secondary batteries.

The electrochemical reactions in Al-air batteries are described as follows:

Anode:

$$Al + 3OH^- \rightarrow Al(OH)_3 + 3e^-$$

Cathode:

$$O_2 + 2H_2O + 4e^- \rightarrow 4OH^-$$

Overall:

$$4Al + 3O_2 + 6H_2O \rightarrow 4Al(OH)_3$$

Although Al-air batteries have a theoretical open-circuit voltage of 2.71 V, they typically operate between 1.0 and 2.0 V. This voltage drop arises primarily from: (1) the native oxide layer on the Al anode, which increases internal resistance, delays the establishment of steady-state voltage, and leads to electrolyte supersaturation with $Al(OH)_3$ precipitates; and (2) parasitic corrosion of the Al anode, which consumes Al metal and produces H_2. Beyond these aqueous-solution corrosion pathways—present both at open circuit and under load—Al-air systems suffer additional issues: accumulation of $Al(OH)_3$ and Al_2O_3 at both electrodes; complex Al^{3+} intercalation chemistry that renders hydrated alumina irreversible; rapid capacity fade; and limited shelf life.

To mitigate capacity loss and improve Coulombic efficiency, on the anode side, rapid aluminum corrosion and HER in aqueous electrolytes cause severe self-discharge [51]. Alloying Al with elements such as Ga, Ti, Sn, In, and especially Zn—which raises the HER overpotential—has proven effective [52]. Representative alloys include Al–1 Mg–0.1 Ga–0.1 Sn (wt %), Al–1 Mg–1 Zn–0.1 Ga–0.1 Sn (wt %) [53], and Al–0.15 Bi–0.15 Pb–0.035 Ga (wt %) [54]. Additionally, inorganic additives (ZnO, $In(OH)_3$, and Na_2SnO_3), hybrid additives (CMC/ZnO, L-cysteine/

ZnO, and CTAB/ZnO), and plant extracts have been used to inhibit Al-anode corrosion and suppress side reactions [55].

In addition to corrosion, electrolyte leakage and capacity fade during cycling also hinder Al-air batteries. Although acidic and neutral aqueous electrolytes have been investigated, alkaline solutions, particularly those containing NaOH and KOH at various molarities, remain the norm due to their high ionic conductivity, efficient oxygen ORR kinetics, low toxicity, and minimal overpotentials 683–685 [56–58]. Ionic liquids (ILs) also offer advantages. Gelman et al. [59] investigated a hydrophilic 1-ethyl-3-methylimidazolium oligo-fluoro-hydrogenate IL with high ionic conductivity (100 $mS{\cdot}cm^{-1}$) and stability in air and water. This enabled an Al-air cell with an Al_4N_7 anode to achieve 2,300 $Wh{\cdot}kg^{-1}$. Polymer and gel electrolytes serve as corrosion inhibitors. For example, Wu et al. [60] developed an alkaline PAA/PVA composite electrolyte for Al-air batteries, and Zhang et al. [61] formulated a gel containing 9 M KOH, 6 wt % PAA, ZnO, and $K_2S_2O_8$, which delivered a power density of 91.13 $mW{\cdot}cm^{-2}$ and a current density of 120 $mA{\cdot}cm^{-2}$. Semisolid electrolytes, such as KOH/PVA/PEO/ZnO/Na_2SnO_3, have also been used to make flexible Al-air devices [62].

Extensive research [63,64] has focused on cathode catalysts, and precious metals such as Pt exhibit high activity for the ORR by facilitating OH^- formation in alkaline media, but their high cost and limited durability are drawbacks. Other noble metals (Pd, Au, and Ag) have also demonstrated good ORR performance; notably, Ag-based catalysts are nearly 50× less expensive than Pt and require only 5–15 wt% loading at the air cathode. Their effectiveness stems from excellent electrical conductivity (6.3×10^7 $S{\cdot}m^{-1}$) and electrochemical stability [65]. Further enhancements in Ag-based electrocatalysts are achieved via alloying [66], incorporation of conductive supports [67], and morphology control [68].

Non-precious metal catalysts of the M–N–C family have emerged as promising alternatives to Pt for the ORR in Al-air batteries. Macrocyclic $M{-}N_x$ complexes, in particular, exhibit exceptional ORR activity with negligible overpotentials [69]. For example, Jin et al. [70] synthesized an $Fe{-}N_x/C$ electrocatalyst via a SiO_2-protective-layer-assisted method and demonstrated ORR performance comparable to Pt/C in 0.1 M KOH. More recently, Lai et al. [71] reported P-doped carbon dots on graphene (P-CD/G), prepared by hydrothermal treatment and annealing, for a flexible solid-state Al-air battery. Optimal annealing at 900°C yields 2.31 at % P, resulting in significantly higher current density and ORR potential than untreated graphene oxide, and achieves a power density of 157.3 $mW{\cdot}cm^{-2}$ slightly surpassing Pt/C (151.5 $mW{\cdot}cm^{-2}$).

5.5 MAGNESIUM-AIR BATTERIES

Mg-air batteries are considered efficient alternatives to Li-based MABs due to their environmental friendliness, cost-effectiveness, and safety. In addition, the non-toxicity of magnesium anodes makes them suitable for biocompatible and biodegradable power applications [72]. Despite these advantages, Mg-air batteries have been less extensively studied. Nevertheless, they exhibit a theoretical energy density of up to 6.8 $kWh{\cdot}kg^{-1}$.

Similar to Al-air systems, Mg-air batteries are not electrochemically rechargeable in aqueous electrolytes because regenerating metallic magnesium from its oxides is thermodynamically prohibitive. Consequently, they are generally implemented as mechanically rechargeable devices through periodic anode replacement, rather than as fully secondary batteries.

The following reactions occur in a typical Mg-air battery:

Anode:

$$Mg + 2OH^- \rightarrow Mg(OH)_2 + 2e^-$$

Cathode:

$$O_2 + 2H_2O + 4e^- \rightarrow 4OH^-$$

Overall:

$$2Mg + O_2 + 2H_2O \rightarrow 2Mg(OH)_2$$

Based on these reactions, the theoretical cell voltage of an Mg-air battery is approximately 3.1 V. During discharge, the magnesium anode is oxidized to Mg^{2+}, while at the cathode oxygen is reduced to OH^-. Despite their high voltage and energy density, Mg-air systems face several challenges that impede practical application: (1) self-corrosion of the Mg anode in alkaline electrolytes, which lowers Coulombic efficiency; (2) high polarization due to accumulation of $Mg(OH)_2$ on the anode surface; and (3) the thermodynamic barrier to regenerating metallic Mg from its oxides, limiting rechargeability [72]. Strategies to address these issues include alloying the Mg anode, employing nanostructured anode materials, adding corrosion-resistant inhibitors to the electrolyte, and incorporating electrocatalysts at the cathode.

Alloying pristine Mg anodes with metallic elements (Al, Mn, Zn, Sn, Ce, Y, and In) effectively suppresses corrosion in Mg-air batteries [73]. Certain Mg-alloy compositions also promote favorable structural transformations in the discharge products, thereby enhancing overall anode utilization and battery efficiency. For example, Zheng et al. [74] demonstrated that an orthogonally structured Mg–Al–Sn–Mn alloy delivers a stable discharge plateau at −1.602 V vs. SCE with minimal corrosion resistance. Similarly, Wang et al. [75] reported that a Mg–Al–Pb–Ce–Y alloy exhibits improved anodic performance in 3.5 wt% NaCl.

Beyond Mg–Al systems, Mg–Li alloys have gained attention for their high specific energy and negative standard potential. Specifically, Mg–14 Li–1 Al–0.1 Ce achieves a specific capacity of 2,076 mAh g^{-1} at 1.272 V, outperforming both pure Mg and AZ31 anodes [76]. Despite these advances, alloying often reduces the operating potential below theoretical values. Consequently, current efforts focus on tailoring the Mg anode's composition, microstructure, and surface morphology. For instance, ultra-high-purity Mg (99.99%)—devoid of Fe, Co, and Ni impurities—outperforms many Mg alloys. Zhao et al. [77] further showed that cross-sectional rolling textures on AZ31 improve both corrosion resistance and electrochemical efficiency relative to conventional rolling surfaces.

On the cathode side, the air-cathode's catalyst layer governs ORR kinetics and thus faradaic efficiency. While 2–4 nm Pt nanoparticles on carbon remain the benchmark ORR catalysts, their cost and scarcity have spurred research into Pt-transition-metal alloys (e.g., Pt_3Y [78], PtCo [73], intermetallic Pt_3Co [78]) and non-precious alternatives. M–N–C catalysts, especially Fe–N_x embedded in carbon, have shown exceptional ORR activity; Cheng et al. [79] recently reported OM–NCNF–FeN_x (ordered mesoporous N-doped carbon nanofibers with Fe–N_x sites), which exhibits an onset potential of 0.905 V and half-wave potential of 0.836 V in alkaline media, delivering a peak current density of 19.2 mA cm^{-2} and power density of 5 mW cm^{-2} in flexible Mg-air cells.

Electrolyte selection critically influences both anode corrosion and overall cell performance. Alkaline electrolytes (NaOH and KOH) remain prevalent for their high conductivity and favorable ORR kinetics, but pH and additive optimization are essential. ILs such as [$P_{6,6,6,14}$][TFSI] and [$P_{6,6,6,14}$][DPP] offer superior Mg-anode corrosion resistance, though their passivating films impede Mg-air operation [80]. To address this, [$P_{6,6,6,14}$][Cl] in water forms an amorphous gel electrolyte that combines adequate ionic conductivity (0.05 mA cm^{-2}) with anode passivation at open-circuit voltage [81]. In the solid-state realm, Jia et al. [82] pioneered a chitosan–choline-nitrate biopolymer film electrolyte, which enabled miniaturized Mg-air cells achieving an open-circuit voltage of 1.80 V and an output power of 118 μW, a level sufficient for use in biomedical implants.

5.6 SUMMARY AND FUTURE PROSPECTS OF METAL-AIR BATTERIES

In this section, we have examined the principles, materials, and architectures of aqueous MABs with sequential attention to Zn-air, Fe-air, Al-air, and Mg-air. We assessed their respective advantages, challenges, and recent advances in electrode design, catalyst development, and electrolyte engineering. We have shown how each system leverages earth-abundant metals and open-air architectures to achieve high theoretical energy densities, and how innovations in anode morphology, bifunctional catalysts, and separator technologies are driving improvements in efficiency, cycle life, and cost. Throughout, we have highlighted the criteria that position MABs as leading candidates for long-duration, GSES, namely material abundance, low capital cost, inherent safety, and the potential for rapid manufacturing scale-up. Finally, we identify key research directions required to unlock their full potential in future renewable-driven power systems.

Despite the remarkable advances in materials and cell architectures, aqueous MABs must overcome several hurdles before they can fully realize their promise for large-scale, stationary energy storage. While nonaqueous systems chase the highest possible energy densities, Zn-air and Fe-air chemistries excel in two critical metrics for grid-scale deployment: low capital cost and abundant, noncritical raw materials. When considered in the context of long-duration energy storage, where capital expenditure, operating costs, and system durability, energy density, and cycle life must all be taken into account, the recyclability and environmental footprint of aqueous MABs offer a compelling advantage.

Yet the very features that make aqueous electrolytes attractive (high ionic conductivity, intrinsic safety, and straightforward cell designs) also demand careful optimization of cell and stack geometries. Narrow interelectrode gaps boost volumetric power density but risk oxygen-transport limitations and polarization; borrowing stack layouts from mature fuel cell technologies may help reconcile these trade-offs at scale.

Most research to date has focused on improving bifunctional electrocatalysts for OER/ORR, but the fundamental challenge associated with reversible metal plating and stripping, which is indispensable for achieving high round-trip efficiency and Coulombic efficiency, has received comparatively little attention. Moving forward, a holistic approach that simultaneously tailors cathode catalysts, anode morphology, electrolyte formulation, separators, and overall cell design will be vital to suppress side reactions, prevent passivation or dendrite growth, and secure consistent performance over hundreds or even thousands of cycles.

Finally, when integrated into real-world grids, aqueous MABs must deliver not only reliable, multiday energy throughput but also seamless control, monitoring, and maintenance protocols. Pilot projects and demonstration plants will be essential to validate system-level strategies, including modular tank-based fuel configurations or hybrid battery–fuel-cell systems, which can capitalize on the intrinsic safety and low cost of water-based metal-air chemistries. Only by resolving these interconnected materials, cell, and system challenges can aqueous MABs earn their place as a powerful, sustainable backbone for our future renewable-powered electricity networks.

REFERENCES

1. Zhu Z, Jiang T, Ali M, et al. Rechargeable batteries for grid scale energy storage. *Chemical Reviews*, 2022, 122(22): 16610–751.
2. Bogomolov K, Ein-Eli Y. Will iron forge the future of metal-air batteries in grid scale energy storage?. *ChemSusChem*, 2025, 18(10): e202402412.
3. Smee A. LIII. On the galvanic properties of the metallic elementary bodies, with a description of a new chemico-mechanical battery. *The London, Edinburgh, and Dublin Philosophical Magazine and Journal of Science*, 1840, 16(103): 315–21.
4. Wang C, Xie N-H, Zhang Y, et al. Silk-derived highly active oxygen electrocatalysts for flexible and rechargeable Zn–air batteries. *Chemistry of Materials*, 2019, 31(3): 1023–9.
5. Peng X, Zhang L, Chen Z, et al. Hierarchically porous carbon plates derived from wood as bifunctional ORR/OER electrodes. *Advanced Materials*, 2019, 31(16): 1900341.
6. Toussaint G, Stevens P, Akrour L, et al. Development of a rechargeable zinc-air battery. *ECS Transactions*, 2010, 28(32): 25.
7. Li Y, Dai H. Recent advances in zinc–air batteries. *Chemical Society Reviews*, 2014, 43(15): 5257–75.
8. Yan X, Ha Y, Wu R. Binder-free air electrodes for rechargeable zinc-air batteries: Recent progress and future perspectives. *Small Methods*, 2021, 5(4): 2000827.
9. Zhang J, Zhou Q, Tang Y, et al. Zinc–air batteries: Are they ready for prime time? *Chemical Science*, 2019, 10(39): 8924–9.
10. Tan P, Chen B, Xu H, et al. Flexible Zn– and Li–air batteries: Recent advances, challenges, and future perspectives. *Energy & Environmental Science*, 2017, 10(10): 2056–80.

11. Fukuda H, Szpunar JA, Kondoh K, et al. The influence of carbon nanotubes on the corrosion behaviour of AZ31B magnesium alloy. *Corrosion Science*, 2010, 52(12): 3917–23.
12. Shu C, Wang E, Jiang L, et al. High performance cathode based on carbon fiber felt for magnesium-air fuel cells. *International Journal of Hydrogen Energy*, 2013, 38(14): 5885–93.
13. Maja M, Orecchia C, Strano M, et al. Effect of structure of the electrical performance of gas diffusion electrodes for metal air batteries. *Electrochimica Acta*, 2000, 46(2): 423–32.
14. Eom S-W, Lee C-W, Yun M-S, et al. The roles and electrochemical characterizations of activated carbon in zinc air battery cathodes. *Electrochimica Acta*, 2006, 52(4): 1592–5.
15. Tomantschger K, Findlay R, Hanson M, et al. Degradation modes of alkaline fuel cells and their components. *Journal of Power Sources*, 1992, 39(1): 21–41.
16. Li H, Sun G, Li N, et al. Design and preparation of highly active Pt–Pd/C catalyst for the oxygen reduction reaction. *The Journal of Physical Chemistry C*, 2007, 111(15): 5605–17.
17. Li G, Jiang L, Jiang Q, et al. Preparation and characterization of Pd_xAg_y/C electrocatalysts for ethanol electrooxidation reaction in alkaline media. *Electrochimica Acta*, 2011, 56(22): 7703–11.
18. Yuan L, Yan Z, Jiang L, et al. Gold-iridium bifunctional electrocatalyst for oxygen reduction and oxygen evolution reactions. *Journal of Energy Chemistry*, 2016, 25(5): 805–10.
19. Lee DU, Xu P, Cano ZP, et al. Recent progress and perspectives on bi-functional oxygen electrocatalysts for advanced rechargeable metal–air batteries. *Journal of Materials Chemistry A*, 2016, 4(19): 7107–34.
20. Guo J, Hsu A, Chu D, et al. Improving oxygen reduction reaction activities on carbon-supported Ag nanoparticles in alkaline solutions. *The Journal of Physical Chemistry C*, 2010, 114(10): 4324–30.
21. Neburchilov V, Wang H, Martin JJ, et al. A review on air cathodes for zinc–air fuel cells. *Journal of Power Sources*, 2010, 195(5): 1271–91.
22. Liu J, Liu J, Song W, et al. The role of electronic interaction in the use of Ag and Mn_3O_4 hybrid nanocrystals covalently coupled with carbon as advanced oxygen reduction electrocatalysts. *Journal of Materials Chemistry A*, 2014, 2(41): 17477–88.
23. Wang Y, Ma X, Lu L, et al. Carbon supported MnO_x–Co_3O_4 as cathode catalyst for oxygen reduction reaction in alkaline media. *International Journal of Hydrogen Energy*, 2013, 38(31): 13611–6.
24. Lamminen J, Kivisaari J, Lampinen MJ, et al. Preparation of air electrodes and long run tests. *Journal of The Electrochemical Society*, 1991, 138(4): 905.
25. Liang Y, Wang H, Diao P, et al. Oxygen reduction electrocatalyst based on strongly coupled cobalt oxide nanocrystals and carbon nanotubes. *Journal of the American Chemical Society*, 2012, 134(38): 15849–57.
26. Gu L, Jiang L, Li X, et al. A Fe-N-C catalyst with highly dispersed iron in carbon for oxygen reduction reaction and its application in direct methanol fuel cells. *Chinese Journal of Catalysis*, 2016, 37(4): 539–48.
27. Guo S, Zhang S, Sun S. Tuning nanoparticle catalysis for the oxygen reduction reaction. *Angewandte Chemie International Edition*, 2013, 52(33): 8526–44.
28. Ferrandon M, Kropf AJ, Myers DJ, et al. Multitechnique characterization of a polyaniline–iron–carbon oxygen reduction catalyst. *The Journal of Physical Chemistry C*, 2012, 116(30): 16001–13.
29. Wang R, Yang Z, Yang B, et al. Superior cycle stability and high rate capability of Zn–Al–In-hydrotalcite as negative electrode materials for Ni–Zn secondary batteries. *Journal of Power Sources*, 2014, 251: 344–50.

30. Zhang Z, Yang Z, Wang R, et al. Electrochemical performance of ZnO/SnO_2 composites as anode materials for Zn/Ni secondary batteries. *Electrochimica Acta*, 2014, 134: 287–92.
31. Yuan YF, Yu LQ, Wu HM, et al. Electrochemical performances of Bi based compound film-coated ZnO as anodic materials of Ni–Zn secondary batteries. *Electrochimica Acta*, 2011, 56(11): 4378–83.
32. Wang T, Yang Z, Huang J, et al. The electrochemical performances of La_2O_3-doped ZnO in Ni–Zn secondary batteries. *Electrochimica Acta*, 2013, 112: 104–10.
33. Zhang XG. Fibrous zinc anodes for high power batteries. *Journal of Power Sources*, 2006, 163(1): 591–7.
34. Masri MN, Mohamad AA. Effect of adding carbon black to a porous zinc anode in a zinc-air battery. *Journal of The Electrochemical Society*, 2013, 160(4): A715.
35. Fu J, Lee DU, Hassan FM, et al. Flexible high-energy polymer-electrolyte-based rechargeable zinc–air batteries. *Advanced Materials*, 2015, 27(37): 5617–22.
36. Tian Q, Guo X. Manufacturing microporous foam zinc materials with high porosity by electrodeposition. *Journal of Wuhan University of Technology-Mater Sci Ed*, 2011, 26(5): 843–6.
37. Yan Z, Wang E, Jiang L, et al. Superior cycling stability and high rate capability of three-dimensional Zn/Cu foam electrodes for zinc-based alkaline batteries. *RSC Advances*, 2015, 5(102): 83781–7.
38. Cheng Y, Lai Q, Li X, et al. Zinc-nickel single flow batteries with improved cycling stability by eliminating zinc accumulation on the negative electrode. *Electrochimica Acta*, 2014, 145: 109–15.
39. Sapkota P, Kim H. Zinc–air fuel cell, a potential candidate for alternative energy. *Journal of Industrial and Engineering Chemistry*, 2009, 15(4): 445–50.
40. Tan WK, Wada Y, Hayashi K, et al. Fabrication of an all-solid-state Zn-air battery using electroplated Zn on carbon paper and KOH-ZrO_2 solid electrolyte. *Applied Surface Science*, 2019, 487: 343–8.
41. Ding C, Huang X, Zhang H, et al. Self-assembled porous Fe_3O_4/C nanoclusters with superior rate capability for advanced lithium-ion batteries. *Journal of Materials Science: Materials in Electronics*, 2018, 29(8): 6491–500.
42. Tan WK, Kawamura G, Muto H, et al. Chapter 3 – Current progress in the development of Fe-air batteries and their prospects for next-generation batteries. Cheong KY, Chen L-C. (eds) *Sustainable Materials for Next Generation Energy Devices*. Elsevier, Amsterdam, Netherlands, 2021: 59–83.
43. Pingale AD, Owhal A, Katarkar AS, et al. Recent researches on Cu-Ni alloy matrix composites through electrodeposition and powder metallurgy methods: A review. *Materials Today: Proceedings*, 2021, 47: 3301–8.
44. Jayaraman K, Kotaki M, Zhang Y, et al. Recent advances in polymer nanofibers. *Journal of Nanoscience and Nanotechnology*, 2004, 4(1–2): 52–65.
45. Ren X, Zhang SS, Tran DT, et al. Oxygen reduction reaction catalyst on lithium/air battery discharge performance. *Journal of Materials Chemistry*, 2011, 21(27): 10118–25.
46. Zoladek S, Blicharska-Sobolewska M, Krata AA, et al. Heteropolytungstate-assisted fabrication and deposition of catalytic silver nanoparticles on different reduced graphene oxide supports: Electroreduction of oxygen in alkaline electrolyte. *Journal of Electroanalytical Chemistry*, 2020, 875: 114694.
47. Lee Y, Suntivich J, May KJ, et al. Synthesis and activities of rutile IrO_2 and RuO_2 nanoparticles for oxygen evolution in acid and alkaline solutions. *The Journal of Physical Chemistry Letters*, 2012, 3(3): 399–404.
48. Cheng F, Shen J, Peng B, et al. Rapid room-temperature synthesis of nanocrystalline spinels as oxygen reduction and evolution electrocatalysts. *Nature Chemistry*, 2011, 3(1): 79–84.

49. Goel P, Dobhal D, Sharma RC. Aluminum–air batteries: A viability review. *Journal of Energy Storage*, 2020, 28: 101287.
50. Chen LD, Nørskov JK, Luntz AC. Al–air batteries: Fundamental thermodynamic limitations from first-principles theory. *The Journal of Physical Chemistry Letters*, 2015, 6(1): 175–9.
51. Ryu J, Park M, Cho J. Advanced technologies for high-energy aluminum–air batteries. *Advanced Materials*, 2019, 31(20): 1804784.
52. Park IJ, Choi S-R, Kim J-G. Aluminum anode for aluminum-air battery – Part II: Influence of In addition on the electrochemical characteristics of Al-Zn alloy in alkaline solution. *Journal of Power Sources*, 2017, 357: 47–55.
53. Moghanni-Bavil-Olyaei H, Arjomandi J. Performance of Al–1 Mg–1 Zn–0.1 Bi–0.02 In as anode for the Al–AgO battery. *RSC Advances*, 2015, 5(111): 91273–9.
54. Wang Q, Miao H, Xue Y, et al. Performances of an Al–0.15 Bi–0.15 Pb–0.035 Ga alloy as an anode for Al–air batteries in neutral and alkaline electrolytes. *RSC Advances*, 2017, 7(42): 25838–47.
55. Lameche-Djeghaba S, Benchettara A, Kellou F, et al. Electrochemical behaviour of pure aluminium and Al–5% Zn alloy in 3% NaCl solution. *Arabian Journal for Science and Engineering*, 2014, 39(1): 113–22.
56. Jingling M, Jiuba W, Hongxi Z, et al. Electrochemical performances of Al–0.5 Mg–0.1 Sn–0.02 in alloy in different solutions for Al–air battery. *Journal of Power Sources*, 2015, 293: 592–8.
57. Müller S, Holzer F, Desilvestro J, et al. Aluminium alloys in sulfuric acid Part II: Aluminium-oxygen cells. *Journal of Applied Electrochemistry*, 1996, 26(12): 1217–23.
58. Smoljko I, Gudić S, Kuzmanić N, et al. Electrochemical properties of aluminium anodes for Al/air batteries with aqueous sodium chloride electrolyte. *Journal of Applied Electrochemistry*, 2012, 42(11): 969–77.
59. Gelman D, Shvartsev B, Ein-Eli Y. Aluminum–air battery based on an ionic liquid electrolyte. *Journal of Materials Chemistry A*, 2014, 2(47): 20237–42.
60. Wu GM, Lin SJ, Yang CC. Alkaline Zn-air and Al-air cells based on novel solid PVA/PAA polymer electrolyte membranes. *Journal of Membrane Science*, 2006, 280(1): 802–8.
61. Zhang Z, Zuo C, Liu Z, et al. All-solid-state Al–air batteries with polymer alkaline gel electrolyte. *Journal of Power Sources*, 2014, 251: 470–5.
62. Chen L, Li B, Zhu L, et al. A PVA/LiCl/PEO interpenetrating composite electrolyte with a three-dimensional dual-network for all-solid-state flexible aluminum–air batteries. *RSC Advances*, 2021, 11(62): 39476–83.
63. Wang Z-L, Xu D, Xu J-J, et al. Oxygen electrocatalysts in metal–air batteries: from aqueous to nonaqueous electrolytes. *Chemical Society Reviews*, 2014, 43(22): 7746–86.
64. Mokhtar M, Talib MZM, Majlan EH, et al. Recent developments in materials for aluminum–air batteries: A review. *Journal of Industrial and Engineering Chemistry*, 2015, 32: 1–20.
65. Zhu C, Fu S, Shi Q, et al. Single-atom electrocatalysts. *Angewandte Chemie International Edition*, 2017, 56(45): 13944–60.
66. Lei Y, Chen F, Jin Y, et al. Ag-Cu nanoalloyed film as a high-performance cathode electrocatalytic material for zinc-air battery. *Nanoscale Research Letters*, 2015, 10(1): 197.
67. Yan S, Xue Y, Li S, et al. Improving catalytic activity of layered lithium transition metal oxides for oxygen electrode in metal-air batteries. *International Journal of Hydrogen Energy*, 2020, 45(3): 1846–56.
68. Van Cleve T, Gibara E, Linic S. Electrochemical oxygen reduction reaction on Ag nanoparticles of different shapes. *ChemCatChem*, 2016, 8(1): 256–61.
69. Gewirth AA, Varnell JA, Diascro AM. Nonprecious metal catalysts for oxygen reduction in heterogeneous aqueous systems. *Chemical Reviews*, 2018, 118(5): 2313–39.

70. Sa YJ, Seo D-J, Woo J, et al. A general approach to preferential formation of active Fe–N_x sites in Fe–N/C electrocatalysts for efficient oxygen reduction reaction. *Journal of the American Chemical Society*, 2016, 138(45): 15046–56.
71. Wang M, Li Y, Fang J, et al. Superior oxygen reduction reaction on phosphorus-doped carbon dot/graphene aerogel for all-solid-state flexible Al–air batteries. *Advanced Energy Materials*, 2020, 10(3): 1902736.
72. Sun Y, Liu X, Jiang Y, et al. Recent advances and challenges in divalent and multivalent metal electrodes for metal–air batteries. *Journal of Materials Chemistry A*, 2019, 7(31): 18183–208.
73. Zhang T, Tao Z, Chen J. Magnesium–air batteries: From principle to application. *Materials Horizons*, 2014, 1(2): 196–206.
74. Zheng T, Hu Y, Zhang Y, et al. Composition optimization and electrochemical properties of Mg-Al-Sn-Mn alloy anode for Mg-air batteries. *Materials & Design*, 2018, 137: 245–55.
75. Feng Y, Xiong W, Zhang J, et al. Electrochemical discharge performance of the Mg–Al–Pb–Ce–Y alloy as the anode for Mg–air batteries. *Journal of Materials Chemistry A*, 2016, 4(22): 8658–68.
76. Ma Y, Li N, Li D, et al. Performance of Mg–14 Li–1 Al–0.1 Ce as anode for Mg-air battery. *Journal of Power Sources*, 2011, 196(4): 2346–50.
77. Yanchun Z, Guangsheng H, Cheng Z, et al. Effect of texture on the performance of Mg-air battery based on rolled Mg-3 Al-1 Zn alloy sheet. *Rare Metal Materials and Engineering*, 2018, 47(4): 1064–8.
78. Greeley J, Stephens IEL, Bondarenko AS, et al. Alloys of platinum and early transition metals as oxygen reduction electrocatalysts. *Nature Chemistry*, 2009, 1(7): 552–6.
79. Cheng C, Li S, Xia Y, et al. Atomic Fe–N_x coupled open-mesoporous carbon nanofibers for efficient and bioadaptable oxygen electrode in Mg–air batteries. *Advanced Materials*, 2018, 30(40): 1802669.
80. Huang P, Latham J-A, Macfarlane DR, et al. A review of ionic liquid surface film formation on Mg and its alloys for improved corrosion performance. *Electrochimica Acta*, 2013, 110: 501–10.
81. Khoo T, Howlett PC, Tsagouria M, et al. The potential for ionic liquid electrolytes to stabilise the magnesium interface for magnesium/air batteries. *Electrochimica Acta*, 2011, 58: 583–8.
82. Jia X, Yang Y, Wang C, et al. Biocompatible ionic liquid–biopolymer electrolyte-enabled thin and compact magnesium–air batteries. *ACS Applied Materials & Interfaces*, 2014, 6(23): 21110–7.

6 The Role of Green H_2 Generation by Water Electrolysis in Grid Energy Storage

Zhenye Kang

6.1 INTRODUCTION

The ongoing global energy transition is characterized by a rapid shift toward renewable energy sources, primarily driven by the urgent need to mitigate climate change and achieve carbon neutrality.[1] While wind and solar power have experienced significant growth and are becoming increasingly cost competitive, their intermittent nature presents a major challenge for grid stability and reliable energy supply.[2] This has led to a growing interest in energy storage solutions that can effectively balance the supply and demand of electricity, ensuring a smooth integration of renewables into the grid.[3]

Among various energy storage technologies, green hydrogen produced through water electrolysis stands out as a promising candidate due to its unique properties and potential for large-scale deployment,[4–7] as shown in Figure 6.1. This chapter delves into the role of green hydrogen generated by water electrolysis in grid energy storage, exploring its advantages, challenges, and applications.

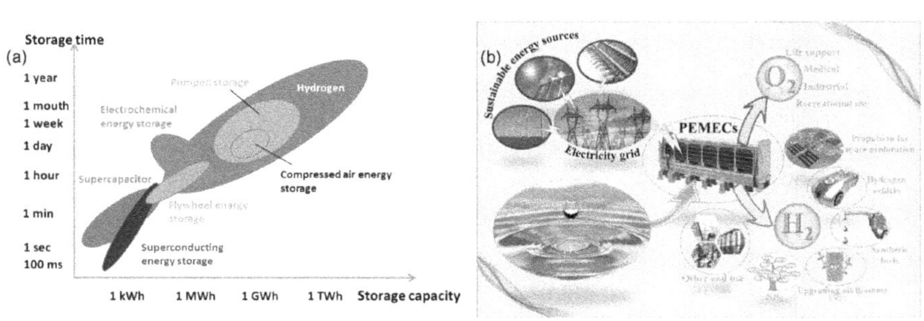

FIGURE 6.1 Schematic of (a) the various energy storage ways, (b) the water electrolysis for grid energy storage and further applications (PEMEC represents the proton exchange membrane electrolyzer cell that produce hydrogen at the cathode and oxygen at the anode).

DOI: 10.1201/9781003492511-6

6.1.1 Green Hydrogen in Grid Energy Storage

Green hydrogen, derived from renewable electricity and water, offers several key advantages that make it a compelling option for grid energy storage (Figure 6.1)[8]:

Renewable and clean: The production of green hydrogen is entirely reliant on renewable energy sources, resulting in minimal greenhouse gas emissions and a significantly reduced carbon footprint compared to conventional hydrogen production methods.[9]

High energy density and long-term storage: Hydrogen has a high energy density, allowing for large amounts of energy to be stored in a compact form. Additionally, hydrogen can be stored for extended periods, making it suitable for long-duration energy storage applications.

Versatility and flexibility: Green hydrogen can be utilized in various applications across multiple sectors, including transportation, heating, and industry.[8] This versatility allows for a diverse range of energy storage solutions and integration options within the grid.

Scalability and decentralization: Water electrolysis plants can be designed and scaled to meet specific energy storage requirements, offering flexibility in deployment. Furthermore, the potential for decentralized hydrogen production and storage systems enables localized energy solutions and enhances grid resilience.

6.1.2 Challenges and Opportunities

Despite its numerous advantages, the widespread adoption of green hydrogen for grid energy storage faces several challenges. First, the current cost of producing and storing green hydrogen remains relatively high compared to other energy storage technologies. This is primarily due to the high capital costs of electrolyzers, as well as the costs associated with electricity, hydrogen compression, storage, and transportation. Second, the conversion process of converting electricity into hydrogen and back into electricity through fuel cells involves energy losses, impacting the overall efficiency of the system. Third, the existing energy infrastructure is not yet optimized for hydrogen-based energy storage and distribution. This necessitates significant investments in new infrastructure and technology development.

Overcoming these challenges and realizing the full potential of green hydrogen for grid energy storage requires collaborative efforts from various stakeholders, including researchers, industry, policymakers, and investors. Advancements in electrolyzer technology, cost reductions through economies of scale, and supportive policy frameworks are crucial for accelerating the deployment of green hydrogen in grid energy storage applications.

In the following sections, we will delve deeper into the fundamentals of water electrolysis, the advantages and limitations of green hydrogen as an energy storage medium, its role in grid energy storage, practical applications and case studies, and future trends and innovations. By exploring these aspects, we aim to provide a comprehensive understanding of the potential of green hydrogen in shaping a sustainable and reliable energy future.

6.2 WATER ELECTROLYSIS FOR HYDROGEN PRODUCTION

6.2.1 Principles of Water Electrolysis

6.2.1.1 Comparison of Various Water Electrolysis Technologies

Hydrogen production technology mainly includes alkaline electrolytic water (ALK), proton exchange membrane electrolysis water (PEMWE), solid oxide electrolysis cell (SOEC), anion exchange membrane electrolysis water (AEMWE). From the perspective of technology maturity, according to the division of the U.S. Department of Energy, ALK, PEMWE, SOEC and AEMWE are 8–9, 8–9, 5–6, and 3–4, respectively. In other words, at present, ALK and PEMWE are relatively mature, SOEC is close to maturity, and AEMWE is still in the early stage. The detailed comparison between the four technologies is shown in Table 6.1.

Among the four technologies, PEMWE has the capability to fit the fluctuating or intermittent renewable electricity, which makes it the most promising way for grid energy storage.[10] The fundamental principle of PEMWE involves the splitting of H_2O into H_2 and O_2 powered by electricity.[11] This core reaction can be presented by Equation (6.1):

$$2H_2O \xrightarrow{\text{Anode}} O_2 + 4H^+ + 4e^-; \quad 4H^+ + 4e^- \xrightarrow{\text{Cathode}} 2H_2 \tag{6.1}$$

When operating the PEMWE, circulating pure water is injected from the anode-side inlet before the application of direct current power to the cathode and anode. Water electrolysis occurs once the external voltage surpasses the thermoneutral potential at ambient temperature. Subsequently, H_2O is oxidized at the anode, leading to the production of O_2, protons, and electrons, which is the process of the oxygen evolution reaction (OER). Meanwhile, the protons migrate through the PEM from anode to cathode and are reduced to H_2 by the electrons transferred from the external circuit, which is the process of hydrogen evolution reaction (HER), as shown in Figure 6.2.

TABLE 6.1
The Comparison Between Four Water Electrolysis Technologies

	ALK	PEMWE	AEMWE	SOEC
Electrolyte	30% KOH/PPS	PEM	AEM	YSZ
Current density (A cm^{-2})	<0.8	1–4	1–2	0.2–0.4
Efficiency (kWh Nm^{-3})	4.2–5.5	3.9–5.2	4.3–5.5	3.0–4.0
Operating temperature (C°)	70–90	50–80	40–60	600–1,000
Hydrogen purity	≥99.8%	≥99.99%	≥99.99%	≥99.99%
Hydrogen pressure (MPa)	≤1	1–10	1–3	≤1
System volume ratio	1	1/3	1/3	–
Operating property	Slow start-stop	Quick start-stop	Quick start-stop	Slow start-stop
Technology maturity	8–9	8–9	3–4	5–6
Cost ($/kW)	160–300	450–900	300–600	1,000–1,400

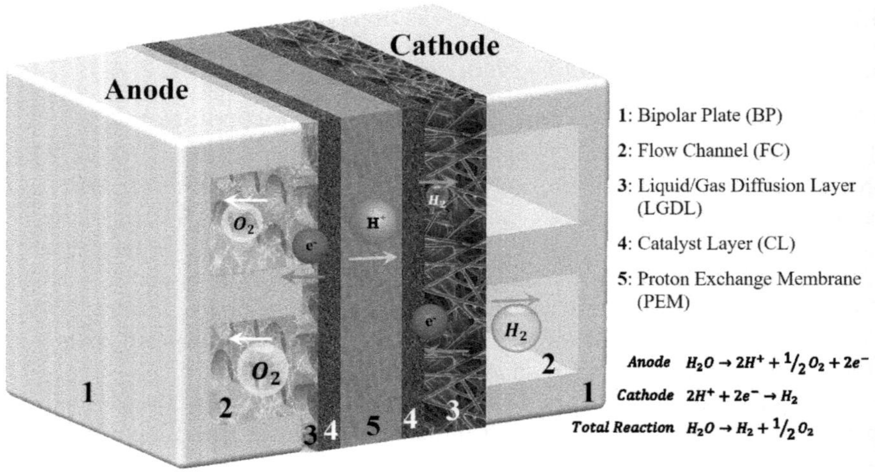

FIGURE 6.2 Schematic of the three-dimensional geometrical proton exchange membrane electrolysis water (PEMWE). (Reproduced with permission from Ref. 12.)

As the PEM separates the two half-reactions into two compartments, the generated H_2 is released through the cathode outlet, while the generated O_2 is emitted from the anode outlet, along with pure water. High-purity H_2 and O_2 can be collected following the drying process. In industrial applications, PEMWE operates under pressure to provide compressed H_2 or O_2 for subsequent applications.[13]

6.2.1.2 Electrochemical Process of Water Splitting

The electrochemical performance can be expressed by the polarization curve, which is the relation between the current density and cell voltage.[12] The cell voltage (V_{cell}) of a PEMWE consists of reversible losses (V_{rev}) and irreversible losses that include ohmic overpotential (η_{Ohm}), activation overpotential (η_{act}), and diffusion overpotential (η_{diff}),[14] which is shown as Equation (6.2):

$$V_{cell} = V_{rev} + \eta_{Ohm} + \eta_{act} + \eta_{diff} \quad (6.2)$$

Water splitting is a non-spontaneous process that necessitates external energy (reaction enthalpy, ΔH_R) to drive it. According to thermodynamic principles, the reaction enthalpy can be expressed as in Equation (6.3):

$$\Delta H_R = \Delta G_R + T\Delta S_R \quad (6.3)$$

Here, ΔG_R represents the Gibbs free energy of the reaction and can be interpreted as the electrical energy demand for reaction. ΔS_R is the change in entropy of reaction, and T represents the temperature. $T\Delta S_R$ is known as the entropy term, corresponding to the thermal energy demand of reaction.[15] Consequently, from the perspective of energy input, the reaction enthalpy of water electrolysis is contributed by both electrical and thermal energy.

At standard conditions ($T=298.15$ K, $p=1$ atm), the formation enthalpies (ΔH_F) of the elemental H_2 and O_2 are both zero. As a result, the reaction enthalpy of water splitting is equal to the formation enthalpy of water. Since the standard value of ΔS_R for the water splitting reaction is 49 kJ mol^{-1}, the value of ΔG_R can be calculated as 236.483 kJ mol^{-1}. It is noteworthy that in PEMWE systems operating at low temperatures (<373.15 K), electrical energy input plays a prominent role, whereas the contribution of thermal energy input is comparatively minor.[16]

In the case of a reversible reaction, its potential difference on the electrode is called the reversible cell voltage (V_{rev}). When an adequate supply of thermal energy is available, the reaction exhibits minimal electrical work requirements, and the corresponding V_{rev} can be determined by Equation (6.4):

$$V_{rev} = \frac{\Delta G_R}{zF} = 1.229 \text{ V} \tag{6.4}$$

Where z represents the number of charges transferred in the reaction, and F denotes the Faraday constant, which is 96,485 C mol^{-1}. Conversely, once the reaction proceeds without any thermal input, the total energy demand (ΔH_R) is solely provided by electrical energy. At this time, the cell voltage is designated as the thermoneutral voltage (V_{th}), which can be calculated using Equation (6.5):

$$V_{th} = \frac{\Delta H_R}{zF} = 1.481 \text{ V} \tag{6.5}$$

In practical application, the minimum operational voltage required by PEMWE is closer to V_{th}, attributed to the relatively restricted influence of thermal energy during low-temperature electrolysis. It is important to note that ΔG_R and ΔH_R are not only functions of temperature but also pressure.

The ohmic overpotential is due to the ohmic resistances of both electron and proton transport within the PEMWE, and also the interfaces between components. It follows the Ohm's law and shows a linear relationship between voltage and current. The resistivity of each component and the interfacial contact resistance (ICR) between each component are critical parameters that determine the ohmic overpotential.[17] Using high-conductivity materials and enhancing the interfacial contact to reduce the ICR are the main methods to reduce this loss.

Given the assumption of unrestricted reactant supply and product removal at the reaction interface during water splitting, the reaction kinetics governed by electron transfer follows Faraday's law.[18] To facilitate the electron transfer process involved in water splitting, an external potential is applied to overcome the activation energy barrier, corresponding to η_{act}. It consists of anodic activation overpotential ($\eta_{act,an}$) and cathodic activation overpotential ($\eta_{act,cat}$), which can be improved by applying HER and OER electrocatalysts, respectively. Given that OER is a four-electron transfer process and exhibits sluggish kinetics, $\eta_{act,an}$ primarily contributes to activation losses.[19] The activation overpotential can be calculated using Equation (6.6).

$$\eta_{act} = \eta_{act,\,an} + \eta_{act,\,cat} = \frac{RT_{an}}{\alpha_{an} F} \arcsin h\left(\frac{i}{2si_{0,an}}\right) + \frac{RT_{cat}}{\alpha_{cat} F} \arcsin h\left(\frac{i}{2i_{0,cat}}\right) \tag{6.6}$$

In this equation, T_{an} and T_{cat} are the anode and cathode temperatures, respectively, and s is the liquid saturation at the interface between catalyst layer (CL) and porous transport layer (PTL), which is defined as the ratio of liquid water volume to total pore volume in PTLs.[20] α_{an} and α_{cat} are the charge transfer coefficients of the anode and cathode, respectively. $i_{0,an}$ and $i_{0,\,cat}$ are the exchange current densities of the anode and cathode, respectively.[21] Their values are greatly affected by the activity of the catalyst materials and the operating temperature.[22]

The diffusion overpotential represents the effects of mass and ion transport processes. It is calculated using the following Equation (6.7).[14]

$$\eta_{diff} = \frac{RT_{an}}{2F}\ln\left(\frac{C_{O_2}}{C_{O_2,0}}\right) + \frac{RT_{cat}}{2F}\ln\left(\frac{C_{H_2}}{C_{H_2,0}}\right) + \zeta \tag{6.7}$$

C_{O_2} and C_{H_2} are the oxygen and hydrogen concentrations at the interface of CL and PEM, respectively. $C_{O_2,0}$ and $C_{H_2,0}$ indicate the reference values. ζ is the loss from other impacts, such as wettability.

6.2.1.3 Role of Critical Components

In a typical PEMWE, the core component is the membrane electrode assembly (MEA), which consists of a piece of proton exchange membrane (or named polymer electrolyte membrane, PEM), CLs, and PTLs.[23] The MEA is sandwiched by anode and cathode flow field plates (FFPs), current collectors, and endplates, respectively.[24] In PEMWE stacks, the flow field plates also act as bipolar plates (BPs), which work as the anode on one side and cathode on the other.[25] Among these components, current collectors are used to introduce the electrical power into the PEMWE for water electrolysis and also provide connecting points for the voltage lead. It is usually made of cheap metals, such as copper and aluminum, and with protective coatings on its surface to reduce the ICR between current collector and FFP, and also enhance the corrosion resistivity of itself to fulfill a long-term stability. In some cases, the BPs and current collectors are combined into a single component. Endplates are actually the most outside components of both the PEMWE single cell and stack, which is made of metals with strong mechanical properties to provide a uniform compression of the whole hardware. In addition, endplates also have flow inlets/outlets, connecting the interior flow fields for pumping reactant water and cooling water, and also collecting product hydrogen/oxygen.

MEA located in the center of the PEMWE, directly determines the performance and stability.[26] It has two categories based on the substrate of CLs, that is catalyst-coated membrane (CCM) and porous transport electrode (PTE), as shown in Figure 6.3. In CCM, the CL is adhered onto the surface of the membrane, which can be directly deposited or decal transferred from other substrates. Currently, the most widely used deposition techniques include ultrasonic spray, air-spray, rod coating, blade coating, slot-die coating, and gravure coating.[27] The swelling of the membrane is a critical concern when directly depositing catalyst onto the PEM by using catalyst inks.[28] Therefore, a vacuum plate or tensioning instrument is usually applied to inhibit this issue. In addition, the compositions and properties of the catalyst ink also

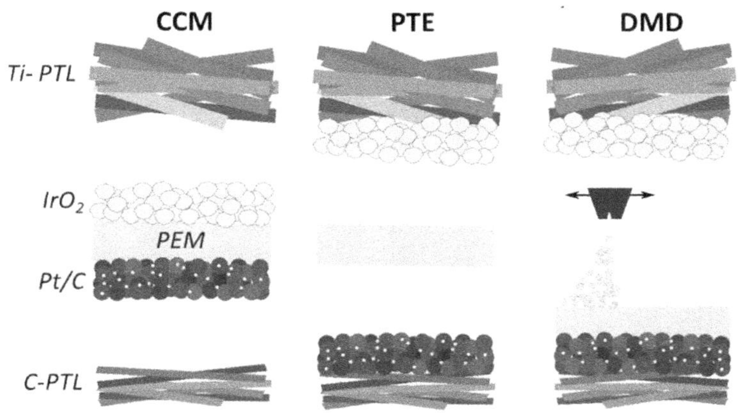

FIGURE 6.3 Schematic of the PTL and its function in PEMWE. (Reproduced with permission from Ref. 32.)

affect the morphology, performance, and stability of the CL and MEA. When applying the decal transfer method, some patterned or gradient-ordered CL structures can be achieved, which could significantly improve the catalyst mass activity.[29–31]

PTE has a totally different structure compared to CCM, and the CL is usually directly deposited onto the PTL. Besides the abovementioned deposition methods used for CCM fabrication, chemical vapor deposition and physical vapor deposition can be easily applied for PTE, and the ionomer-free thin-film CL is developed. CL in PTE commonly has a larger surface area than in CCM due to the porous structure of the PTL, which improves the performance and catalyst utilization. While, the porous structure also results in some wastes of the materials when it is located in the deep pores of the PTLs. Thus, microporous layers with smaller pore sizes are developed to make sure the catalyst materials are deposited on the surface of the porous structure. Between CCM and PTE, CCM has better interfacial contact between the PEM and CL, while PTE has better interfacial contact between CL and PTL. Aiming at improving the contact, hot pressing is widely used during the MEA assembling and a thin ionomer layer can be sprayed on top of the PTE before it is assembled with PEM.[33] Recently, an integrated MEA has been developed, which is fabricated layer by layer, also named like direct membrane deposition (DMD).[32] The integrated MEA can significantly reduce the interfacial contact between each part, having the advantages of both CCM and PTE. This contributes to a good performance, which will probably be a promising direction for next-generation MEA.[34]

PTL, also named as liquid/gas diffusion layer, is a critical component and has a direct influence on PEMWE performance and stability.[35] It not only provides efficient transport paths for reactant water and product oxygen at the anode, or product oxygen and diffused water at cathode, but also acta as a conductor for electrons and heat, as shown in Figure 6.4.[36] Therefore, the PTL is a multifunctional component, which should also sustain a good mechanical support for the PEM and CL under the compression of the system or the differential pressure.[37] In a single PEMWE device,

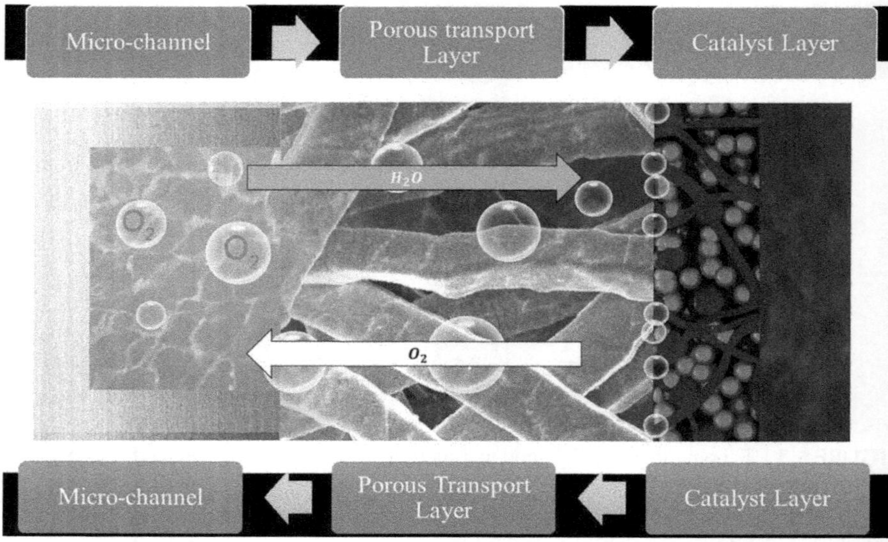

FIGURE 6.4 Schematic of the PTL and its function in PEMWE. (Reproduced with permission from Ref. 44.)

Ti-based materials (like Ti felt, sintered Ti powders, Ti mesh, etc.) are widely used on the anode side because of the harsh environment, such as high humidity, high overpotential, oxygen-rich conditions, which results in severe corrosion with most materials, like stainless steel.[38–41] Ti with high corrosion resistivity, should also be surface protected with coatings to prolong its life. Recently, some newly structured PTLs have been proposed and demonstrated, like pore-graded or hierarchically structured PTLs,[42] thin/tunable PTLs.[35] The pore-graded or hierarchically structured PTLs can improve the transport and interfacial contact.[43] The thin/tunable PTL has a unique structure, including a planar surface, straight-through pores, and small thickness, which contributes to better interfacial contact, mass transport, electron and thermal transport, and improved performance at last. At the cathode, carbon paper is widely used because of its high electrical conductivity and good stability. Around the PTL, there is usually a gasket, which is used to seal the MEA and maintain a good compression.

FFPs distribute the reactants/products and conduct the current and heat, which are indispensable components in PEMWE.[45] Therefore, the flow field pattern is the most concerned aspect, which mainly includes parallel flow channel, (single/double/triple/multi) serpentine flow channel, pin cushion flow channel, and chamber flow filed.[46] The flow field pattern mainly impacts the flow uniformity and reactant/pressure/thermal distribution, which could affect the performance of the PEMWE.[47] As reported, the most widely used materials for FFP include titanium, stainless steel, and graphite.[48–50] At anode side, due to the harsh environment, FFP is usually made of metal with protective coatings, such as Pt. Recently, nitridation-processed FFP has been proposed for the surface protection, which also shows good interfacial contact and corrosion resistivity.[51] In a stack, BPs accounts for 40%–50% the cost, which is mainly due to the high cost of materials and manufacturing process.[25] BPs are usually manufactured via molding and cutting processes conventionally, and additive manufacturing has also been applied to investigate the new flow field design and integrated devices. For example, Gao et al. have used the selective laser melting technique to fabricate the FFP with stainless steel powder, and they successfully fabricated the FFP together with the 3D-printed PTL forming an integrated structure.[45]

6.2.2 Advances in Electrolysis Technology

Recent advancements in electrolysis technology have significantly improved the efficiency, performance, and durability of water electrolyzers, paving the way for the widespread adoption of green hydrogen for grid energy storage.[52] The catalyst materials and PTLs play a crucial role in determining the efficiency and cost of water electrolysis, as shown in Figure 6.5. Ongoing research and development efforts are focused on improving the activity, stability, and cost-effectiveness of these materials. This section delves into the key technological advancements in water electrolysis, focusing on innovations in catalyst materials and PTLs, performance and durability enhancements, and developments in system integration and automation.

6.2.2.1 Innovations in Catalysts

The performance and durability of PEMWE systems are heavily reliant on the catalyst materials and electrodes used. While conventional catalysts like platinum group metals (PGMs) offer high activity, their high cost and susceptibility to degradation pose significant challenges for widespread adoption.[54] Currently, Ir-based catalysts are the most widely used OER electrocatalysts, and Pt-based catalysts are the most accepted HER electrocatalysts.[55] This section explores the latest innovations in catalyst materials and electrodes, focusing on alternative materials, nanotechnology, and advanced manufacturing techniques that aim to enhance performance, reduce costs, and improve long-term stability.

Currently, a lot of high-performing and durable PGM catalysts have been developed, and some of them can meet industrial operation requirements,[56] like Ir nanocatalyst,[53] $IrRuO_x$ nanowires,[57] Ta-doped Ru catalyst,[58] etc. Ir-based catalysts have

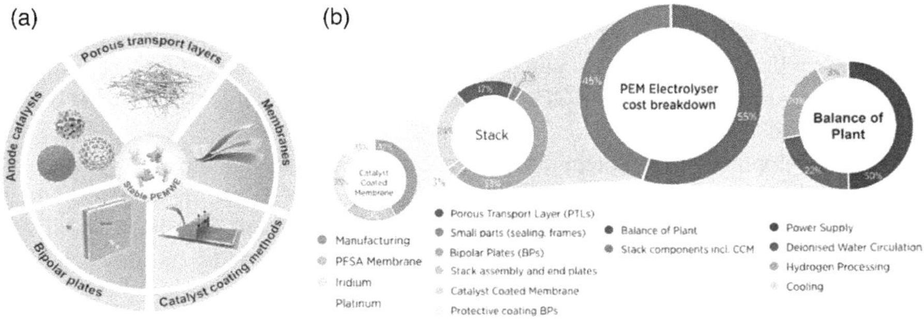

FIGURE 6.5 (a) Key materials research for PEMWE. (b) Total cost breakdown for 1 MW PEMWE system. (Reproduced with permission from Ref. 53.)

emerged as the top choice for OER in PEMWEs due to their intrinsic activity and stability in acidic environments.[59] However, the scarcity and high cost of Ir necessitate significant advancements to reduce its usage while maintaining or enhancing catalytic performance.[60]

Doping Ir with other elements is a promising strategy to optimize its electronic structure and the adsorption/desorption energy of oxygen species, leading to enhanced OER activity.[54] This strategy also effectively reduces the cost of the catalyst by incorporating less expensive elements.[61] Other doping strategies involve the use of transition metals, such as Ce, Ni and Co, which can further enhance the OER performance of Ir- or Ru-based catalysts.[62] These doped catalysts often exhibit heightened intrinsic activity and a high specific electrochemical surface area, making them particularly efficient.[63]

The nanostructure of Ir catalysts plays a crucial role in determining overall performance.[64] Nanostructural modifications can increase the electrochemical surface area and improve mass transfer capabilities under high current densities, which are critical factors for OER performance. Many nanostructured Ir-based catalysts have been developed, including nanorods, nanoparticles, and nanosheets. These nanomaterials offer another pathway to lower Ir usage without compromising durability.

Supported catalysts with enhanced electronic conductivity and optimized Ir density are another promising strategy.[65] Innovations such as atomically dispersed Ir sites, Ir-based alloys, and Ir clusters on supports have demonstrated promising electroactivity for OER in both neutral and acidic conditions. Supports, such as TiO_2,[66] Nb_2O_5,[67] Ta_2O_5[6], and MnO_2,[68] have been explored to enhance the stability and activity of Ir-based catalysts.

Besides the developed PGM electrocatalysts, the promising direction of catalyst development includes non-PGM catalysts, composite catalysts, and metal-organic frameworks (MOFs).[69] Research efforts are intensifying to develop non-PGM catalysts that offer comparable activity and stability to PGMs but at a significantly lower cost. Transition metal-based catalysts, such as cobalt (Co), are promising candidates due to their high activity for the OER. However, challenges remain in achieving the desired activity and stability while minimizing the loading of these materials.

Combining PGMs with non-PGMs in composite catalysts can offer a synergistic effect, leading to improved activity and stability. MOFs are porous coordination polymers with a high surface area and tunable properties.[70] They offer a promising platform for developing efficient and cost-effective catalysts for the OER. Research is ongoing to explore the use of MOFs as standalone catalysts or as supports for PGMs and non-PGMs.

The development of efficient CLs is crucial for enhancing the performance of PEMWEs.[71] The CL is responsible for catalyzing the OER and must possess high electrochemical activity, good mass transport properties, and long-term stability.[72] Several innovations have been made in CL development, focusing on improving the structure, composition, and deposition techniques.

Porous CLs have been shown to enhance mass transport and improve catalyst utilization. Techniques such as spray drying and evaporation-driven self-assembly have been used to engineer porous hierarchical structures in Ir-based catalysts. These structures enhance the accessibility of reactants and products to nanoparticle surfaces, significantly boosting catalytic performance.

Direct deposition techniques, such as roll-to-roll processing and electrochemical deposition, offer advantages in terms of scalability and cost-effectiveness. These techniques allow for the uniform deposition of catalysts on the MEA, ensuring good contact between the catalyst and the membrane. Roll-to-roll processing, in particular, is a promising technology for the mass production of PEMWEs.

Composite CLs, consisting of a mix of different catalysts or catalyst supports, have been developed to enhance the overall performance, as shown in Figure 6.6. These composites can combine the advantages of different catalysts, such as high activity and stability, leading to improved OER performance. For instance, composite CLs containing both Ir-based and Ru-based catalysts have been shown to exhibit enhanced activity and durability.

Furthermore, the structure or morphology also shows a great impact on PEMWE performance and durability.[74–76] Figure 6.7 presents a thought-provoking figure that delves into the intricacies of CL structures and their impact on performance.[77] The central focus is on the conventional porous electrode, surrounded by inquiries about ideal electrode characteristics and comparisons between dense and porous thin-film electrodes. Unlike conventional gas diffusion electrodes with deep porous structures that underutilize the catalyst due to limited triple-phase boundary conditions, the flat CLs with straight-through pores overcome this restriction.[78]

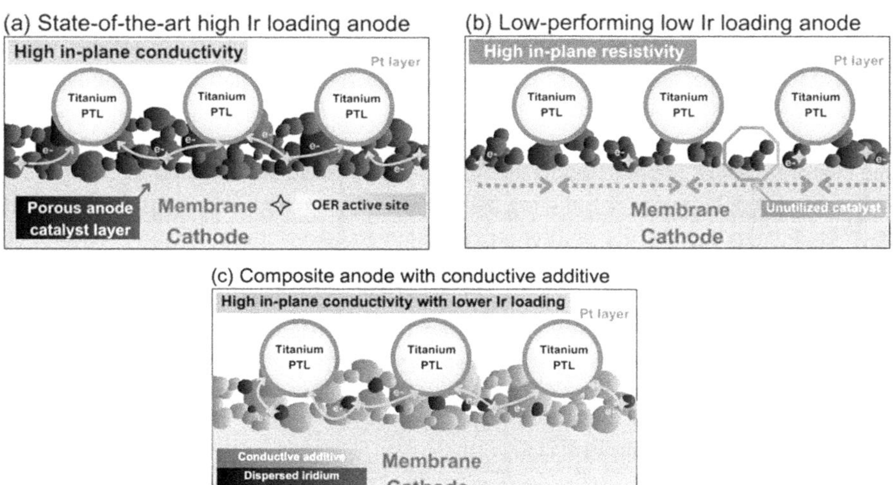

FIGURE 6.6 Schematic of the catalyst layer (CL) in PEMWE showing the improved structure using a composite anode with a conductive additive from a cross-sectional view. (a) State-of-the-art high Ir loading PEMWE anode with sufficient thickness and high in-plane conductivity, but the high Ir loading leads to high cost. (b) Typical low-loaded IrOx anode, leading to a thin electrode, electrically disconnected and unutilized catalyst, and low performance. (c) Composite anode using a conductive additive to increase in-plane electrical conductivity and electrode thickness while enabling low Ir loadings. (Reproduced with permission from Ref. 73.)

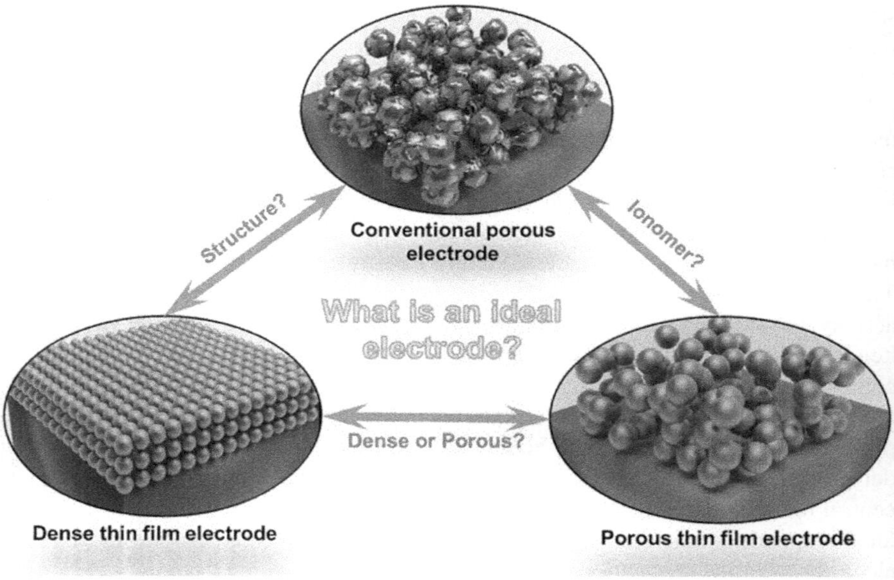

FIGURE 6.7 Schematic of the three types of CLs. (Reproduced with permission from Ref. 77.)

Despite significant advancements in catalyst and CL development, several challenges remain. One of the primary challenges is the high cost of Ir-based catalysts, which necessitates further reductions in Ir loading while maintaining or enhancing catalytic performance.[10] Alternative catalyst systems, such as Ru-based and noble-metal-free catalysts, offer potential solutions but still require significant advancements to match the performance of Ir-based catalysts.[79] Long-term operations can lead to catalyst dissolution and agglomeration, reducing the active Ir content and impairing overall stability.[80] Understanding the dissolution mechanisms and developing strategies to mitigate these effects is crucial for the long-term viability of PEMWEs.[81]

In conclusion, innovations in catalyst and CL development are crucial for the commercialization of PEMWEs for green hydrogen generation.[82] By addressing the challenges of cost, performance, and durability, researchers can pave the way for the widespread adoption of PEMWEs in grid energy storage systems.

6.2.2.2 Advancements in PTLs

Historically, conventional PTL designs in PEMWE systems have relied on materials like Ti powder sintered plates and Ti fiber felts, with manufacturing processes such as sintering and incorporating metal foams.[41] While these approaches have been widely used, they often result in suboptimal pore structures and inadequate interfacial contact between the PTL and other components. This compromises both the performance and durability of the electrolyzer, hindering its efficiency and longevity. Recognizing the limitations of conventional PTL designs, researchers have embarked on a journey of innovation and exploration, seeking to enhance the performance and efficiency of PEMWE systems.

While titanium-based materials continue to dominate due to their excellent corrosion resistance, researchers are actively investigating lower-cost alternatives to reduce the overall cost of PTLs. One promising avenue is the use of stainless steel, which can be treated with novel passivation strategies to improve its durability and resistance to degradation. Additionally, incorporating hydrophilic and hydrophobic additives into PTL materials can optimize their wettability, leading to improved water and gas transport within the layer. This enhances the efficiency of mass transfer and the overall performance of the electrolyzer.

Innovative structural designs have emerged as a key focus in PTL development.[43] Spatially gradient PTLs, achieved through microporous layers or dual-layer configurations, offer tailored pore structures that optimize water and gas transport.[42] This mitigates mass transport limitations and reduces ohmic losses, leading to improved efficiency. Furthermore, surface/interface optimization techniques like laser treatment and the creation of patterned through-pores enhance interfacial contact and facilitate bubble detachment.[39] This further improves the overall performance and efficiency of PEMEC (proton exchange membrane electrolytic cell) systems.

While the advancements in PTL development have been significant, challenges remain in achieving scalability for mass production. The wet-chemical etching processes used to create intricate pore structures are precise but can be environmentally hazardous and difficult to scale up.[35] This poses a barrier to the widespread adoption of these advanced PTL designs for large-scale hydrogen production. In response to the scalability challenge, recent research efforts have explored a novel mechanical manufacturing methodology for the PTLs with straight-through pores (SP-PTLs). This innovative approach involves mechanical patterning, stacking arrangement, welding, and cutting to create PTLs with straight-through pores. These SP-PTLs offer several advantages, including enhanced catalyst utilization, efficient gas removal, and reduced voltage losses. Importantly, this mechanical manufacturing process is suitable for industrial deployment, addressing the scalability concerns associated with wet-chemical etching while maintaining performance attributes. The SP-PTLs can also be easily implemented with novel FFPs and other adjacent components, forming integrated 6-in-1 BP (Uni-BPP) for PEMWE stacks, as shown in Figure 6.8.

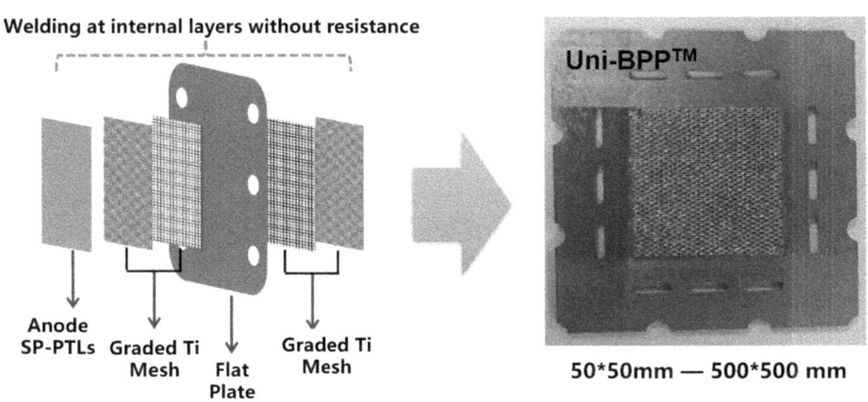

FIGURE 6.8 Schematic and photo of the integrated 6-in-1 Uni-BPP for PEMWE stack.

The advancements in PTL development present a promising future for PEMWE systems. The exploration of new materials, innovative structural designs, and the development of scalable manufacturing processes are driving the evolution of PTL technology. These advancements will contribute to improved efficiency, durability, and cost-effectiveness of PEMWE systems, paving the way for the widespread adoption of green hydrogen production for grid energy storage and other applications. As research and development continue to push the boundaries of PTL technology, we can expect further breakthroughs that will unlock the full potential of PEMWE systems, contributing to a sustainable and clean energy future.

6.2.2.3 Developments in System Integration

The integration of water electrolysis systems into the broader energy infrastructure is a crucial aspect of advancing green hydrogen production for grid energy storage. Recent developments in system integration have focused on optimizing stack design, enhancing control systems, facilitating close integration with renewable energy sources, and leveraging digitalization and artificial intelligence (AI) to improve efficiency and reliability.

Stack design optimization is a fundamental aspect of improving the overall performance and cost-effectiveness of water electrolysis systems. One key trend in stack design is the adoption of modular architectures. Modular stacks allow for greater flexibility in system configuration, enabling operators to tailor the system's capacity to specific application requirements. This approach not only minimizes waste but also facilitates easier maintenance and upgrades. In addition to modular design, the integration of components into a single, compact unit has become increasingly important. By reducing the number of interconnections and simplifying the overall system layout, manufacturers can achieve higher power densities and lower energy losses, as shown in Figure 6.8. These improvements are critical for reducing the levelized cost of hydrogen, which is a key metric for assessing the economic viability of green hydrogen production.

Advanced control or diagnostic systems play a pivotal role in optimizing the operation of water electrolysis systems.[83] These systems continuously monitor key parameters such as current density, voltage, temperature, and pressure.[84] By collecting and analyzing this data in real time, operators can make informed decisions about adjusting operating conditions to maximize efficiency and minimize energy losses.[16] For example, the *in situ* characterization methods applying unique thin wires embedded in MEA[85] (Figure 6.9a) or sensors in PEMWE stacks (Figure 6.9b) for monitoring the losses and degradations can provide insight and meaningful data for the PEMWE analysis and operation monitor.[86] Moreover, advanced control systems can incorporate predictive analytics and machine learning algorithms to anticipate potential system failures and preemptively address them. This proactive approach to maintenance not only reduces downtime but also extends the lifespan of critical components, such as membranes and catalysts.

The integration of water electrolysis systems with renewable energy sources is crucial for achieving sustainable and economically viable green hydrogen production. Renewable energy sources, such as wind and solar power, are inherently intermittent and variable, posing significant challenges to the consistent operation of electrolyzers.

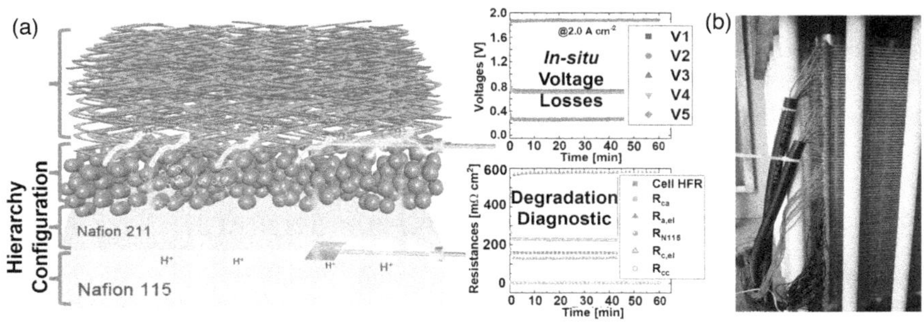

FIGURE 6.9 (a) *In situ* voltage loss and degradation diagnostic methods for PEMWE. (Reproduced with permission from Ref. 85.) (b) Sensors for 1 MW PEMWE system.

To mitigate these challenges, advanced power management systems and energy storage solutions have been devised. These systems effectively balance the intermittent nature of renewable energy with the continuous demands of electrolyzers by storing excess energy during periods of high generation and releasing it during periods of low generation. This ensures the optimal operation of the electrolyzer, regardless of fluctuations in renewable energy availability. Additionally, employing a combination of PEMWE and ALK technologies can strike a balance between cost and efficiency. PEMWE can operate over a wide power range of 5%–200%, while ALK operates at a stable status, thus optimizing production by the coupled system. Furthermore, the development of hybrid renewable energy systems that integrate multiple renewable energy sources can further bolster the reliability and cost-effectiveness of green hydrogen production. For instance, combining solar and wind energy can create a more stable and predictable energy supply, minimizing the reliance on extensive energy storage solutions.

The application of digitalization and AI has revolutionized the operation and maintenance of water electrolysis systems. AI-based optimization algorithms can analyze vast amounts of data to identify patterns and trends that are not easily detectable by human operators. These algorithms can then optimize operating conditions in real time, ensuring that the system operates at its maximum efficiency. Moreover, AI can predict system failures with high accuracy, enabling operators to take preemptive action to prevent downtime and extend the lifespan of critical components. This predictive maintenance approach not only reduces operating costs but also enhances system reliability and safety. In addition to optimization and predictive maintenance, AI can also automate routine maintenance tasks, such as cleaning and inspecting system components. This automation reduces the need for human intervention, lowering labor costs and improving overall system performance.

Developments in system integration have played a crucial role in advancing the commercialization of green hydrogen production for grid energy storage. By optimizing stack design, enhancing control systems, facilitating seamless integration with renewable energy sources, and leveraging digitalization and AI, manufacturers have achieved significant improvements in efficiency, reliability, and cost-effectiveness. As research and development continue to push the boundaries of water electrolysis technology, we can expect further breakthroughs that will unlock the full potential of green hydrogen as a sustainable and cost-effective energy storage solution. These advancements will pave the way for a more resilient and sustainable energy future, enabling us to harness the power of renewable energy and reduce our dependence on fossil fuels.

6.3 GREEN HYDROGEN IN GRID ENERGY STORAGE

Renewable energy sources, notably solar and wind, are inherently intermittent due to their reliance on environmental factors such as sunlight and wind speed. These natural phenomena vary significantly based on weather conditions and the time of day, leading to fluctuations in energy production. For instance, solar power generation peaks during sunny days and drops to zero at night, while wind power can be abundant during certain seasons or times of the day but scant in others. This variability poses significant challenges for grid operators, who must maintain a delicate balance between electricity supply and demand to ensure reliable and stable power distribution. To mitigate these challenges, grid energy storage systems have emerged as a crucial technology. These systems allow for the capture and storage of excess energy generated during periods of high renewable production.[87] When production falls and demand remains high, the stored energy can be released back into the grid, thereby bridging the gap between supply and demand. Various storage technologies are employed for this purpose, each with its own set of advantages and limitations.[26] For example, batteries can provide fast response times and are well-suited for short-term storage, while pumped hydro storage can offer large-scale capacity but requires specific geographical conditions.

Figure 6.10 clearly shows the role of green hydrogen in grid energy storage, emphasizing its potential to revolutionize the energy sector. As a renewable and clean energy source, green hydrogen offers a sustainable alternative to traditional fossil fuels.[88] The figure illustrates a comprehensive energy system that integrates various sources of green power, including solar, wind, hydroelectricity, and biomass, to generate electricity. Electrolyzers play a crucial role by producing hydrogen, which serves as an essential input for several industries. The chemical industry uses hydrogen for ammonia production, while agriculture benefits from it through the manufacturing of e-fertilizers. Additionally, the petrochemical industry relies on hydrogen for synthesizing e-methanol. In terms of mobility, the rail, road, aviation, and marine sectors all depend on green power-generated electricity. Rail systems are directly powered by electricity, whereas road vehicles can be fueled by either electricity or hydrogen produced via electrolysis. Aviation and marine transportation primarily rely on hydrogen-based fuels derived from electrolysis. Furthermore, the figure highlights the importance of the steel and cement industries, which consume significant amounts of energy. These industries utilize both direct electrical inputs and indirect energy derived from natural gas grids and further industrial processes. Overall, this intricate network showcases how different sectors interconnect and leverage sustainable energy solutions to minimize environmental impact while meeting diverse energy demands across various domains. It is reasonable to say that green hydrogen has the potential to play a pivotal role in the transition to a low-carbon, sustainable energy future.[89]

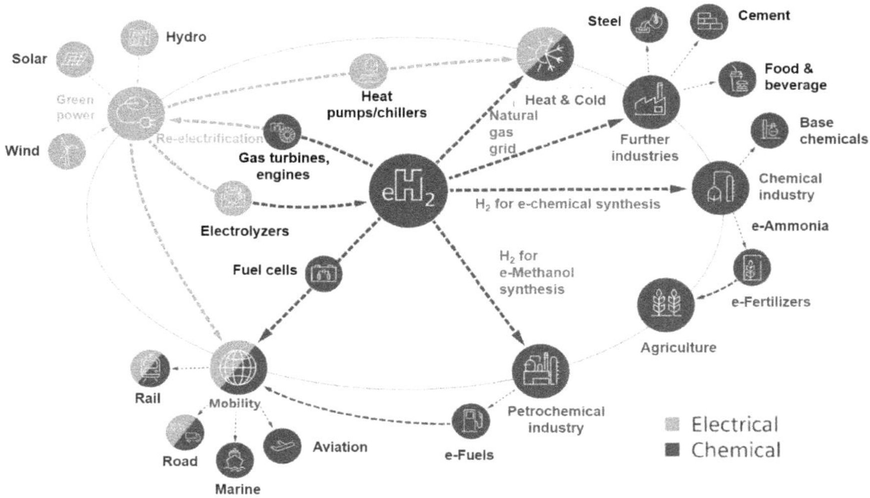

FIGURE 6.10 Hydrogen as a core in energy supply and demand chains. (Reproduced with permission from Ref. 90.)

6.3.1 Role of Green Hydrogen

Green hydrogen stands out as a pioneering energy carrier in the quest for sustainable development. Its production process, which involves the electrolysis of water powered by renewable energy sources, epitomizes the transition from fossil-fuel-dependent systems to clean, renewable alternatives. The water electrolysis, driven by solar, wind, or hydroelectric power, not only harnesses the abundant supply of water on Earth but also leverages the intermittent nature of renewable energy sources effectively. During electrolysis, direct current electricity is passed through the device, causing the chemical bonds between hydrogen and oxygen atoms to break, thereby releasing hydrogen gas at the cathode and oxygen gas at the anode. This method ensures that the production of green hydrogen does not emit greenhouse gases, in stark contrast to gray hydrogen, which relies on natural gas or coal steam reforming and contributes substantially to climate change.

The scalability of green hydrogen production is a significant advantage. It can be tailored to meet the varying capacities of renewable energy installations. For instance, solar panels installed in sun-drenched regions can generate electricity for a long time, which, when in excess, can be directed toward electrolysis units to produce hydrogen. Similarly, wind farms in windy areas can convert the energy of wind into electrical power, which is then used to split water molecules. This flexibility allows for optimal utilization of renewable resources and ensures a continuous supply of green hydrogen. When electricity demand spikes or renewable production wanes, the stored hydrogen can be converted back into electricity using fuel cells. Alternatively, it can be burned in hydrogen turbines to generate power, offering a clean and efficient alternative to fossil fuels. This capability allows grid operators to harness the full potential of renewable energy sources, even during periods of low production, thereby enhancing grid stability and reliability. Moreover, green hydrogen's role extends beyond mere energy storage. It can also serve as a valuable resource for balancing the grid during periods of peak demand. By injecting hydrogen into the natural gas grid, grid operators can increase the calorific value of the gas and boost electricity production in combined cycle gas turbines. Additionally, dedicated hydrogen-powered turbines can be deployed to provide additional capacity during high-demand scenarios. This flexibility enables grid operators to meet peak demand without resorting to fossil fuels, thereby promoting a more sustainable and resilient energy system.

Storage and transportation of green hydrogen further enhance its appeal. Once produced, hydrogen can be stored in compressed gas form, liquefied, or bound to chemical carriers like ammonia. These storage solutions facilitate the distribution of green hydrogen to remote locations where renewable energy sources might be scarce but energy demand is high. For example, hydrogen produced in sunny or windy regions can be transported to urban centers for use in fuel cell vehicles, thus reducing reliance on traditional fuels and cutting down emissions from the transportation sector. Moreover, the versatility of green hydrogen extends beyond transportation. In industrial processes, it can serve as a reducing agent in the production of chemicals such as ammonia and methanol, replacing fossil-based feedstocks. Additionally, green hydrogen can be used in fuel cells to generate electricity, providing a clean and reliable power source for homes, businesses, and even large-scale grid applications. This multifaceted utility underscores green hydrogen's potential as a cornerstone of a decarbonized energy system.

Countries and regions that embrace green hydrogen production are poised to make substantial strides in reducing their carbon footprints. By integrating green hydrogen into their energy mix, they can diversify their energy sources, enhance energy security, and mitigate the environmental impacts associated with fossil fuel extraction and combustion. Governments and private sector alike are investing in research and development to improve electrolysis efficiency, reduce costs, and scale up production capacities.

In conclusion, green hydrogen offers a transformative and multifaceted clean energy solution, positioning it as a pivotal player in the global transition to a sustainable energy future. Its renewable nature, scalability, and versatility, combined with its storage capabilities, enable the capture and preservation of excess energy from intermittent renewable sources and enhance grid resilience during peak demand periods. As technology advances and both production and storage costs decline, green hydrogen is increasingly viable and poised to become a cornerstone for countries seeking to decrease their dependency on fossil fuels, combat climate change, and ensure a sustainable and reliable energy future.

6.3.2 Practical Applications of Green Hydrogen

Green hydrogen, as a renewable and clean energy source, is emerging as a pivotal player in various practical applications, showcasing its transformative potential in the energy sector. By leveraging its unique properties as a storable and versatile energy carrier, green hydrogen can be utilized in a multitude of ways to enhance grid resilience, reduce emissions, and promote sustainability. This section delves into the practical applications of green hydrogen, exploring how it is being integrated into power-to-gas (P2G) and power-to-liquid (P2L) systems, hydrogen-based microgrids and distributed energy systems, and successful implementations in grid energy storage. As technology advances and production costs decline, the practical applications of green hydrogen are expanding, positioning it as a cornerstone for countries seeking to decrease their dependency on fossil fuels and ensure a sustainable and reliable energy future.

6.3.2.1 Power-to-Gas and Power-to-Liquid Systems

Power-to-X (P2X) refers to a range of technologies that convert surplus renewable energy into various forms of energy storage and reconversion pathways, enabling the decoupling of power from the electricity sector for use in sectors such as transport, chemicals, and more.[91] The term, widely used in Germany, encompasses various conversion processes including power-to-ammonia, power-to-chemicals, power-to-fuel, P2G (such as hydrogen and methane), P2L (synthetic fuel), power-to-food, and power-to-heat.[92] These schemes are considered flexibility measures, particularly useful in energy systems with high shares of renewable generation and strong decarbonization targets. In 2016, the German government funded a significant research project into P2X options. Among them, P2G and P2L systems represent cutting-edge technologies designed to convert renewable electricity into storable fuels, such as hydrogen and synthetic fuels, respectively. These innovative systems offer a highly promising solution for addressing the challenges associated with storing excess renewable energy and balancing supply and demand within the electrical grid.[93]

In a P2G system, the process begins with the utilization of renewable electricity to split water molecules into hydrogen and oxygen through a process known as electrolysis. Once produced, the hydrogen can be injected directly into the existing natural gas grid, where it can be used as a fuel source, or it can be stored for later use, as shown in Figure 6.11.[94] This capability allows grid operators to harness excess renewable energy and convert it into a form that can be stored and utilized as needed, thereby enhancing grid stability and reliability.[95] On the other hand, P2L systems operate in a similar manner to P2G systems but with a focus on producing synthetic fuels such as methane, methanol, or diesel. These synthetic fuels offer several advantages, including their compatibility with existing transportation infrastructure and their potential to significantly reduce greenhouse gas emissions in the transportation sector. By converting renewable electricity into these fuels, P2L systems provide a viable pathway to decarbonizing transportation and achieving broader sustainability goals.

The feasibility and benefits of P2G and P2L systems have been demonstrated through several notable case studies. Currently, Iceland boasts one operational P2L plant, while Europe has approximately 40 pilot and demonstration facilities for P2G/P2L technologies. In November 2016, an EU project was launched to establish and assess a P2-methanol pilot plant in Luleå, Sweden, which aimed to utilize carbon dioxide-rich blast furnace gas from SSAB's (Swedish Steel AB) steel production and renewable hydrogen from intermittent electricity generation. The project's goal was to explore and propose potential P2G/P2L solutions in Norrbotten, considering the

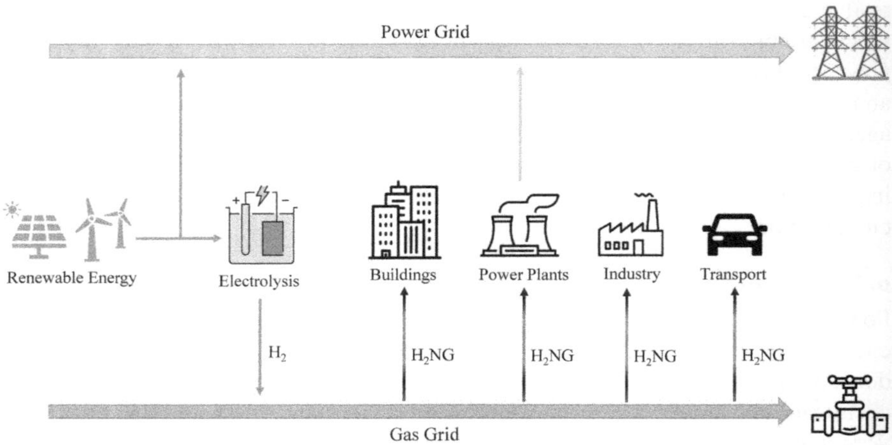

FIGURE 6.11 Schematic view of P2G system with hydrogen blending into the gas grid. (Reproduced with permission from Ref. 94.)

regional electricity market and hydrogen demand, with Piteå's biorefinery infrastructure serving as the foundation. The analysis, which considers both current and future scenarios, builds upon a 2015 study by ÅF that identified Piteå-Luleå-Norrbotten as a prime location for demonstrating P2G/P2L in Sweden.

The status of P2G is currently in a phase of development and demonstration, with several projects around the world showcasing its potential. The technology is seen as a promising solution to the challenge of integrating renewable energy sources into the existing energy grid, as it allows for the storage of energy in a form that can be dispatched on demand.

One of the key examples of P2G in practice is the H2FUTURE project in Germany. This project, funded by the German Federal Ministry for Economic Affairs and Energy, aims to demonstrate the technical and economic feasibility of integrating hydrogen produced via electrolysis into the existing natural gas grid. The hydrogen is produced using renewable energy sources and is then injected into the gas grid, where it can be used to generate power or heat in conventional gas-fired power plants or boilers. Another notable example is the Power-to-Gas Demonstration Plant in Austria, which is operated by a consortium of companies. This plant uses wind power to produce hydrogen via electrolysis, which is then injected into the natural gas grid. The project aims to showcase the potential of P2G for storing renewable energy and reducing greenhouse gas emissions. In addition to these examples, there are several other P2G projects ongoing around the world, including in Japan, South Korea, and the U.S. These projects are exploring different approaches to P2G, such as using methane as the storage medium instead of hydrogen, or integrating P2G with other energy systems such as district heating networks.

Overall, the status of P2G is promising, with ongoing research and development efforts aimed at improving its efficiency, reducing costs, and demonstrating its feasibility in real-world applications. As renewable energy sources become increasingly important in the global energy mix, P2G has the potential to play a significant role in facilitating their integration into the existing energy grid. However, challenges such as the need for large-scale storage facilities and the high costs of electrolysis equipment remain to be addressed before P2G can be widely adopted. As these technologies continue to advance, they have the potential to play a crucial role in achieving broader sustainability goals and fostering a more resilient and sustainable energy future.

6.3.2.2 Hydrogen-Based Microgrids and Distributed Energy Systems

Hydrogen-based microgrids and distributed energy systems are emerging as key technologies to enhance grid resilience and reliability while reducing emissions.[96] These systems leverage hydrogen as a clean fuel source for electricity generation and storage, providing communities with the ability to operate independently of the traditional grid during outages or emergencies.[97]

In a hydrogen-based microgrid, renewable electricity is used to produce hydrogen through the process of electrolysis. The hydrogen is then stored in tanks or converted into other forms such as ammonia or methanol for long-term storage. When electricity is needed, the hydrogen can be converted back into electricity using fuel cells or burned in a turbine to generate power. This process ensures that critical infrastructure and communities receive continuous and reliable electricity, even during outages. Therefore, the hydrogen-based microgrids usually contain three steps: (1) Electrolysis: Surplus renewable energy, such as solar or wind power, is used to electrolyze water into hydrogen and oxygen. (2) Storage: The produced hydrogen is stored in high-pressure tanks or underground caverns for later use. (3) Conversion: When needed, the stored hydrogen can be converted back into electricity via fuel cells or used as an input for combined cooling, heating, and power (CCHP) systems.

Distributed energy systems involve the deployment of small-scale renewable energy sources such as solar panels and wind turbines in close proximity to the point of consumption. By integrating hydrogen-based energy storage systems with these distributed energy sources, communities can further reduce their reliance on the traditional grid and achieve greater energy independence. This integration also helps to balance supply and demand, ensuring that renewable energy is used efficiently and effectively.

Figure 6.12 presents a schematic diagram of the Energy Storage as a Service (ESaaS) mode.[98] This diagram illustrates the interaction between multiple energy microgrids. The microgrids have diverse energy demands, including electricity, heating energy, and cooling energy. To fulfill these demands, each microgrid is equipped with a renewable energy generation system and a CCHP system. They are also connected to the utility grid, natural gas pipelines, and a P2G system facilitated by the ESaaS operator. The ESaaS operator plays a crucial role in this system. They are responsible for investing in and operating the advanced P2G system, which converts surplus renewable energy into green hydrogen for storage. This hydrogen can then be used as an energy carrier and traded with the microgrids as needed. The energy trading process is represented by dotted lines in the diagram.

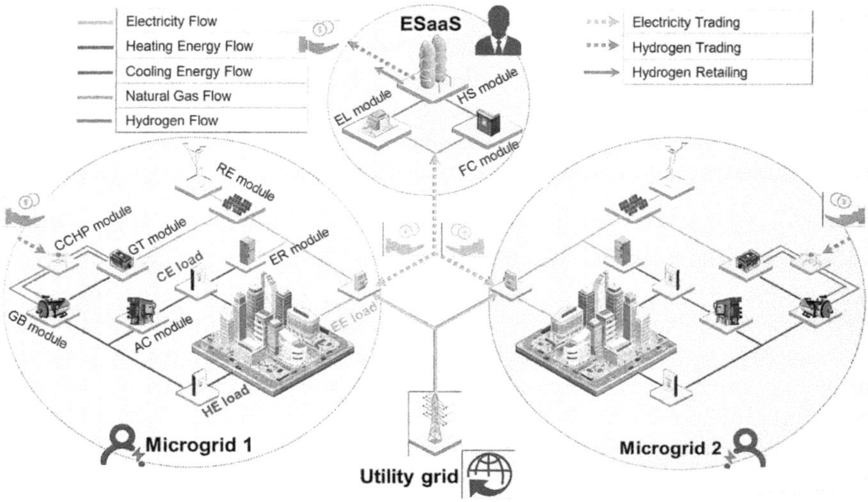

FIGURE 6.12 Schematic diagram of the Energy Storage as a Service (ESaaS) mode. (Reproduced with permission from Ref. 98.)

Overall, hydrogen-based microgrids and distributed energy systems offer a promising solution to enhance grid resilience and reliability while reducing emissions. As these technologies continue to develop and mature, they will play an increasingly important role in shaping the future of energy systems around the world.

6.3.2.3 Successful Implementations of Green Hydrogen in Grid Energy Storage

Several successful implementations of green hydrogen in grid energy storage have demonstrated the feasibility and benefits of this technology. This part will briefly introduce these examples to show the current status of the green hydrogen in grid energy storage.

i. H2RES (Offshore Wind to Hydrogen Demonstration Project) project in Denmark

   The H2RES project, led by Ørsted, utilizes two 3.6 MW offshore wind turbines at Avedøre Holme to power a 2 MW electrolysis facility, making it the world's first electrolyzer directly connected to offshore wind turbines. The project aims to produce up to 1,000 kg of green hydrogen daily, which will be used to fuel zero-emission road transport in the Greater Copenhagen area. Additionally, it will generate knowledge and insights into further optimization and cost-reduction of green hydrogen production, serving as a critical stepping stone for future green hydrogen projects. This innovative project aligns with the EU's ambitious hydrogen strategy, targeting significant expansions in renewable hydrogen capacity by 2024 and 2030, and demonstrates the potential of green hydrogen in achieving decarbonization targets.

ii. HyBalance project in Denmark

The HyBalance project, a pioneering endeavor focused on demonstrating the concept of multi-sectoral hydrogen end-use in Denmark. The project leverages the abundant wind resources in the country to produce hydrogen through Power-to-Hydrogen (PtH2) technology, utilizing cheap renewable electricity in times of surplus. The primary objective is to showcase the feasibility of large-scale electrolyzers in providing grid-balancing services while simultaneously supplying hydrogen to various end-use sectors, including industry and transportation. The HyBalance plant, located in Hobro, Nordjylland, will be designed for combined operation, aiming to identify potential revenue streams under current and future constraints. It will also employ recent developments in hydrogen distribution and storage, with relevant applications such as hydrogen refueling stations for fuel cell vehicles and buses, supply to local industry, and potential hydrogen storage in salt caverns.

iii. Advanced Clean Energy Storage project in U.S.

The Advanced Clean Energy Storage project, a cutting-edge facility in Delta, Utah, that combines 220 MW of alkaline electrolysis with two massive 4.5-million-barrel salt caverns for long-term, seasonal energy storage. The project aims to capture excess renewable energy, store it as hydrogen, and deploy it as fuel for a hydrogen-capable gas turbine combined cycle power plant. The facility is expected to play a crucial role in grid stabilization, reduction of curtailment of renewable energy, and contribute to significant carbon dioxide emissions reductions annually. By using salt caverns for hydrogen storage and deploying large-scale electrolyzers, the project represents a significant innovation in clean energy storage solutions.

iv. P2G project in Germany

The P2G technology project in Germany initiative focuses on converting excess renewable energy into hydrogen through electrolysis. Key players in this field have developed the "proWindgas" tariff, reinvesting the premiums to further technology advancements. A notable project involves Audi, EWE, and SolarFuel GmbH constructing a pilot facility to produce "e-gas" from renewable energy surpluses, aiming to fuel Audi's TCNG vehicle model range. ENERTRAG, another company, operates a hybrid power station that transforms wind energy into hydrogen for CO_2-free mobility. Additionally, the Deutsche Energie-Agentur (dena) has established the P2G Strategy Platform to assess the integration of renewable energy sources into the energy mix and facilitate the practical implementation of P2G technology. Collaborations between industry stakeholders and research institutions are crucial for the development and commercialization of this innovative energy storage solution. The P2G technology project highlights Germany's commitment to leveraging renewable energy sources for sustainable mobility and energy storage, with significant investments and partnerships driving technological advancements and commercial applications.

Apart from the abovementioned projects, there are also lots of successful implementation of the green hydrogen in grid energy storage. For example, in the U.S., there are 67 green hydrogen projects planned before 2029, with a total investment of $26 billion. States like Texas, Louisiana, California, and Wyoming are leading the green hydrogen development, reflecting a nationwide commitment to this clean energy source. These projects hold promise for reducing carbon emissions and advancing the U.S. toward net-zero emissions. Several significant green hydrogen projects are set to begin operations in 2025, marking a substantial advancement in the U.S. clean energy landscape. Some notable projects include: (1) St. Gabriel Green Hydrogen Plant, planned to be one of the largest green hydrogen facilities in the world, leveraging Louisiana's industrial infrastructure and renewable energy resources. (2) Sauk Valley Green Hydrogen Plant, aiming to produce low-carbon hydrogen by utilizing renewable energy sources and capturing CO_2 emissions. (3) Kingsland Green Hydrogen Plant, a project that will supply green hydrogen to existing infrastructure and potentially export it globally. (4) Casa Grande Green Hydrogen Plant, strategically located in Arizona to benefit from abundant solar resources, making it a significant player in the state's clean energy transition. (5) Donaldsonville Green Hydrogen Project, focusing on the production of green ammonia as well as hydrogen, with the potential to create jobs and drive economic growth in the region. These projects underscore the U.S.'s commitment to decarbonizing its energy sector through green hydrogen production.

These successful implementations demonstrate the potential for green hydrogen to play a pivotal role in the transition to a low-carbon, sustainable energy future. By leveraging the unique properties of hydrogen as a storable and versatile energy carrier, grid operators can enhance grid resilience and reliability while reducing emissions and promoting sustainability.

6.4 SUMMARY

The role of green hydrogen in grid energy storage has emerged as a pivotal element in the transition to a sustainable energy future. Green hydrogen, produced through the process of water electrolysis powered by renewable energy sources, offers a clean, renewable, and versatile alternative to traditional fossil fuels. This clean energy solution stands out due to its unique properties and potential for large-scale deployment, making it a compelling option for grid energy storage.

Water electrolysis involves splitting water molecules into hydrogen and oxygen using electricity. This process can be powered by renewable energy sources such as wind and solar power, making the production of green hydrogen entirely reliant on renewable resources. Recent advancements in electrolysis technology, including catalyst materials and PTLs, have significantly improved the efficiency, performance, and durability of water electrolyzers. These improvements have paved the way for the widespread adoption of green hydrogen for grid energy storage.

The advantages of green hydrogen in grid energy storage are numerous. Firstly, green hydrogen is renewable and clean, producing minimal greenhouse gas emissions during its production. Secondly, hydrogen has a high energy density, allowing for large amounts of energy to be stored in a compact form. This makes it suitable for long-duration energy storage applications. Additionally, green hydrogen is versatile and flexible, capable of being utilized in various applications across multiple sectors, including transportation, heating, and industry. This versatility allows for a diverse range of energy storage solutions and integration options within the grid. Furthermore, water electrolysis plants can be designed and scaled to meet specific energy storage requirements, offering flexibility in deployment. The potential for decentralized hydrogen production and storage systems enables localized energy solutions and enhances grid resilience.

Implications for future research and development in the field of green hydrogen in grid energy storage are vast. As technology advances, improvements in electrolysis efficiency, cost reductions, and the scaling up of production capacities are essential to enhance the viability of green hydrogen as a mainstream energy carrier. Furthermore, research into more efficient and cost-effective storage solutions, as well as the integration of green hydrogen into existing energy grids, is necessary to facilitate its widespread adoption. Collaboration between industry stakeholders and research institutions is crucial for the development and commercialization of this innovative energy storage solution.

In conclusion, green hydrogen offers a transformative and multifaceted clean energy solution, contributing significantly to the broader goal of achieving a sustainable energy future. Its renewable nature, scalability, and versatility, combined with its storage capabilities, enable the capture and preservation of excess energy from intermittent renewable sources. As research and development continue to push the boundaries of water electrolysis technology and system integration, we can expect further breakthroughs that will unlock the full potential of green hydrogen as a sustainable and cost-effective energy storage solution. These advancements will pave the way for a more resilient and sustainable energy future, enabling us to harness the power of renewable energy and reduce our dependence on fossil fuels.

REFERENCES

1. Turner, J. A., A realizable renewable energy future. *Science* **1999**, *285* (5428), 687–689.
2. Turner, J. A., Sustainable hydrogen production. *Science* **2004**, *305* (5686), 972–974.
3. Chen, M.; Kitiphatpiboon, N.; Feng, C.; Abudula, A.; Ma, Y.; Guan, G., Recent progress in transition-metal-oxide-based electrocatalysts for the oxygen evolution reaction in natural seawater splitting: A critical review. *eScience* **2023**, *3* (2), 100111.
4. Guo, J.; Zheng, Y.; Hu, Z.; Zheng, C.; Mao, J.; Du, K.; Jaroniec, M.; Qiao, S.-Z.; Ling, T., Direct seawater electrolysis by adjusting the local reaction environment of a catalyst. *Nature Energy* **2023**, *8* (3), 264–272.
5. Ge, S.; Xie, R.; Huang, B.; Zhang, Z.; Liu, H.; Kang, X.; Hu, S.; Li, S.; Luo, Y.; Yu, Q.; Wang, J.; Chai, G.; Guan, L.; Cheng, H.-M.; Liu, B., A robust chromium–iridium oxide catalyst for high-current–density acidic oxygen evolution in proton exchange membrane electrolyzers. *Energy & Environmental Science* **2023**, *16* (9), 3734–3742.
6. Wang, Y.; Zhang, M.; Kang, Z.; Shi, L.; Shen, Y.; Tian, B.; Zou, Y.; Chen, H.; Zou, X., Nano-metal diborides-supported anode catalyst with strongly coupled $TaO_{(x)}$/$IrO_{(2)}$ catalytic layer for low-iridium-loading proton exchange membrane electrolyzer. *Nature Communications* **2023**, *14* (1), 5119.
7. Sun, J.-P.; Zhao, Z.; Li, J.; Li, Z.-Z.; Meng, X.-C., Recent advances in electrocatalytic seawater splitting. *Rare Metals* **2023**, *42* (3), 751–768.

8. Pivovar, B., Catalysts for fuel cell transportation and hydrogen related uses. *Nature Catalysis* **2019**, *2*, 562–565.
9. Kang, Z.; Wang, M.; Yang, Y.; Wang, H.; Liu, Y.; Mo, J.; Li, J.; Deng, P.; Jia, C.; Tian, X., Performance improvement induced by membrane treatment in proton exchange membrane water electrolysis cells. *International Journal of Hydrogen Energy* **2022**, *47* (9), 5807–5816.
10. Lee, J. K.; Anderson, G.; Tricker, A. W.; Babbe, F.; Madan, A.; Cullen, D. A.; Arregui-Mena, J. D.; Danilovic, N.; Mukundan, R.; Weber, A. Z.; Peng, X., Ionomer-free and recyclable porous-transport electrode for high-performing proton-exchange-membrane water electrolysis. *Nature Communications* **2023**, *14* (1), 4592.
11. Lin, C.; Li, J.-L.; Li, X.; Yang, S.; Luo, W.; Zhang, Y.; Kim, S.-H.; Kim, D.-H.; Shinde, S. S.; Li, Y.-F.; Liu, Z.-P.; Jiang, Z.; Lee, J.-H., In-situ reconstructed Ru atom array on α-MnO_2 with enhanced performance for acidic water oxidation. *Nature Catalysis* **2021**, *4*, 1012–1023.
12. Kang, Z.; Mo, J.; Yang, G.; Li, Y.; Talley, D. A.; Han, B.; Zhang, F.-Y., Performance modeling and current mapping of proton exchange membrane electrolyzer cells with novel thin/tunable liquid/gas diffusion layers. *Electrochimica Acta* **2017**, *255*, 405–416.
13. Zheng, J.; Kang, Z.; Han, B.; Mo, J. Three-dimensional numerical simulation of the performance and transport phenomena of oxygen evolution reactions in a proton exchange membrane water electrolyzer. *Materials* **2023**, *16* (3), 1310.
14. Kang, Z.; Alia, S. M.; Young, J. L.; Bender, G., Effects of various parameters of different porous transport layers in proton exchange membrane water electrolysis. *Electrochimica Acta* **2020**, *354*, 136641.
15. Mench, M. M., *Fuel Cell Engines*. John Wiley & Sons: New York, 2008.
16. Kang, Z.; Wang, H.; Liu, Y.; Mo, J.; Wang, M.; Li, J.; Tian, X., Exploring and understanding the internal voltage losses through catalyst layers in proton exchange membrane water electrolysis devices. *Applied Energy* **2022**, *317*, 119213.
17. Lopata, J.; Kang, Z.; Young, J.; Bender, G.; Weidner, J. W.; Shimpalee, S., Effects of the transport/catalyst layer interface and catalyst loading on mass and charge transport phenomena in polymer electrolyte membrane water electrolysis devices. *Journal of The Electrochemical Society* **2020**, *167* (6), 064507.
18. Kai, J.; Saito, R.; Terabaru, K.; Li, H.; Nakajima, H.; Ito, K., Effect of temperature on the performance of polymer electrolyte membrane water electrolysis: Numerical analysis of electrolysis voltage considering gas/liquid two-phase flow. *Journal of The Electrochemical Society* **2019**, *166* (4), F246.
19. Han, B.; Steen, S. M.; Mo, J.; Zhang, F.-Y., Electrochemical performance modeling of a proton exchange membrane electrolyzer cell for hydrogen energy. *International Journal of Hydrogen Energy* **2015**, *40* (22), 7006–7016.
20. Han, B.; Mo, J.; Kang, Z.; Zhang, F.-Y., Effects of membrane electrode assembly properties on two-phase transport and performance in proton exchange membrane electrolyzer cells. *Electrochimica Acta* **2016**, *188*, 317–326.
21. Lopata, J.; Kang, Z.; Young, J.; Bender, G.; Weidner, J. W.; Cho, H.-S.; Shimpalee, S., Resolving anodic current and temperature distributions in a polymer electrolyte membrane water electrolysis cell using a pseudo-two-phase computational fluid dynamics model. *Journal of The Electrochemical Society* **2021**, *168*, 054518.
22. Su, H.; Linkov, V.; Bladergroen, B. J., Membrane electrode assemblies with low noble metal loadings for hydrogen production from solid polymer electrolyte water electrolysis. *International Journal of Hydrogen Energy* **2013**, *38* (23), 9601–9608.
23. Kang, Z.; Tian, X., Chapter 9: Degradations in PEMWE. In: *Green Hydrogen Production by Water Electrolysis*. Hou, J.; Yang, M., Eds. CRC Press, Boca Raton, FL, 2024; pp. 190–213.

24. Kang, Z.; Schuler, T.; Chen, Y.; Wang, M.; Zhang, F.-Y.; Bender, G., Effects of interfacial contact under different operating conditions in proton exchange membrane water electrolysis. *Electrochimica Acta* **2022**, *429*, 140942.
25. Yang, G.; Mo, J.; Kang, Z.; List, F. A.; Green, J. B.; Babu, S. S.; Zhang, F.-Y., Additive manufactured bipolar plate for high-efficiency hydrogen production in proton exchange membrane electrolyzer cells. *International Journal of Hydrogen Energy* **2017**, *42* (21), 14734–14740.
26. Bodard, A.; Chen, Z.; ELJarray, O.; Zhang, G., Green hydrogen production by low-temperature membrane-engineered water electrolyzers, and regenerative fuel cells. *Small Methods* **2024**, *8* (12), 2400574.
27. Kang, Z.; Fan, Z.; Zhang, F.; Zhang, Z.; Tian, C.; Wang, W.; Li, J.; Shen, Y.; Tian, X., Studying performance and kinetic differences between various anode electrodes in proton exchange membrane water electrolysis cell. *Materials* **2022**, *15* (20), 7209.
28. Yang, G.; Yu, S.; Mo, J.; Li, Y.; Kang, Z.; Bender, G.; Pivovar, B. S.; Green, J. B.; Cullen, D. A.; Zhang, F.-Y., Impacts of catalyst nanolayers on water permeation and swelling of polymer electrolyte membranes. *Journal of Power Sources* **2020**, *448*, 227582.
29. Lee, C.; Kort-Kamp, W. J. M.; Yu, H.; Cullen, D. A.; Patterson, B. M.; Arman, T. A.; Komini Babu, S.; Mukundan, R.; Borup, R. L.; Spendelow, J. S., Grooved electrodes for high-power-density fuel cells. *Nature Energy* **2023**, *8*, 685–694.
30. Dong, S.; Zhang, C.; Yue, Z.; Zhang, F.; Zhao, H.; Cheng, Q.; Wang, G.; Xu, J.; Chen, C.; Zou, Z., Overall design of anode with gradient ordered structure with low iridium loading for proton exchange membrane water electrolysis. *Nano Letters* **2022**, *22* (23), 9434–9440.
31. Debe, M.; Hendricks, S.; Vernstrom, G.; Meyers, M.; Brostrom, M.; Stephens, M.; Chan, Q.; Willey, J.; Hamden, M.; Mittelsteadt, C. K., Initial performance and durability of ultra-low loaded NSTF electrodes for PEM electrolyzers. *Journal of The Electrochemical Society* **2012**, *159* (6), K165–K176.
32. Holzapfel, P.; Bühler, M.; Van Pham, C.; Hegge, F.; Böhm, T.; McLaughlin, D.; Breitwieser, M.; Thiele, S., Directly coated membrane electrode assemblies for proton exchange membrane water electrolysis. *Electrochemistry Communications* **2020**, *110*, 106640.
33. Bühler, M.; Holzapfel, P.; McLaughlin, D.; Thiele, S., From catalyst coated membranes to porous transport electrode based configurations in PEM water electrolyzers. *Journal of The Electrochemical Society* **2019**, *166* (14), F1070.
34. Liu, H.; Kang, X.; Zhao, T.; Zhang, Z.; Ge, S.; Hu, S.; Luo, Y.; Yang, F.; Li, S.-H.; Sun, C.; Yu, Q.; Cheng, H.-M.; Liu, B., Engineering membrane electrode assembly for advanced polymer electrolyte water electrolyzer. *Science China Materials* **2022**, *65* (12), 3243–3272.
35. Kang, Z.; Mo, J.; Yang, G.; Retterer, S. T.; Cullen, D. A.; Toops, T. J.; Green Jr, J. B.; Mench, M. M.; Zhang, F.-Y., Investigation of thin/well-tunable liquid/gas diffusion layers exhibiting superior multifunctional performance in low-temperature electrolytic water splitting. *Energy & Environmental Science* **2017**, *10* (1), 166–175.
36. Stiber, S.; Balzer, H.; Wierhake, A.; Wirkert, F. J.; Roth, J.; Rost, U.; Brodmann, M.; Lee, J. K.; Bazylak, A.; Waiblinger, W.; Gago, A. S.; Friedrich, K. A., Porous transport layers for proton exchange membrane electrolysis under extreme conditions of current density, temperature, and pressure. *Advanced Energy Materials* **2021**, *11* (33), 2100630.
37. Kang, Z.; Mo, J.; Yang, G.; Li, Y.; Talley, D. A.; Retterer, S. T.; Cullen, D. A.; Toops, T. J.; Brady, M. P.; Bender, G.; Pivovar, B. S.; Green, J. B.; Zhang, F.-Y., Thin film surface modifications of thin/tunable liquid/gas diffusion layers for high-efficiency proton exchange membrane electrolyzer cells. *Applied Energy* **2017**, *206*, 983–990.

38. Lee, J. K.; Schuler, T.; Bender, G.; Sabharwal, M.; Peng, X.; Weber, A. Z.; Danilovic, N., Interfacial engineering via laser ablation for high-performing PEM water electrolysis. *Applied Energy* **2023**, *336*, 120853.
39. Weber, C. C.; Wrubel, J. A.; Gubler, L.; Bender, G.; De Angelis, S.; Buchi, F. N., How the porous transport layer interface affects catalyst utilization and performance in polymer electrolyte water electrolysis. *ACS Applied Materials & Interfaces* **2023**, *15* (29), 34750–34763.
40. Zhu, X.; Zhang, J.; Du, S.; Dang, J.; Yang, M.; Liu, B.; Ma, J.; Yang, F.; Ouyang, M., Exploring the influence mechanism of porous transport layer structure and type on performance in proton exchange membrane electrolyzer. *Journal of The Electrochemical Society* **2023**, *170* (12), 124510.
41. Yuan, X.-Z.; Shaigan, N.; Song, C.; Aujla, M.; Neburchilov, V.; Kwan, J. T. H.; Wilkinson, D. P.; Bazylak, A.; Fatih, K., The porous transport layer in proton exchange membrane water electrolysis: Perspectives on a complex component. *Sustainable Energy & Fuels* **2022**, *6* (8), 1824–1853.
42. Schuler, T.; Ciccone, J. M.; Krentscher, B.; Marone, F.; Peter, C.; Schmidt, T. J.; Büchi, F. N., Hierarchically structured porous transport layers for polymer electrolyte water electrolysis. *Advanced Energy Materials* **2020**, *10* (2), 1903216.
43. Lettenmeier, P.; Kolb, S.; Sata, N.; Fallisch, A.; Zielke, L.; Thiele, S.; Gago, A. S.; Friedrich, K. A., Comprehensive investigation of novel pore-graded gas diffusion layers for high-performance and cost-effective proton exchange membrane electrolyzers. *Energy & Environmental Science* **2017**, *10* (12), 2521–2533.
44. Mo, J.; Dehoff, R. R.; Peter, W. H.; Toops, T. J.; Green, J. B.; Zhang, F.-Y., Additive manufacturing of liquid/gas diffusion layers for low-cost and high-efficiency hydrogen production. *International Journal of Hydrogen Energy* **2016**, *41* (4), 3128–3135.
45. Yang, G.; Mo, J.; Kang, Z.; Dohrmann, Y.; List, F. A.; Green, J. B.; Babu, S. S.; Zhang, F.-Y., Fully printed and integrated electrolyzer cells with additive manufacturing for high-efficiency water splitting. *Applied Energy* **2018**, *215*, 202–210.
46. Li, Y.; Kang, Z.; Mo, J.; Yang, G.; Yu, S.; Talley, D. A.; Han, B.; Zhang, F.-Y., In-situ investigation of bubble dynamics and two-phase flow in proton exchange membrane electrolyzer cells. *International Journal of Hydrogen Energy* **2018**, *43* (24), 11223–11233.
47. Fiedler, L.; Ma, T.-C.; Fritsch, B.; Risse, J. H.; Lechner, M.; Dworschak, D.; Merklein, M.; Mayrhofer, K. J. J.; Hutzler, A., Stability of bipolar plate materials for proton-exchange membrane water electrolyzers. *ChemElectroChem* **2023**, *10* (20), e202300373.
48. Rojas, N.; Sánchez-Molina, M.; Sevilla, G.; Amores, E.; Almandoz, E.; Esparza, J.; Cruz Vivas, M. R.; Colominas, C., Coated stainless steels evaluation for bipolar plates in PEM water electrolysis conditions. *International Journal of Hydrogen Energy* **2021**, *46* (51), 25929–25943.
49. Wang, H.; Turner, J., Reviewing metallic PEMFC bipolar plates. *Fuel Cells* **2010**, *10* (4), 510–519.
50. Young, J. L.; Kang, Z.; Ganci, F.; Madachy, S.; Bender, G., PEM electrolyzer characterization with carbon-based hardware and material sets. *Electrochemistry Communications* **2021**, *124*, 106941.
51. Yang, G.; Chen, Z.; Huang, Y.; Mo, J.; Kang, Z.; Zhang, F.-Y., Thermal analysis during metallic additive manufacturing. In: *Advanced Materials for Multidisciplinary Applications*. Wu, M.; Gao, W.; Li, L.; Lu, Y.; Liu, J. L., Eds. Springer International Publishing, Cham, 2024; pp. 237–264.
52. Pham, C. V.; Escalera-López, D.; Mayrhofer, K.; Cherevko, S.; Thiele, S., Essentials of high performance water electrolyzers – From catalyst layer materials to electrode engineering. *Advanced Energy Materials* **2021**, *11*, 2101998.

53. Song, Y.; Chen, H.; Wang, X.; Weng, C.; Zou, K.; Wang, C.; Yuan, Y.; Ma, Y.; Yang, X.; Lin, W., Engineering Ir-based catalysts for high current density applications in proton exchange membrane water electrolyzers. *Energy & Environmental Science* **2025**, *18* (1), 130–154.
54. Wu, Q.; Wang, Y.; Zhang, K.; Xie, Z.; Sun, K.; An, W.; Liang, X.; Zou, X., Advances and status of anode catalysts for proton exchange membrane water electrolysis technology. *Materials Chemistry Frontiers* **2023**, *7* (6), 1025–1045.
55. Xie, Z.; Liang, X.; Kang, Z.; Zou, Y.; Wang, X.; Wu Yimin, A.; King, G.; Liu, Q.; Huang, Y.; Zhao, X.; Chen, H.; Zou, X., High-porosity, layered iridium oxide as an efficient, durable anode catalyst for water splitting. *CCS Chemistry* **2024**, *7* (1), 1–24.
56. Gao, T.; Jiao, D.; Wang, L.; Ge, X.; Wen, X.; Zhang, L.; Zheng, L.; Zou, X.; Zhang, W.; Zheng, W.; Fan, J.; Cui, X., Switchable acidic oxygen evolution mechanisms on atomic skin of ruthenium metallene oxides. *Journal of the American Chemical Society* **2024**, *147* (5), 4159–4166.
57. Pang, B.; Feng, S.; Xu, Y.; Chen, H.; Li, J.; Yuan, Y.; Zou, X.; Tian, X.; Kang, Z., Ru doped Ir nanowires for high-efficient and durable proton exchange membrane water electrolyzers. *Advanced Functional Materials* **2024**, *34* (52), 2411062.
58. Zhang, J.; Fu, X.; Kwon, S.; Chen, K.; Liu, X.; Yang, J.; Sun, H.; Wang, Y.; Uchiyama, T.; Uchimoto, Y.; Li, S.; Li, Y.; Fan, X.; Chen, G.; Xia, F.; Wu, J.; Li, Y.; Yue, Q.; Qiao, L.; Su, D.; Zhou, H.; Goddard, W. A.; Kang, Y., Tantalum-stabilized ruthenium oxide electrocatalysts for industrial water electrolysis. *Science* **2025**, *387* (6729), 48–55.
59. Wang, C.; Yang, F.; Feng, L., Recent advances in iridium-based catalysts with different dimensions for the acidic oxygen evolution reaction. *Nanoscale Horizons* **2023**, *8* (9), 1174–1193.
60. Gao, G.; Sun, Z.; Chen, X.; Zhu, G.; Sun, B.; Yamauchi, Y.; Liu, S., Recent advances in Ru/Ir-based electrocatalysts for acidic oxygen evolution reaction. *Applied Catalysis B: Environmental* **2024**, *343*, 123584.
61. Han, X.; Shi, L.; Chen, H.; Zou, X., Key role of subsurface doping in optimizing active sites of IrO_2 for the oxygen evolution reaction. *Chemical Communications* **2024**, *60* (25), 3453–3456.
62. Bai, J.; Cheng, L.; Liu, S.; Zhang, H.; Lian, Y.; Deng, Y.; Zhou, Q.; Tang, Y.; Su, Y., Unravel the mechanism of D-orbital modulation and oxygen vacancy in cerium-doped RuO_2 catalysts for acidic oxygen evolution reaction. *Applied Surface Science* **2024**, *642*, 158613.
63. Qin, R.; Chen, G.; Feng, X.; Weng, J.; Han, Y., Ru/Ir-based electrocatalysts for oxygen evolution reaction in acidic conditions: From mechanisms, optimizations to challenges. *Advanced Science* **2024**, *11* (21), 2309364.
64. Wang, L.; Shi, L.; Liu, Q.; Huang, Y.; Yan, W.; Liang, X.; Zhao, X.; Chen, H.; Zou, X., Structurally robust honeycomb layered strontium iridate as an oxygen evolution electrocatalyst in acid. *ACS Catalysis* **2023**, *13* (11), 7322–7330.
65. Hua, K.; Li, X.; Rui, Z.; Duan, X.; Wu, Y.; Yang, D.; Li, J.; Liu, J., Integrating atomically dispersed ir sites in $MnCo_2O_{4.5}$ for highly stable acidic oxygen evolution reaction. *ACS Catalysis* **2024**, *14* (5), 3712–3724.
66. Pham, C. V.; Bühler, M.; Knöppel, J.; Bierling, M.; Seeberger, D.; Escalera-López, D.; Mayrhofer, K. J. J.; Cherevko, S.; Thiele, S., IrO_2 coated TiO_2 core-shell microparticles advance performance of low loading proton exchange membrane water electrolyzers. *Applied Catalysis B: Environmental* **2020**, *269*, 118762.
67. Audichon, T.; Mayousse, E.; Napporn, T. W.; Morais, C.; Comminges, C.; Kokoh, K. B., Elaboration and characterization of ruthenium nano-oxides for the oxygen evolution reaction in a proton exchange membrane water electrolyzer supplied by a solar profile. *Electrochimica Acta* **2014**, *132*, 284–291.

68. He, Y.; Kang, Z.; Li, J.; Li, Y.; Tian, X., Recent progress of manganese dioxide based electrocatalysts for the oxygen evolution reaction. *Industrial Chemistry & Materials* **2023**, *1*, 312–331.
69. Sun, L.; Feng, M.; Peng, Y.; Zhao, X.; Shao, Y.; Yue, X.; Huang, S., Constructing oxygen vacancies by doping Mo into spinel Co_3O_4 to trigger a fast oxide path mechanism for acidic oxygen evolution reaction. *Journal of Materials Chemistry A* **2024**, *12* (15), 8796–8804.
70. Chen, J.; Liu, J.; Xu, S.; Wu, Y.; Ye, Y.; Qian, J., Bimetallic ZnCo-MOF derived porous Ir-doped cobalt oxides for water oxidation with improved activity and stability. *Inorganic Chemistry Frontiers* **2024**, *11* (15), 4876–4885.
71. Kang, Z.; Chen, Y.; Wang, H.; Alia, S. M.; Pivovar, B. S.; Bender, G., Discovering and demonstrating a novel high-performing 2D-patterned electrode for proton-exchange membrane water electrolysis devices. *ACS Applied Materials & Interfaces* **2022**, *14* (1), 2335–2342.
72. Ferner, K. J.; Park, J.; Kang, Z.; Mauger, S. A.; Ulsh, M.; Bender, G.; Litster, S., Morphological analysis of iridium oxide anode catalyst layers for proton exchange membrane water electrolysis using high-resolution imaging. *International Journal of Hydrogen Energy* **2024**, *59*, 176–186.
73. Ferner, K. J.; Litster, S., Composite anode for PEM water electrolyzers: Lowering iridium loadings and reducing material costs with a conductive additive. *ACS Appllied Energy Materials* **2024**, *7* (18), 8124–8135.
74. Lv, H.; Wang, S.; Sun, Y.; Chen, J.; Zhou, W.; Zhang, C., Anode catalyst layer with hierarchical pore size distribution for highly efficient proton exchange membrane water electrolysis. *Journal of Power Sources* **2023**, *564*, 232878.
75. Lin, X.; Seow, J. Z. Y.; Xu, Z. J., A brief introduction of electrode fabrication for proton exchange membrane water electrolyzers. *Journal of Physics: Energy* **2023**, *5* (3), 034003.
76. Zhang, K.; Liang, X.; Wang, L.; Sun, K.; Wang, Y.; Xie, Z.; Wu, Q.; Bai, X.; Hamdy, M. S.; Chen, H.; Zou, X., Status and perspectives of key materials for PEM electrolyzer. *Nano Research Energy* **2022**, *1*, 9120032.
77. Kang, Z.; Yang, G.; Mo, J., Development of an ultra-thin electrode for the oxygen evolution reaction in proton exchange membrane water electrolyzer. *Renewable Energy* **2024**, *224* (5428), 120159.
78. Kang, Z.; Yang, G.; Mo, J.; Li, Y.; Yu, S.; Cullen, D. A.; Retterer, S. T.; Toops, T. J.; Bender, G.; Pivovar, B. S.; Green, J. B.; Zhang, F.-Y., Novel thin/tunable gas diffusion electrodes with ultra-low catalyst loading for hydrogen evolution reactions in proton exchange membrane electrolyzer cells. *Nano Energy* **2018**, *47*, 434–441.
79. Hou, L.; Gu, X.; Cui, X.; Tang, J.; Li, Z.; Liu, X.; Cho, J., Strategies for the design of ruthenium-based electrocatalysts toward acidic oxygen evolution reaction. *EES Catalysis* **2023**, *1* (5), 619.
80. Lickert, T.; Fischer, S.; Young, J. L.; Klose, S.; Franzetti, I.; Hahn, D.; Kang, Z.; Shviro, M.; Scheepers, F.; Carmo, M.; Smolinka, T.; Bender, G.; Metz, S., Advances in benchmarking and round robin testing for PEM water electrolysis: Reference protocol and hardware. *Applied Energy* **2023**, *352*, 121898.
81. Liu, C.; Shviro, M.; Bender, G.; Gago, A. S.; Morawietz, T.; Dzara, M. J.; Biswas, I.; Gazdzicki, P.; Kang, Z.; Zaccarine, S. F.; Pylypenko, S.; Friedrich, K. A.; Carmo, M.; Lehnert, W., Degradation effects at the porous transport layer/catalyst layer interface in polymer electrolyte membrane water electrolyzer. *Journal of The Electrochemical Society* **2023**, *170* (3), 034508.
82. Doo, G.; Park, J.; Park, J.; Heo, J.; Jung, J.; Lee, D. W.; Bae, H.; Hyun, J.; Oh, E.; Kwen, J.; Kim, K. M.; Kim, H.-T., Contact problems of IrO_x anodes in polymer electrolyte membrane water electrolysis. *ACS Energy Letters* **2023**, *8*, 2214–2220.

83. Kang, Z.; Pak, M.; Bender, G., Introducing a novel technique for measuring hydrogen crossover in membrane-based electrochemical cells. *International Journal of Hydrogen Energy* **2021**, *46* (29), 15161–15167.
84. Liu, H.; Tao, H. B.; Liu, B., Kinetic insights of proton exchange membrane water electrolyzer obtained by operando characterization methods. *Journal of Physical Chemistry Letters* **2022**, *13* (28), 6520–6531.
85. He, Y.; Feng, S.; Chen, H.; Liu, Y.; Shi, X.; Rao, P.; Li, J.; Wu, X.; Huang, S.; Li, K.; Wang, H.; Tian, X.; Kang, Z., Direct kinetic loss analysis with hierarchy configuration of catalyst coated membrane in proton exchange membrane water electrolysis cell. *Fuel* **2025**, *379*, 133028.
86. Kang, Z.; Alia, S. M.; Carmo, M.; Bender, G., In-situ and in-operando analysis of voltage losses using sense wires for proton exchange membrane water electrolyzers. *Journal of Power Sources* **2021**, *48*, 229012.
87. Wang, W.; Xie, Z.; Li, K.; Yu, S.; Ding, L.; Zhang, F.-Y., Recent progress in in-situ visualization of electrochemical reactions in electrochemical energy devices. *Current Opinion in Electrochemistry* **2022**, *35*, 101088.
88. Riedmayer, R.; Paren, B. A.; Schofield, L.; Shao-Horn, Y.; Mallapragada, D., Proton exchange membrane electrolysis performance targets for achieving 2050 expansion goals constrained by iridium supply. *Energy & Fuels* **2023**, *37* (12), 8614–8623.
89. Uzunoglu, M.; Onar, O.; Alam, M., Modeling, control and simulation of a PV/FC/UC based hybrid power generation system for stand-alone applications. *Renewable Energy* **2009**, *34* (3), 509–520.
90. Mehr, A. S.; Phillips, A. D.; Brandon, M. P.; Pryce, M. T.; Carton, J. G., Recent challenges and development of technical and technoeconomic aspects for hydrogen storage, insights at different scales; A state of art review. *International Journal of Hydrogen Energy* **2024**, *70*, 786–815.
91. Palys, M. J.; Daoutidis, P., Power-to-X: A review and perspective. *Computers & Chemical Engineering* **2022**, *165*, 107948.
92. Sorrenti, I.; Harild Rasmussen, T. B.; You, S.; Wu, Q., The role of power-to-X in hybrid renewable energy systems: A comprehensive review. *Renewable and Sustainable Energy Reviews* **2022**, *165*, 112380.
93. Hu, J.; Zou, Y.; Zhao, Y., Robust operation of hydrogen-fueled power-to-gas system within feasible operating zone considering carbon-dioxide recycling process. *International Journal of Hydrogen Energy* **2024**, *58*, 1429–1442.
94. Lo Basso, G.; Pastore, L. M.; Sgaramella, A.; Mojtahed, A.; de Santoli, L., Recent progresses in H2NG blends use downstream Power-to-Gas policies application: An overview over the last decade. *International Journal of Hydrogen Energy* **2024**, *51*, 424–453.
95. Wu, X.; Zhang, H.; Yang, M.; Jia, W.; Qiu, Y.; Lan, L., From the perspective of new technology of blending hydrogen into natural gas pipelines transmission: Mechanism, experimental study, and suggestions for further work of hydrogen embrittlement in high-strength pipeline steels. *International Journal of Hydrogen Energy* **2022**, *47* (12), 8071–8090.
96. Handique, A. J.; Peer, R.; Haas, J.; Osorio-Aravena, J. C.; Reyes-Chamorro, L., Distributed hydrogen systems: A literature review. *International Journal of Hydrogen Energy* **2024**, *85*, 427–439.
97. Qamar, H. G. M.; Guo, X.; Ahmad, F., Intelligent energy management system of hydrogen based microgrid empowered by AI optimization technique. *Renewable Energy* **2024**, *237*, 121738.
98. Qiu, R.; Zhang, H.; Wang, G.; Liang, Y.; Yan, J., Green hydrogen-based energy storage service via power-to-gas technologies integrated with multi-energy microgrid. *Applied Energy* **2023**, *3507*, 121738.

7 The General Aspects of Fuel Cells Applied in Large-Scale Electricity Generation

Zhengpei Miao and Quan Zhang

7.1 INTRODUCTION

7.1.1 The Role of Fuel Cells in Grid-Scale Energy Storage and Generation

Electrochemical energy storage and conversion technologies are pivotal in modern efforts to decarbonize power grids. Among these, fuel cells can directly convert chemical energy into electricity with high efficiency and minimal greenhouse gas emissions.[1,2] Unlike batteries, which store energy in chemical form and must undergo periodic recharging, fuel cells can continuously generate electricity as long as a constant supply of fuel (such as hydrogen, natural gas, or biogas) is provided.[3] This makes them uniquely suited for large-scale, dispatchable power generation, particularly when integrated with renewable energy sources like wind and solar, which suffer from intermittency.[4–6] Below, we expand on these attributes, contextualizing their significance in modern energy systems and their potential to reshape grid infrastructure.[7] For grid applications, fuel cells offer three critical advantages:

7.1.1.1 Scalability: From Distributed Power to Utility-Scale Plants

Fuel cell systems are inherently modular in design, which enables their deployment across a broad spectrum of power capacities and application scales. This feature is rooted in their architecture, including repeatable electrochemical unit cells stacked in series or parallel configurations. The structural scalability enables precise tuning of power output from watts to megawatts, making them adaptable to diverse applications such as portable electronics, distributed energy systems, and grid-scale stabilization.[8] Central to this modularity is the decoupling of active materials, such as catalyst-coated membranes in proton exchange membrane fuel cells (PEMFCs) or ceramic electrolytes in solid oxide fuel cells (SOFCs), from the balance-of-plant components. This separation allows for the independent optimization of both the electrochemical interfaces and the mechanical subsystems. Advanced manufacturing techniques, including roll-to-roll fabrication of membrane electrode assemblies

 DOI: 10.1201/9781003492511-7

(MEAs) and laser-sintered metallic bipolar plates, further enhance reproducibility across scales. Crucially, the modular design permits phased capacity expansion with minimal efficiency losses and has a material advantage over conventional combustion systems. Such inherent scalability positions fuel cells as critical enablers for renewable integration, where rapid response and partial-load efficiency (maintained above 60% even at 30% rated power) are essential for mitigating solar/wind intermittency.

Scalability is further enhanced by the ability to stack individual fuel cell units. For example, a single SOFC module typically generates between 10 and 100kW, while clusters composed of hundreds of modules can be combined to deliver outputs on the order of gigawatt-hours. This modular approach reduces both financial and technical risks, as capacity can be incrementally expanded to meet grid demand.[9] At the kilowatt scale, PEMFCs are used for residential or commercial backup power, providing uninterrupted electricity during grid outages. In contrast to large-scale batteries, which are constrained by limitations in energy density and cycle life, fuel cells do not experience the scaling penalty observed in lithium-ion systems, where efficiency decreases as storage duration increases.

7.1.1.2 Fuel Flexibility: Bridging Legacy and Renewable Infrastructure

A defining strength of fuel cells is their compatibility with diverse fuels, enabling integration with existing energy infrastructure while supporting the transition to renewables. While hydrogen is the ideal fuel for zero-emission operation, most commercial systems today utilize natural gas, biogas, or synthetic hydrocarbons, leveraging established pipelines and storage networks. High-temperature fuel cells, such as SOFCs and molten carbonate fuel cells (MCFCs), can internally reform methane (CH_4) into hydrogen, bypassing the need for external reformers. This capability allows operators to blend hydrogen with natural gas, reducing carbon emissions incrementally as green hydrogen production scales up.[10] The PEMFCs, employing Pt-based catalysts and sulfonated polymer electrolytes, can integrate with high-purity hydrogen streams derived from water electrolysis, while modified membranes (e.g., phosphoric acid-doped polybenzimidazole) allow intermediate-temperature operation compatible with methanol reformate. Emerging fuels like ammonia (NH_3) and synthetic methane are also gaining traction. Ammonia, which is easier to store and transport than hydrogen, can be directly fed into SOFCs, where it decomposes into nitrogen and hydrogen.[11] Japan's "Green Ammonia Consortium" aims to use ammonia-fed fuel cells for 10% of its power generation by 2050.[12] Similarly, biogas from landfills or wastewater treatment plants provides a carbon-neutral feedstock for fuel cells, which convert biogas into electricity, heat, and hydrogen for vehicles.[9] This flexibility ensures that fuel cells remain relevant throughout the decades-long transition to fully renewable power grids. In contrast, combustion turbines are restricted to specific fuels, and batteries rely entirely on grid-charged electricity, which may be generated from fossil fuel sources.

Furthermore, hybrid configurations coupling high-temperature fuel cells with gas turbines exploit waste heat for combined-cycle efficiencies exceeding 60%, reducing transitional carbon footprints. By decoupling fuel processing from electrochemical pathways through tailored electrode architectures and defect-tolerant ceramic electrolytes, fuel cells emerge as adaptive nodes within heterogeneous energy grids, accelerating decarbonization without requiring a full infrastructure overhaul.

7.1.1.3 Low Environmental Impact: Efficiency and Emissions Reductions

Fuel cells achieve efficiencies unmatched by conventional thermal power plants. While combined-cycle gas turbines (CCGTs) peak at ~60% efficiency, SOFCs can reach 60%–65% electrical efficiency in standalone operation and up to 85% in CHP mode, reducing CO_2 emissions compared to combustion turbines.[13] This is because fuel cells bypass the Carnot cycle limitations of heat engines, converting chemical energy directly to electricity. For every megawatt-hour generated, SOFCs emit 30%–50% less CO_2 than CCGTs when using natural gas, and near-zero emissions with green hydrogen. High-temperature fuel cells also excel at carbon capture. MCFCs, for instance, concentrate CO_2 at the cathode as part of their electrochemical process, enabling cost-effective carbon capture and utilization. The U.S. Department of Energy's (DOE) "Hydrogen Shot" initiative targets $1 kg^{-1} green hydrogen by 2030, which would make fuel cell power generation cheaper and cleaner than fossil alternatives.[14]

Moreover, fuel cells eliminate criteria pollutants like NO_x and SO_x, which are by-products of combustion. Emissions reductions are equally transformative: PEMFCs operating on green hydrogen produce only water by-products, while SOFCs utilizing methane achieve near-zero NO_x/SO_x emissions through inherently low-temperature oxidation (<800°C) compared to combustion (>1,300°C). Recent advances in platinum group metal (PGM)-free catalysts, such as Fe–N–C cathodes, further mitigate lifecycle environmental impacts by reducing reliance on scarce metals.[15] Moreover, reversible fuel cells have recently demonstrated round-trip efficiencies exceeding 60% in energy storage applications, surpassing the performance of battery–electrolysis hybrid systems. These attributes position fuel cells as critical enablers of deep decarbonization, particularly when integrated with carbon capture strategies during hydrocarbon reforming. Net-negative emissions can be achieved through the electrochemical separation of CO_2 using oxygen-ion-conducting membranes.

7.1.1.4 Integration with Renewables: Stabilizing the Grid

The intermittent nature of wind and solar power creates a demand for technologies that can balance supply and demand. Fuel cells address this challenge through two pathways: (1) Hybrid systems: Pairing fuel cells with electrolyzers allows excess renewable energy to produce hydrogen via water electrolysis. This hydrogen is stored and later used by fuel cells during periods of low renewable generation.[16] (2) Grid-forming capabilities: Advanced fuel cell inverters can regulate voltage and frequency, mimicking the inertia of traditional generators. This "grid-forming" ability is critical as solar and wind displace synchronous machines, which historically provided grid stability.[17] Critical to grid stability is the electrochemical impedance matching between variable renewable inputs and fuel cell stacks, achieved through materials innovations. The advent of SOFC with Y-doped $BaCeO_3$ electrolytes enables direct methanation of excess CO_2 using renewable H_2, effectively converting transient electrical surges into chemically stored energy.[18] This materials-mediated symbiosis not only mitigates the duck curve dilemma but also elevates renewable penetration limits by providing >95% capacity firming through adaptive oxygen reduction/evolution reactivity at triple-phase boundaries (TPBs), as governed by defect chemistry in mixed ionic-electronic-conducting cathodes. As more new energy systems emerge,

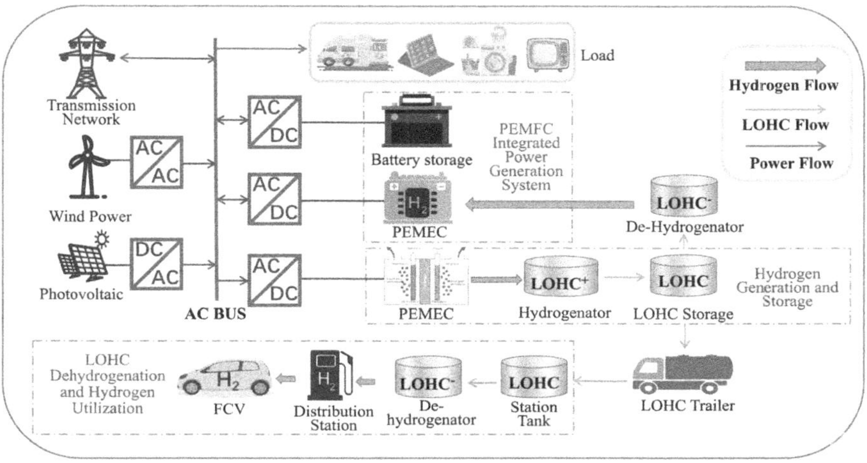

FIGURE 7.1 Power grids based on new energy systems are equipped with hydrogen fuel cells. (Reproduced with permission from Ref. 19.)

it is expected that more will be integrated into the existing power grid in the future (Figure 7.1). The integration of additional renewable energy sources into the existing power grid helps to compensate for fluctuations within the network.[19]

7.1.2 Historical Context and Evolution

The fuel cells technology traces its roots to the 19th-century experiments of Sir William Grove, who demonstrated the first gas battery in 1839. Grove's seminal "gas voltaic battery" demonstrated the electrochemical recombination of hydrogen and oxygen to generate electricity and water, thereby laying the theoretical foundation for modern fuel cells.[20] In 1889, Charles Langer and Ludwig Mond, in their research on fuel cells using gas as a fuel, coined the term "fuel cell."[21] However, practical applications remained elusive for over a century, constrained by material limitations and the dominance of combustion-based power systems. A transformative leap occurred in the mid-20th century with the U.S. space program, where alkaline fuel cells (AFCs), valued for their high efficiency and potable water by-product, were deployed aboard National Aeronautics and Space Administration (NASA's) Apollo and Gemini missions.[17] These systems supplied both onboard power and potable water for astronauts, demonstrating the reliability of fuel cells in mission-critical environments and sparking interest in terrestrial applications. During the 1990s, there was a pivotal transition from aerospace applications to grid-connected energy systems, driven by increasing environmental concerns and significant advances in materials science. Early stationary fuel cell installations, such as the 200-kW phosphoric acid fuel cells (PAFCs) units deployed by UTC Power in hospitals and universities, demonstrated the feasibility of decentralized, low-emission power generation.[22] By the 2010s, MCFCs and SOFCs had matured sufficiently for utility-scale deployment. A landmark project, the 2.8 MW FuelCell Energy plant in Bridgeport, Connecticut (2013), showcased

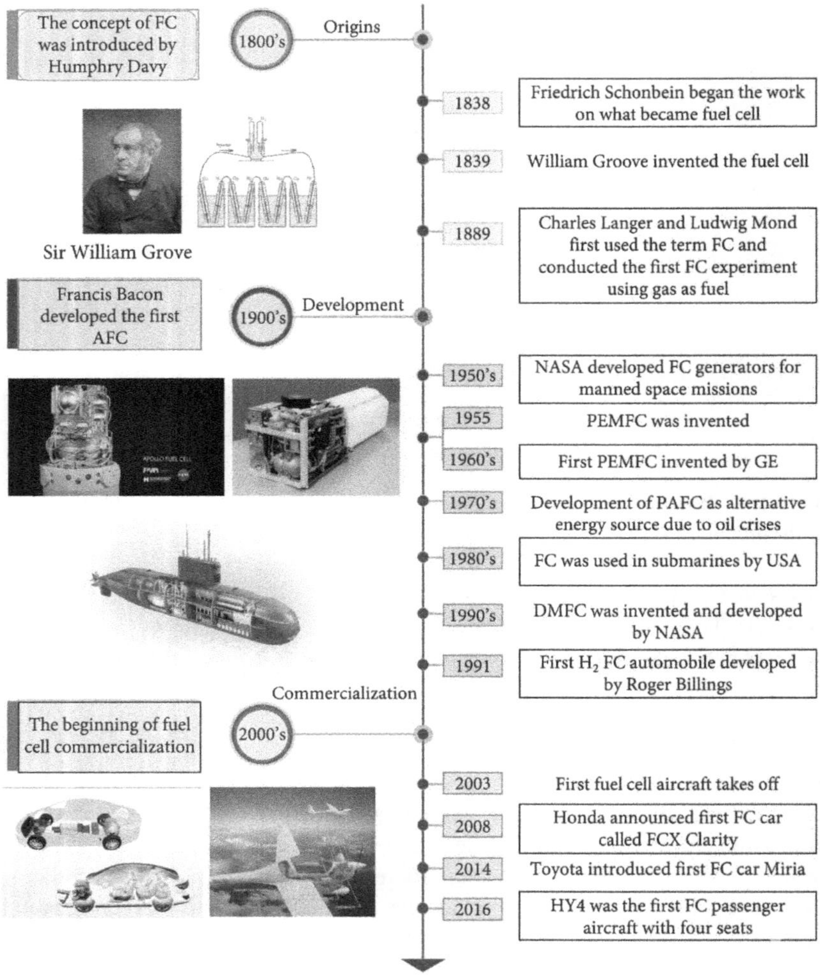

FIGURE 7.2 Timeline of the historical development of fuel cells. (Reproduced with permission from Ref. 23.)

MCFCs operating on biogas from wastewater treatment, achieving 47% electrical efficiency while reducing landfill methane emissions, a model for sustainable urban energy infrastructure. Figure 7.2 summarizes the main historical milestones in the development of fuel cells.[23,24]

Global momentum accelerated in the 2020s as nations prioritized hydrogen economies to meet decarbonization targets. South Korea emerged as a leader, commissioning the 59 MW Daesan Fuel Cell Power Plant in 2020, one of the world's largest fuel cell facilities.[25] Utilizing by-product hydrogen from petrochemical industries, the Daesan plant supplies baseload power to 135,000 households, offsetting 250,000 tons of CO_2 annually.[26] Similarly, Japan's ENE-FARM program has installed over

400,000 residential SOFC units, achieving 52% electrical efficiency through cogeneration.[27] These initiatives reflect strategic policy frameworks, such as South Korea's Renewable Portfolio Standard and Japan's Basic Hydrogen Strategy, which mandate grid integration of clean energy technologies and subsidize hydrogen infrastructure. This historical progression highlights the unique capacity of fuel cells to address both longstanding energy challenges, such as dependence on fossil fuels, and emerging grid requirements, including renewable intermittency and distributed generation. As the technology transitions from policy-driven niches to mainstream markets, its integration into global energy systems epitomizes the convergence of electrochemical innovation and sustainable infrastructure.

7.2 FUNDAMENTALS OF FUEL CELLS

The deployment of fuel cells for grid-scale electricity generation demands a comprehensive understanding of how fundamental electrochemical principles scale and interact within large, engineered systems. At the core of fuel cell operation is the integration of thermodynamic, kinetic, and mass transport principles that govern electrochemical energy conversion. For large-scale electricity generation, these principles must be evaluated within the context of challenges related to scalability, efficiency, and system durability.

7.2.1 Core Principles of Electrochemical Energy Conversion

The theoretical foundation of fuel cell operation is rooted in the interplay between thermodynamics and electrochemical kinetics, which together dictate both the idealized performance and practical limitations of these systems. While the Gibbs free energy (ΔG) defines the theoretical maximum voltage through the relationship: $E° = -\Delta G/nF$, where n is the number of electrons transferred (2 for H_2/O_2), and F is Faraday's constant (96,485 C mol^{-1}). Under standard conditions, this yields $E° = 1.23$ V. However, there is a significant difference between the theoretical and practical performance of fuel cells. When the circuit generates current, electrical energy can be obtained from fuel cells, but irreversible losses caused by various reasons result in the actual potential of fuel cells being lower than the theoretical equilibrium potential. These irreversible losses are commonly referred to as polarization or overpotential, mainly including activated polarization, Ohmic polarization, and concentration. A typical polarization curve of PEMFC is shown in Figure 7.3.[28]

7.2.1.1 Activation Polarization

This phenomenon arises from the inherently slow reaction kinetics within PEMFCs, which are determined by the reaction rates of the two half-reactions as well as the operating conditions of the PEMFC. The voltage loss in the low current density region is particularly pronounced and can be quantitatively described by the Butler–Volmer equation. As a result, a certain overpotential is necessary to overcome the activation energy barrier associated with the electrochemical reaction. This energy loss can be mitigated by increasing the electrocatalytic activity of the redox reaction catalyst, thereby enhancing the reaction kinetics.

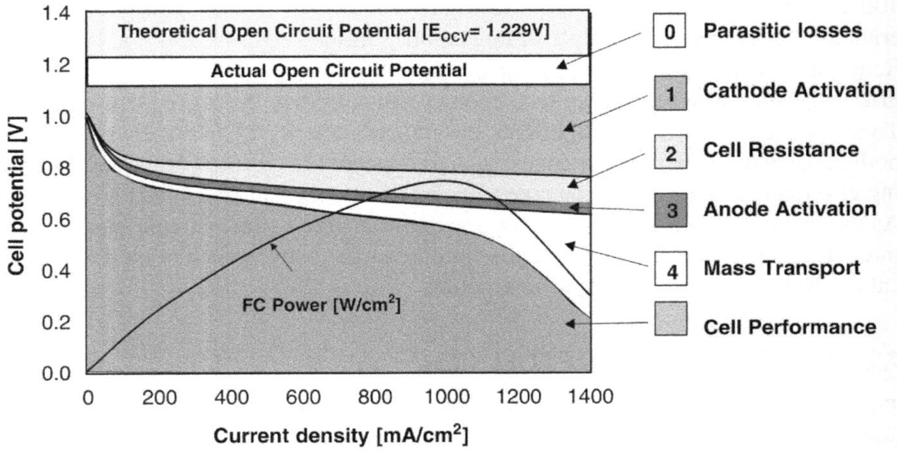

FIGURE 7.3 Typical proton exchange membrane fuel cell (PEMFC) polarization curve with different losses. (Reproduced with permission from Ref. 28.)

7.2.1.2 Ohmic Polarization

In MEA, Ohmic loss arises from the combined effects of ionic resistance and electronic resistance. The proton conductivity of the electrolyte membrane influences the ionic resistance, whereas the electronic resistance is governed by the conductivity of the electrode material, particularly that of the catalyst support. Therefore, using electrolyte membranes with high ionic conductivity and catalyst supports with high conductivity during MEA preparation can reduce the overall Ohmic loss of the fuel cell, as this loss is proportional to the operating current density of PEMFCs.

7.2.1.3 Concentration Polarization

Concentration polarization arises due to the decrease in reactant concentration at the electrode surface compared to the bulk concentration, which limits the active sites' ability to access sufficient reactants, leading to a voltage loss. This decrease in reactant concentration is caused by slow diffusion of reactants to the electrode and inadequate removal of reaction products from the electrode surface. This phenomenon predominantly occurs in high-current-density/low-voltage operating regions.

As shown in the efficiency loss graph of the polarization curve for the PEMFC in Figure 7.3, voltage losses vary across different current density regions. Among these, the loss attributable to cathode activation constitutes the largest proportion. This is due to the extremely slow kinetics of cathode oxygen reduction reaction (ORR) and its complex multi-electron transfer steps, which are six orders of magnitude lower than the reaction kinetics of the anode hydrogen oxidation reaction (HOR). The ORR kinetics at the cathode of PEMFCs require an overpotential of 300–400 mV, which necessitates the use of high-performance catalysts, such as those with a Pt loading of 0.125–0.350 mg Pt cm^{-2}. These catalysts accelerate reaction kinetics and enhance the energy conversion efficiency of PEMFCs.

In grid-scale deployments, these losses compound due to the need to scale individual cells into multi-megawatt stacks. Uniform reactant distribution across hundreds of cells, mitigation of thermal gradients, and humidity control in PEMFCs introduce systemic engineering challenges. Furthermore, transient grid demands accelerate degradation mechanisms—such as catalyst sintering, carbon corrosion, and electrolyte delamination—necessitating advanced materials and adaptive control strategies. Emerging solutions include machine learning-driven system optimization, sulfur-tolerant catalysts for impurity resilience, and redox-stable perovskites for SOFC electrodes. By contextualizing core electrochemical principles within the framework of scalability and durability, it underscores the multidisciplinary innovations required to align fuel cell technology with the rigorous demands of large-scale grid integration.

7.2.2 Efficiency Metrics

Fuel cell efficiency (η) is a key parameter for evaluating fuel cell performance, especially in the context of large-scale electricity generation. It is defined as the ratio of electrical energy output to total energy input, typically expressed on the basis of the fuel's higher heating value (HHV). The HHV accounts for the total energy content of hydrogen, including the latent heat of vaporization, providing a comprehensive measure of the fuel's energy potential. State-of-the-art SOFCs have demonstrated efficiencies exceeding 60% on a lower heating value basis, surpassing the efficiency of CCGTs, which typically operate in the range of 55%–60%. This high efficiency is a key advantage of SOFCs, positioning them as a promising technology for grid-scale energy storage and electricity generation.

In addition to overall fuel cell efficiency, voltage efficiency (η_v) is a refined metric that highlights the electrochemical performance of individual cells. This parameter represents the efficiency of the fuel cell at actual operating voltage, compared to the theoretical maximum voltage (such as the Nernst voltage). Voltage efficiency is one of the most important indicators for measuring the performance of fuel cells, which reflects the degree of voltage drop caused by various losses (such as activation polarization, Ohmic polarization, and concentration polarization) during actual operation of the cell. Operating voltages in fuel cells typically range from 0.6 to 0.8 V per cell, depending on the type of fuel cell and operating conditions. This voltage range translates to a voltage efficiency of approximately 50%–65%, indicating the proportion of the theoretical maximum voltage (typically around 1.23 V for hydrogen fuel cells) that is effectively utilized in practical operation. Voltage efficiency is influenced by factors such as overpotentials arising from activation losses, ohmic losses, and mass transport limitations, which are inherent to electrochemical systems.

A thorough understanding of these efficiency metrics is essential for optimizing fuel cell systems for grid applications. While overall fuel cell efficiency provides a broad assessment of energy conversion effectiveness, voltage efficiency offers more detailed insights into the electrochemical processes occurring within the cell. Together, these metrics inform the design and operation of fuel cell systems to maximize energy output, minimize losses, and enhance both the economic and environmental viability of fuel cells in large-scale electricity generation.

7.2.3 Performance Parameters

Power density is a fundamental performance parameter in fuel cell technology, reflecting the amount of power generated per unit area of the electrode surface. It is a critical metric for evaluating the suitability of fuel cells for large-scale electricity generation, as it directly impacts the size, weight, and cost of the system. Among the various fuel cell technologies, PEMFCs exhibit superior power density, often exceeding 1.0 W cm^{-2} under optimal operating conditions. This high-power density makes PEMFCs particularly well-suited for applications requiring compact and lightweight systems, such as transportation and portable power. However, PEMFCs often face challenges related to durability and long-term stability, which can limit their lifecycle performance in stationary applications.

In contrast, SOFCs typically operate at significantly lower power densities, often less than 0.3 W cm^{-2}. Despite this limitation, SOFCs excel in terms of longevity and operational stability, making them highly attractive for stationary power generation and grid-scale applications. Their ability to operate at high temperatures (typically 600°C–1,000°C) enables the use of a wider range of fuels, including natural gas and biogas, and contributes to their enhanced durability. Furthermore, SOFCs benefit from lower material costs and reduced sensitivity to catalyst degradation compared to PEMFCs, which often require expensive platinum-based catalysts.

The trade-offs between power density and longevity underscore the importance of selecting the appropriate fuel cell technology based on the specific requirements of the application. While PEMFCs offer high power density and rapid startup times, SOFCs provide long-term reliability and fuel flexibility, making them complementary technologies in the broader landscape of electrochemical energy storage and generation. Understanding these performance parameters is essential for optimizing the design, operation, and integration of fuel cells into the grid, ensuring that they meet the demanding efficiency, durability, and cost criteria of large-scale electricity generation.

7.3 FUEL CELL TECHNOLOGIES FOR GRID-SCALE APPLICATIONS

Fuel cells are electrochemical devices that convert chemical energy directly into electrical energy through redox reactions. The fundamental principle involves the oxidation of a fuel at the anode and the reduction of an oxidant at the cathode, with an electrolyte facilitating ion transport between the electrodes.[29] Fuel cells consist of two electrodes: the anode, where fuel oxidation occurs, and the cathode, which mediates the reduction of the oxidizing agent. An electrolyte is positioned between the electrodes, enabling proton transfer while physically separating them and thereby blocking electron flow. The electrodes are supported by a porous framework containing catalyst materials, which serve to enhance reaction kinetics.[19] Depending on the application and operating conditions, the electrolyte may exist in solid, liquid, or molten states. Fuel cells are classified based on operating temperature, efficiency, intended application, and cost. Their classification is primarily determined by the type of electrolyte used, resulting in six major types: AFCs, PAFCs, SOFCs, MCFCs, and PEMFCs.[30] Different fuel cell technologies offer varying advantages and limitations for grid applications. PEMFCs are suitable for mobile and stationary applications due to their low operating temperatures and fast start-up times. SOFCs operate at high temperatures and can use a variety of fuels, making them well-suited for distributed power generation. MCFCs and PAFCs are well-suited for large-scale stationary power plants; MCFCs operate at similarly high temperatures, while PAFCs provide high efficiency and reliability. The following classifications are organized according to the operating temperature ranges of the various fuel cell types.

7.3.1 High-Temperature Fuel Cells

7.3.1.1 Solid Oxide Fuel Cells

The SOFCs are high-temperature fuel cells that utilize metal oxide solid ceramic electrolytes. The SOFCs are composed of three primary layers: a porous nickel-based anode, a dense ceramic electrolyte (typically yttria-stabilized zirconia (YSZ)), and a lanthanum-based perovskite cathode (e.g., $La_{0.6}Sr_{0.4}Co_{0.2}Fe_{0.8}O_{3-\delta}$).[31] Figure 7.4 shows a schematic diagram of SOFC, which typically uses a mixture of hydrogen and carbon monoxide generated from the internal reforming of hydrocarbon fuels and air as the oxidant for the fuel cell.[32] YSZ is the most widely used electrolyte for SOFCs because of its high chemical and thermal stability, as well as its purely ionic conductivity. In the reduction reaction at the cathode (7.2), oxygen is oxidized at 1,000°C, while fuel oxidation occurs at the anode (7.1). The anode is porous to facilitate fuel diffusion and to transport oxidation products away from the anode–electrolyte interface.

$$(\text{Oxidation})(1/2)O_2(g) + 2e^- \rightarrow O^{2-}(s) \tag{7.1}$$

$$(\text{Reduction})O^{2-}(s) + H_2(g) \rightarrow H_2O(g) + 2e^- \tag{7.2}$$

The SOFCs have been well applied in large-scale distributed power generation systems with a capacity of several hundred megawatts. The SOFCs are distinguished by their high operating temperatures (600°C–1,000°C) and unique capacity for direct electrochemical conversion of hydrocarbon fuels, positioning them as a cornerstone technology for large-scale electricity generation and combined heat and power (CHP) systems.[34] The by-product heat generated by SOFCs is typically used to drive gas turbines to produce additional electricity, thereby increasing the CHP efficiency to 70%–80%.[35] The SOFCs achieve electrical efficiencies of 45%–60% in standalone operation and up to 85% in CHP mode by recovering waste heat for residential or industrial heating. The SOFC systems are known for their reliability, modularity, fuel flexibility, and low emissions of harmful gases (such as NO_x and SO_x). The SOFCs excel in natural gas integration through internal reforming, where methane is converted to hydrogen and carbon monoxide within the anode, bypassing external reformers. These characteristics make them suitable for local power generation in rural areas that are connected to the public grid.[36] Additionally, SOFCs also have noise-free operation and low maintenance costs. However, long startup and cooling times, as well as various

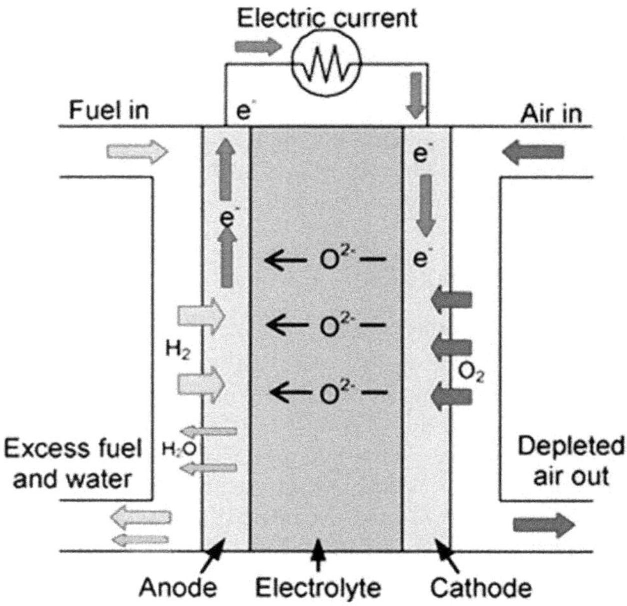

FIGURE 7.4 Schematic diagram of solid oxide fuel cells. (Reproduced with permission from Ref. 33.)

mechanical and chemical compatibility issues, limit their widespread application. The policy incentives (e.g., U.S. Inflation Reduction Act) and material innovations (e.g., PGM-free catalysts) are driving commercialization. Future research targets gigawatt-scale systems and hybrid renewable-SOFCs networks for grid resilience. Current research is focused on developing solutions to reduce operating temperatures. If successful and sustainable measures are established, SOFCs could become a large-scale technology for the next generation of energy devices.[37]

7.3.1.2 Molten Carbonate Fuel Cells

The MCFCs operate at elevated temperatures (~650°C) and are a high-temperature fuel cell that uses a mixture of molten carbonates as the electrolyte, suspended in a porous, chemically inert ceramic matrix of β-alumina solid electrolyte. Their unique electrochemical dynamics and carbon utilization capabilities make them particularly suited for industrial and utility-scale power generation.[38] The schematic diagram of MCFCs is shown in Figure 7.5. In the MCFCs, the electrolyte facilitates ion transport via carbonate ions (CO_3^{2-}) migration from the cathode to the anode. At the cathode, oxygen and carbon dioxide react to form CO_3^{2-} ions, which are then consumed at the anode during hydrogen oxidation, releasing CO_2 for recirculation. This intrinsic CO_2 dependency enables MCFCs to synergize with carbon capture systems.[39]

The following reactions (7.3) and (7.4) occur at the anode, where the fed methane (CH_4) and water (H_2O) are converted to hydrogen (H_2), carbon monoxide (CO), and carbon dioxide (CO_2):

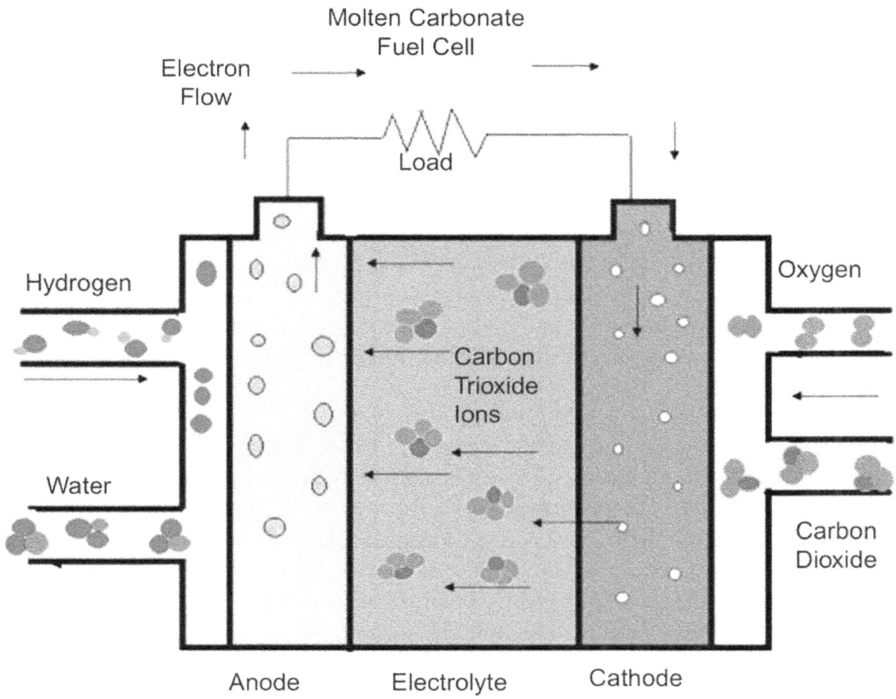

FIGURE 7.5 Schematic diagram of molten carbonate fuel cells. (Reproduced with permission from Ref. 39.)

$$(\text{Conversion reaction 1})\, CH_4 + H_2O \rightarrow CO + 3H_2 \tag{7.3}$$

$$(\text{Conversion reaction 2})\, CO + H_2O \rightarrow CO_2 + H_2 \tag{7.4}$$

Simultaneously, two electrochemical reactions consume H_2 and CO, generating electrons at the anode. Both reactions (7.5) and (7.6) utilize the CO_3^{2-} available.

$$(\text{Oxidation 1})\, H_2 + CO_3^{2-} \rightarrow H_2O + CO_2 + 2e^- \tag{7.5}$$

$$(\text{Oxidation 2})\, CO + CO_3^{2-} \rightarrow 2CO_2 + 2e^- \tag{7.6}$$

At the cathode, a reduction reaction occurs (Reaction 7.7), generating new carbonate ions from oxygen (O_2) and CO_2. These newly formed carbonate ions at the cathode are transported to the anode through the electrolyte. The fuel cell current and voltage can be collected at the electrodes.

$$(\text{Reduction})(1/2)O_2 + CO_2 + 2e^- \rightarrow CO_3^{2-} \tag{7.7}$$

MCFCs are currently employed in natural gas and coal-fired power plants for electricity generation, as well as in various industrial and military applications. MCFCs can directly utilize hydrogen, carbon monoxide, natural gas, and propane as fuels, without the need for precious metal catalysts or the development of additional installation infrastructure. However, they have a slow response time, requiring a long period to reach operating temperature and generate electricity.[40,41] However, MCFCs still face several challenges. Key limitations include carbonate evaporation and nickel cathode oxidation, which are being addressed through the implementation of pulsed operation strategies and the application of protective coatings. Despite capital costs (~\$3,500 kW^{-1}), MCFCs are gaining traction in carbon-intensive sectors due to policy incentives (e.g., the EU Emissions Trading System) and scalability. Future research targets gigawatt-scale deployments and integration with blue hydrogen production.[42]

7.3.1.3 Phosphoric Acid Fuel Cells

A fuel cell system based on liquid phosphate (H_3PO_4) electrolyte using carbon paper electrodes (CP) for both the anode and the cathode has been developed. This type of fuel cell, known as the phosphate type fuel cells, operates efficiently within a temperature range of 150°C–220°C. At lower temperatures, such as 40°C, the system faces challenges due to the low ionic conductivity of phosphoric acid, which hinders its operation. The working principle of PAFCs involves the transfer of protons (H^+ or H_3O^+) through the electrolyte from the anode to the cathode, while electrons are transferred through an external circuit from the cathode to the anode, generating electricity.[43] As shown in Figure 7.6, on the cathode side, due to the reaction among electrons, protons, and oxygen, water is generated at an accelerated rate in the presence of a platinum catalyst. The discharged water is usually used for heating applications. At a temperature of 40°C, the continuous operation and system start-up of PAFCs are problematic due to the solidification of phosphoric acid at this temperature. As shown in Figure 7.6, the oxidation reaction (7.8) occurs at the anode, and hydrogen gas is oxidized to generate 4 protons and 4 electrons; at the cathode, a reduction reaction (7.9) occurs, in which 4 protons and 4 electrons combine with oxygen to form water. The anode and cathode reactions in the PAFCs system are as follows:

$$(\text{Oxidation})2H_2 \rightarrow 4H^+ + 4e^- \tag{7.8}$$

$$(\text{Reduction})O_2 + 4H^+ + 4e^- \rightarrow 4H_2O \tag{7.9}$$

Electrons and protons generate current and heat through external circuits and electrolytes, respectively. Heat is usually used to heat water or generate steam at atmospheric pressure; however, steam reforming reactions produce some CO around the electrodes, which may poison fuel cells and affect PAFC performance. The solution to reduce the absorption of CO is to improve the temperature tolerance of the anode. A higher tolerance to CO means a higher temperature tolerance at the anode. At elevated temperatures, CO is desorbed from the cathode during the electrocatalytic reaction. Unlike other acidic electrolytes that require water for ionic conductivity, the concentrated phosphoric acid electrolyte used in PAFCs enables operation at temperatures above the boiling point of water.[44]

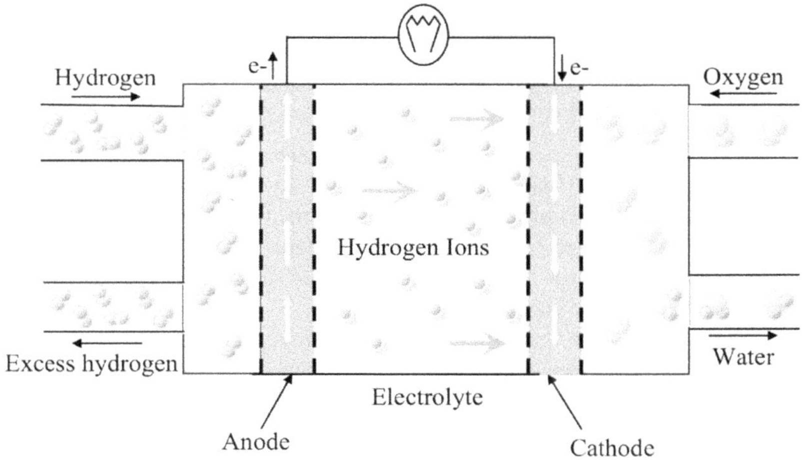

FIGURE 7.6 Schematic diagram of phosphoric acid fuel cells. (Reproduced with permission from Ref. 43.)

The operation of PAFCs does not require pure oxygen, as CO_2 does not adversely affect the performance of the electrolyte or the fuel cell. PAFCs operate using air and can readily utilize reformed fossil fuels. In addition, H_3PO_4 exhibits low volatility and long-term stability. The use of air containing ~21% oxygen instead of pure oxygen in PAFCs results in a threefold reduction in current density. Consequently, PAFCs are designed with stacked bipolar plates to increase electrode surface area and thereby generate more power, which implies high initial costs for the technology. Currently, PAFC systems are commercially available with capacities of up to 200 kW, and higher-capacity, megawatt-scale systems have also been demonstrated in testing. The manufacturing cost of PAFCs is elevated due to the requirement for finely dispersed platinum catalyst coatings on the electrodes. Unlike AFCs, hydrogen steam impurities (such as CO_2) do not affect PAFCs. The electrical efficiency of this type of fuel cell ranges between 40% and 50%, with CHP efficiency reaching approximately 85%. They are typically used in on-site stationary applications.[45]

7.3.2 Low-Temperature Fuel Cells

7.3.2.1 Proton Exchange Membrane Fuel Cells

In PEMFCs, hydrogen is activated by a catalyst to form protons and ions, generating electrons at the anode. Protons migrate through the proton exchange membrane, while electrons flow through an external circuit to the cathode, producing electricity. Then the electrons react with oxygen and proton ions to form water.[46] The PEMFC is illustrated in Figure 7.7. The chemical reactions occurring at each electrode are given in Equations (7.10) and (7.11), with the overall reaction shown in Equation (7.12):

$$(\text{Anode}): \text{H}_2(\text{g}) \rightarrow 2\text{H}^+ + 2\text{e}^- \tag{7.10}$$

$$(\text{Cathode}): (1/2)O_2(g) + 2H^+ + 2e^- \rightarrow H_2O(L) \tag{7.11}$$

$$(\text{Overall Reaction}): H_2(g) + (1/2)O_2(g) \rightarrow H_2O(L) \tag{7.12}$$

The PEMFC is essentially composed of bipolar plates and a MEA. The MEA consists of a dispersed catalyst layer (CL), a carbon cloth or gas diffusion layer (GDL), and the membrane. The membrane transports protons from the anode to the cathode while blocking the passage of electrons and reactants. The GDL ensures uniform distribution of the fuel, and the electrons generated at the anode flow through the external circuit to produce electricity.[47]

The PEMFC is a low-temperature fuel cell with an operating temperature ranging from 60°C to 100°C. It is a lightweight and compact system with a rapid start-up process. However, the manufacturing cost remains relatively high, and the total cost of a vehicle equipped with a PEMFC system is approximately ten times greater than that of a conventional internal combustion engine vehicle. The total cost of a PEMFC includes the costs of assembly processes, bipolar plates, catalytic layer electrodes, proton exchange membranes, and peripheral equipment.[48] From an efficiency perspective, the higher operating temperatures lead to higher efficiencies due to increased reaction rates. Nevertheless, operating temperatures above 100°C can cause water evaporation, leading to membrane dehydration and a subsequent reduction in the proton conductivity of the membrane. The power generation efficiency of PEMFCs ranges from 40% to 50%, with output power capacities reaching up to 250 kW. The PEMFC systems are commonly used in portable, stationary, and transportation applications. Among these, grid-scale energy storage and generation

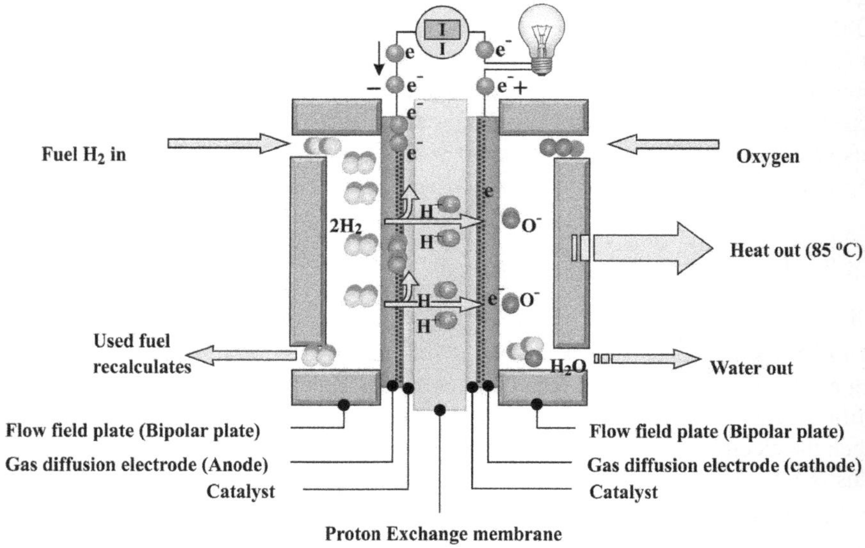

FIGURE 7.7 Schematic diagram of PEMFCs. (Reproduced with permission from Ref. 32.)

represent one of the most promising applications of PEMFCs systems, as their high efficiency and power density can provide a continuous electrical power supply. Additionally, maintenance requirements are minimal due to the absence of moving parts in the PEMFCs power generation stack.[49] The application of fuel cells depends on the type of fuel cell to be used. Various types of fuel cell technologies are suitable for different specific applications. Fuel cells can generate a wide range of power outputs, from 1 to 10 MW, making them applicable to virtually any scenario requiring electricity. For small-scale power applications, fuel cells can be used in low-power devices and personal electronic devices, such as mobile phones and personal computers. Medium-scale power applications include fuel cell vehicles, household appliances, military applications, and public transportation. Finally, in large-scale power applications (1–10 MW), fuel cells can be utilized in distributed power systems and large energy systems, such as power plants.[50]

The PEMFCs are pivotal for grid-scale energy systems due to their rapid start-up times, modular design, and compatibility with hydrogen—an ideal energy carrier for renewable integration. However, their scalability and infrastructure dependencies pose challenges that must be addressed to realize their full potential in decarbonizing power grids. While PEMFCs are inherently modular, scaling to multi-megawatt installations requires overcoming material and system-level limitations. The MEA comprising proton exchange membranes, CLs, and GDLs, remains a cost driver, particularly due to platinum-based catalysts and perfluorinated sulfonic acid (PFSA) membranes.[51] The PEMFCs rely on high-purity hydrogen, necessitating robust production, storage, and distribution networks. Current infrastructure gaps, particularly for green hydrogen (produced via renewable-powered electrolysis), limit scalability. Despite progress, achieving Japan's ambitious targets (6 kW L^{-1} by 2030) demands further optimization of water and thermal management to prevent localized flooding or drying. The PEMFCs excel in stabilizing grids dominated by intermittent renewables. Their sub-minute response times enable dynamic load-following, making them ideal for peak shaving. Advanced inverters with grid-forming capabilities allow PEMFCs to regulate voltage and frequency, replacing the inertia traditionally provided by fossil-fuel turbines. To overcome scalability barriers in the future directions, research focuses on PGM-free catalysts and AI-driven MEA optimization.

7.3.2.2 Alkaline Fuel Cells

AFCs use an alkaline electrolyte, potassium hydroxide (KOH), to generate electricity in aqueous solution. The existing hydroxide ions (OH^-) pass through the electrolyte to form a closed electrical circuit and generate electrical energy. Figure 7.8 illustrates a schematic diagram of an AFC. At the anode, two hydrogen molecules react with four hydroxide ions to produce four water molecules and four electrons, as shown in Reaction (7.13):

$$(\text{Oxidation})\,2H_2 + 4OH^- \rightarrow 4H_2O + 4e^- \tag{7.13}$$

These released electrons flow through the external circuit to the cathode, where they are combined with water molecules and oxygen to produce hydroxide ions. The reduction reaction at the cathode is represented in Reaction (7.14):

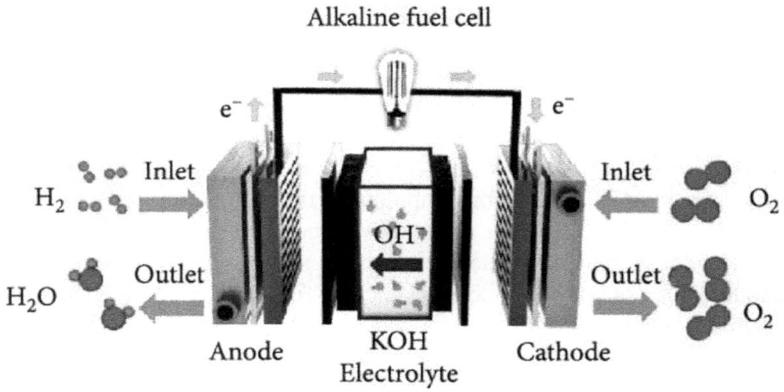

FIGURE 7.8 Schematic diagram of alkaline fuel cells. (Reproduced with permission from Ref. 52.)

$$(\text{Reduction})\, O_2 + 2H_2O + 4e^- \rightarrow 4OH^- \quad (7.14)$$

AFCs typically operate at temperatures between 60°C and 90°C. Recent advancements have enabled operation at low temperatures between 23°C and 70°C. These cells are designed for low-temperature applications and utilize low-cost catalysts, such as nickel-based catalysts, to accelerate reactions at both electrodes. The efficiency of AFCs is approximately 60%, while those integrated with heat recovery systems achieve efficiencies exceeding 80%. They can generate up to 20kW of electricity.[53]

The U.S. NASA has successfully utilized AFCs for spacecraft, providing both drinking water and power for space missions. Currently, they are deployed in submarines, ships, forklifts, and various transport applications. AFCs are recognized for their cost-effectiveness, as they employ potassium hydroxide (KOH) as the electrolyte. Nickel catalysts, used in the electrodes, are relatively inexpensive compared to other catalyst types. The absence of bipolar plates simplifies the system design and enables the simultaneous generation of potable water, heat, and electricity. The by-product water generated by this process is potable, making it highly valuable for use in spacecraft and space missions. In addition, the system operates with high efficiency and does not produce greenhouse gas emissions. Despite their advantages, AFCs are susceptible to carbon dioxide poisoning. The use of the aqueous alkaline solution (KOH) electrolyte absorbs CO_2, converting it into potassium carbonate (K_2CO_3), which reduces the lifespan of the fuel cell. To address this challenge, spacecraft and aircraft frequently use purified oxygen or nitrogen, which increases operational costs.[54] These challenges necessitate ongoing research and development to enhance the performance and reliability of AFCs.

7.3.3 Emerging and Hybrid Systems

Emerging fuel cell technologies and hybrid configurations are reshaping grid-scale energy storage and generation, addressing challenges of intermittency, efficiency, and decarbonization. This section explores recent advancements in reversible fuel cells and hybrid systems, emphasizing their integration into modern grids. Reversible solid oxide cells (rSOCs) operate bidirectionally, functioning as SOFCs for power generation and electrolyzers (SOECs solid oxide electrolytic cell) for hydrogen production. This duality enables energy storage by converting surplus renewable electricity into hydrogen, which can be re-electrified during demand peaks. Recent breakthroughs include in situ material optimization, such as nanostructured gadolinium-doped ceria electrolytes, which reduce operating temperatures to 500°C–700°C while maintaining high ionic conductivity.[55] For instance, the GRHYD project in France demonstrated a 1 MW rSOC system achieving 64% round-trip efficiency, a 15% improvement over standalone electrolyzer–fuel cell setups. However, challenges persist, including electrode degradation under cyclic operation and the need for cost-effective catalysts. Innovations like sulfur-tolerant anodes (e.g., Sr-doped $LaCrO_3$) and cobalt-free cathodes are mitigating these issues.[56]

Hybrid systems combine the steady power output of fuel cells with the rapid response capabilities of batteries or supercapacitors, thereby optimizing grid stability. For example, wind–SOEC–battery hybrid systems utilize SOECs to store excess wind energy as hydrogen, while lithium-ion batteries are employed to manage short-term fluctuations. A 2024 study highlighted a 10 MW hybrid microgrid in Germany combining PEMFCs with vanadium redox flow batteries, achieving 95% reliability in islanded mode and reducing levelized costs by 20%.[57]

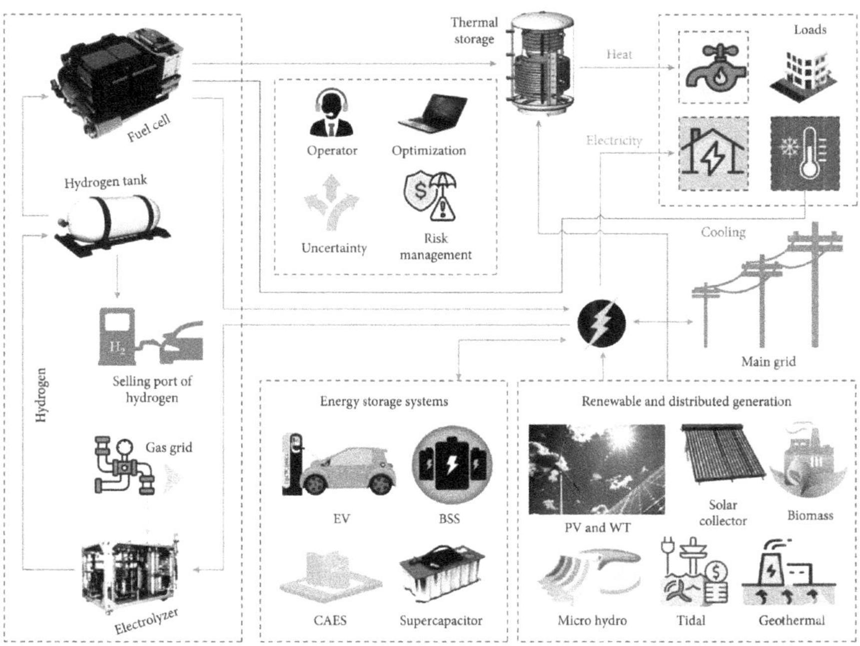

FIGURE 7.9 Typical FC-based hybrid energy system. (Reproduced with permission from Ref. 59.)

In the field of renewable buffering, the SOECs coupled with solar/wind farms store excess energy as hydrogen, later used for peak shaving or synthetic fuel production.[58] In the context of grid resilience, hybrid systems offer black-start capabilities and frequency regulation, thereby serving as replacements for fossil-fuel peaker plants. In the field of carbon utilization, high-temperature SOECs enable co-electrolysis of CO_2 and steam, producing syngas for industrial feedstocks while sequestering emissions. Although some progress has been made, there are still challenges ahead. The scalability remains hindered by hydrogen infrastructure gaps and material costs. Future research targets gigawatt-scale rSOC deployments and advanced thermal integration, such as utilizing industrial waste heat to enhance SOEC efficiency by 16%.[58] As shown in Figure 7.9, fuel cells play the role of cogeneration units in energy storage systems. Electrolyzers use electricity from the power grid and renewable energy sources such as photovoltaic, wind, hydro, biomass, geothermal, and tidal energy to produce hydrogen. Hydrogen is stored in hydrogen tanks, which can provide electricity and heat during peak demand periods. If the fuel supply is insufficient or the electrolyzers malfunction, the hydrogen storage tank can provide hydrogen to the fuel cell to generate electricity and heat. Excess hydrogen can be sold as a by-product at hydrogen sales points, increasing the flexibility of fuel cells.[59]

7.4 THE COUPLING RELATIONSHIP BETWEEN HYDROGEN FUEL CELLS AND THE GRID

Hydrogen's role as an energy vector and a storage medium is pivotal. The U.S. DOE Hydrogen Shot initiative targets $1 kg^{-1} hydrogen production via electrolysis, which would make fuel cell-generated electricity cost-competitive with natural gas peaker plants. In China, Sinosynergy and SinoHyKey have advanced PEMFC technologies, reducing MEA costs by 60% through large-scale production, thereby accelerating hydrogen adoption in public transport and stationary power systems.[60]

Hydrogen fuel cells are electrochemical devices that convert hydrogen and oxygen into electricity, heat, and water, offering a sustainable pathway for grid-scale energy storage and generation. Their modular design, high efficiency (40%–60% electrical efficiency in standalone operation), and compatibility with renewable energy sources position them as critical enablers of grid decarbonization and resilience. HFCs contribute to grid stability by providing dispatchable power and ancillary services. HFCs are increasingly recognized as a critical enablers of grid resilience and decarbonization, bridging the gap between intermittent renewable energy sources and stable electricity supply. Their integration into grid systems hinges on bidirectional energy flows, fuel flexibility, and scalability, which collectively enhance energy security and reduce reliance on fossil fuels. A typical HFCs system comprises four core components: membrane electrode materials, hydrogen fuel cell assembly, hydrogen supply system, and balance of plant (BoP).[61] The content of these four sections is presented sequentially as follows.

7.4.1 Materials Engineering

7.4.1.1 Electrocatalyst

The CL is the primary component of the PEMFCs, and the electrocatalyst is a substance that alters the rate of electrochemical reactions occurring at the electrode surface. It is crucial to understand the kinetics of the electrocatalyst in PEMFCs. The Pt catalysts in currently used PEMFCs account for approximately 50% of the total cost of fabrication.[62] Compared with the hydrogen oxidation reaction (HOR) at the anode, the oxygen reduction reaction (ORR) at the cathode has extremely slow kinetics, which is six orders of magnitude (10^6) lower than that of HOR.[63] Therefore, the development of highly efficient ORR electrochemical catalysts is essential to reduce the overpotential and accelerate the reaction rate. Overall, the primary objective for improving the mass activity (MA), durability, and cost-effectiveness of PEMFCs is to design and develop advanced noble metal-based catalysts, non-precious metal-based catalysts, and nonmetallic catalysts. The research progress of these three types of catalysts in PEMFCs is detailed below.

7.4.1.1.1 Noble-Metal-Based Catalysts

In PEMFCs, the most widely used platinum-based catalyst is platinum nanoparticles supported on carbon carriers (Pt/C). However, the application of Pt/C is hindered by several challenges, primarily due to the significant cost increase of Pt and the corrosion of carbon materials under operating conditions caused by electrochemical oxidation of the surface, leading to aggregation and dissolution of Pt nanoparticles.[64]

The following section introduces the ORR catalytic activity of Pt alloys and other noble metals. When Pt is alloyed with other metals, the Pt alloy exhibits lattice contraction, resulting in a favorable Pt–Pt distance that facilitates the adsorption of O_2. Alloy catalysts can exploit synergistic effects and interactions between constituent metals to inhibit Pt aggregation and detachment.[65] Moreover, alloying can enhance the oxidation potential of Pt, promoting the desorption of O_2 from the Pt surface, thereby improving the stability and ORR activity of the catalyst.[66] Li et al. synthesized an ordered $PtCu_3$ alloy catalyst with a particle size of 3.3 nm using $NH_3{\cdot}H_2O$ as a pH regulator. The half-wave potential of the catalyst is 0.945 V, and the MA is 1.061 mA μg_{Pt-1}, which is 11.7 times that of commercial 20% Pt/C. The enhancement of catalytic activity can be attributed to the incorporation of transition metal copper atoms into the catalyst, which leads to the formation of an ordered intermetallic structure. This structure induces compressive strain, lowers the d-band center, and weakens the adsorption of oxygen-containing intermediates, thereby enhancing the catalytic activity for the ORR.[67]

Compared with binary Pt alloys, ternary Pt alloys generally allow greater flexibility in adjusting the electronic structure and catalytic properties of Pt. In recent studies, ternary Pt alloys achieve ordered transitions more easily than binary Pt alloys. For example, the addition of small amounts of Cu can effectively reduce the temperature required for organizational ordering in Pt–Fe alloys.[68] Compared with Pt–Ni alloys, the doping of Co can also enhance the orderliness of Pt-Ni-Co alloys. Unlike traditional alloys primarily based on a single dominant element, high-entropy alloys

(HEAs) contain five or more elements with similar atomic ratios, and nanoscale HEAs have made significant progress. By maximizing their surface-to-volume ratio, the applications of HEAs have been extended to catalysis and energy storage fields. Chen et al. prepared a PtRuCuOsIr nanoporous HEA (np-HEA) catalyst using a top-down alloying–dealloying synthesis technique. The catalyst exhibited a three-dimensional bicontinuous ligament/channel structure with a length of approximately 2.5 nm. Compared with commercial Pt/C catalysts, the MA and specific activity of the np-HEA catalyst were increased by 1.8 and 3.8 times, respectively, and it demonstrated excellent long-term stability.[69]

7.4.1.1.2 PGM-Free Catalysts

In the past decades, researchers have conducted intensive studies on the development of non-noble metal-based catalysts, striving to develop non-noble-metal catalysts with Pt-like ORR catalytic performance. Non-noble-metal catalysts include molecular catalysts, metal nitrides/oxides, and metal-nitrogen-carbon (M-N_x/C) structured catalysts, where M represents earth-abundant transition metals (M = Co, Fe, Ni, Mn, etc.). Among these, M-N_x/C catalysts have garnered attention due to their high catalytic activity and stability in acidic electrolytes, as well as their abundant and cost-effective utilization of materials, and are considered the most promising candidates to replace noble-metal catalysts for ORR.[70] As shown in Figure 7.10, Miao et al. conducted research on ORR catalysts in PEMFCs and explored the preparation, performance improvement, and stability enhancement strategies of Co-N-C and Fe-N-C catalysts, respectively. They regulate the microstructure and composition of materials through strategies such as double-crosslinked hydrogels and metal–organic polymer supramolecules, improve the activity and durability

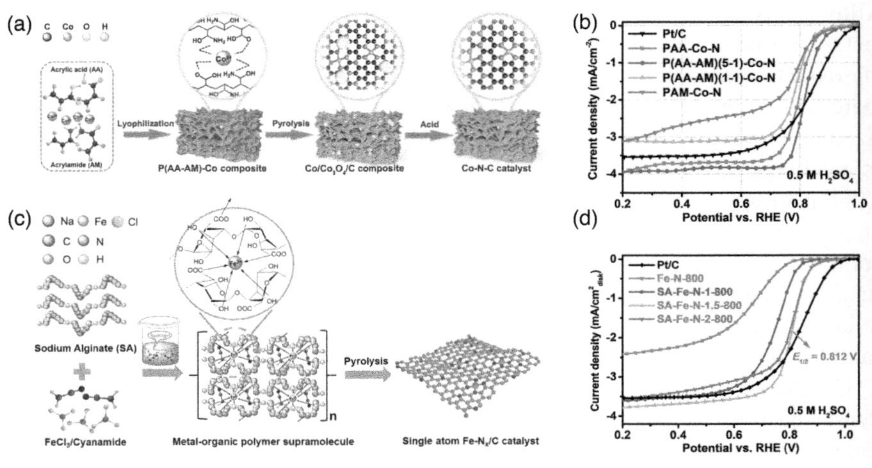

FIGURE 7.10 (a) Schematic illustration of the P(AA-AM)-Co-N catalyst. (b) Oxygen reduction reaction (ORR) steady-state RRDE polarization curves. (Reproduced with permission from Ref. 70.) (c) Schematic illustration of the single atom SA-Fe-N catalyst. (d) ORR steady-state RRDE (Rotating ring disk electrode) polarization curves in 0.5 M H_2SO_4 electrolyte. (Reproduced with permission from Ref. 71.)

of catalysts, and provide potential solutions for the commercial application of PEMFCs.[70,71]

M-N_x/C catalysts are typically prepared by pyrolysis of metal salts, carbon, and nitrogen precursors at high temperatures (>800°C). Bimetallic M_2-N_x/C structures exhibit greater thermodynamic stability compared to monometallic structures. In recent years, a series of bimetallic M_2-N_x/C-structured ORR catalysts with excellent activity and stability in acidic media have been reported.[72] Miao et al. developed a universal strategy to improve the stability of non-precious metal M-N-C catalysts in PEMFCs by regulating the interaction between metal ions (such as iron, cobalt, and manganese) and polymers. Research has found that by changing the structure and composition of polymers, the length and coordination number of metal nitrogen bonds can be regulated, thereby enhancing the stability of metal active centers.[73] Sun et al. prepared a hollow carbon-based (Fe, Co)/N-C catalyst with Fe-Co dual sites by controlling the bonding between Co nodes (hosts) and Fe ions adsorbed within the cavities of metal–organic frameworks. The results showed that the best-performing Fe-Co two-site catalyst has negligible loss of working potential over 100 hours in constant-current operation tests at 600 and 1,000 mA cm^{-2} in PEMFCs.[74]

7.4.1.1.3 Metal-Free Catalysts

In 2006, Ozkane et al. found that ORR catalysis does not necessarily require transition metal substances, as they demonstrated that nitrogen-doped carbon with metal contamination of less than 1 ppm exhibited ORR catalytic activity in acidic electrolytes.[75] In 2009, Dai Liming's team reported in *Science* that nitrogen-doped carbon nanotube arrays exhibited ORR performance in alkaline media comparable to that of commercial Pt/C, highlighting the potential of metal-free ORR catalysts as viable alternatives to commercial Pt/C.[76]

As a result, research on metal-free ORR catalysts has garnered widespread attention. The main components of metal-free catalysts are typically various carbon-based materials, among which graphite, graphene, carbon nanotubes, and ordered mesoporous carbon are the most widely used. The doping of heteroatoms (such as B, N, S, Se, P, and F) into carbon materials fundamentally alters the nature of the carbon-based catalysts, thereby achieving excellent catalytic performance. However, the catalytic mechanisms of carbon-based catalysts doped with various heteroatoms remain highly complex, and definitive conclusions have yet to be established. Various metal-free catalysts based on heteroatom-doped carbon nanomaterials have been studied, including boron doping, nitrogen doping, phosphorus doping, sulfur doping, and dual or multiatom doping involving two or three heteroatoms.[77]

7.4.1.2 Membrane Electrodes

In general, a MEA consists of a GDL and a CL. The electrode serves not only as the center for electrochemical reactions but also as a conduit for electrons to flow from the anode to the cathode. In addition, an appropriate PEMFCs electrode structure should possess three phases: gas-phase reactants, ionic conductors, and solid-phase catalyst particles. The GDL and the CL are described in detail below.

7.4.1.2.1 Gas Diffusion Layer

The GDL is typically located externally to the CL and plays several critical roles during PEMFCs operation, including gas distribution, water transport, thermal conduction, and electrical conduction, while also providing mechanical integrity to the MEA. Essentially, the GDL interconnects the CL and the bipolar plate, serving as a medium for the transport of reactants to the electrocatalytic sites. In addition, the GDL facilitates the drainage of product water and consists of two layers, a hydrophobic agent and a microporous layer (MPL) made of CP or carbon cloth covered with a MPL through which reactant gases diffuse through the pores of these layers. The hydrophobic agent generates open pores within the material, thereby preventing the accumulation of excess water. Functionally, the GDL also helps to remove product water. Otherwise, water blockage at GDL sites may lead to flooding, which restricts the contact of reactants with the catalytic sites and severely degrades the performance of the fuel cell.[78]

The GDL was initially composed of a macroporous substrate, typically made from woven or nonwoven sheets, with pore sizes ranging from 10 to 30 μm. Since the early 2000s, carbon-based materials (such as CP, carbon felt, and carbon cloth) have been widely utilized as substrate materials. However, a few carbon-free GDLs have also been reported, such as titanium and its alloys as well as copper and nickel substrates.[79] To enhance the electrical conductivity and water management properties of GDLs, MPLs have been incorporated into modern GDL designs. These MPLs are primarily composed of carbon powders and hydrophobic agents, such as polytetrafluoroethylene, and are applied to the inner side of the carbon substrate facing the CL. The pore diameters of MPLs typically range from 100 to 500 nm.[80] It has been reported that MPLs promote fuel cell performance due to the excellent electrical properties of carbon powders and enhanced gas transport toward the CL. Additionally, the hydrophobic agent in the MPL facilitates the removal of excess water from the system.

7.4.1.2.2 Catalyst Layer

The CL is the site of the electrode reactions and serves as a critical component of PEMFCs. The CL typically consists of a continuous and interconnected network formed by carbon-supported catalysts (providing active sites and pathways for electron transfer), ionomers (providing pathways for proton transfer and acting as binders to enhance the mechanical stability of the CL), and pores (providing pathways for reactants and water transport). The electrochemical reactions take place at the intersection of these three components, commonly referred to as the TPB. Therefore, not only the activity of the catalyst, but also the accessibility of the active sites to gaseous reactants, protons, and electrons—that is, the formation of the TPB—is crucial for enhancing the performance and durability of PEMFCs, thus reducing costs. The performance, durability, and cost of the CL, as well as the performance of PEMFCs, largely depend on the micro-, meso-, and macrostructures of the CL, which determine the electrochemical surface area (or TPB) and the effective transport properties of protons, electrons, reactants, water, and heat.[81]

One approach to optimizing the micro-, meso-, and macrostructures is to improve the manufacturing processes of traditional microstructures. Another way is to fabricate CL with ordered structures. Traditional CLs are primarily manufactured using catalyst ink processing methods. To create the TPB and mass transport network, catalyst-loaded and ionomer-coated particles are typically dispersed in a liquid chemical solvent called catalyst ink. The catalyst ink is then coated onto a substrate surface to form an ink film, and the solvent in the ink film evaporates during the drying process to form the solid structure of the CL.[82] To mitigate the mass transport issues caused by entangled pathways in traditional CL, Middelman et al. first designed an ordered-structure CL composed of aligned electron conductors, with Pt nanoparticles loaded perpendicular to the membrane and covered by an approximately 10 nm thick ionomer layer. This structure allows for the shortest possible pathways for the transport of protons, electrons, reactants, and water, potentially maximizing the TPB and facilitating mass transport.[83]

Currently, several studies focus on optimizing the formulation and dispersion techniques of catalyst inks, characterizing and modeling catalyst inks, coating technologies, and the potential interactions among solvents, ionomers, and catalysts. To clarify the relationship between the formulation and preparation process of catalyst inks or the CL formation process and the multi-scale structure of the CL, further efforts are needed to enhance the performance and durability of PEM fuel cells. The challenges and future directions for CLs should focus on several key areas: characterization of the unmodified microstructures of catalyst inks, assessment of catalyst ink stability during storage and the reversibility of aged inks, development of high-solid-content catalyst inks, and optimization of the drying process.

7.4.1.3 Proton Exchange Membrane

The membrane is the most critical component of both anion exchange membrane fuel cells (AEMFCs) and PEMFCs, as it allows the transfer of protons and hydroxide ions while preventing the passage of fuel and electrons. PEMFCs can be categorized into two types based on their membranes. Structurally, these two types are identical, but the type of membrane is different. In PEMFCs, the carrier is the hydrogen ion (H^+), while in AEMFCs, the carrier is the hydroxide ion (OH^-). The high-pH operating environment reduces the corrosion risk in AEMFCs, thereby expanding the range of materials suitable for fuel cell components. However, this also has several drawbacks. For example, at higher pH levels, CO_2 can react with OH^- ions during the ORR to form a carbonated environment, which necessitates high-purity air at the cathode and increases the cost of AEMFCs. Compared to AEMFCs, PEMFCs are a relatively mature and more extensively studied technology.[84]

The proton exchange membrane is the core component of PEMFCs and plays a crucial role in cell performance. The first perfluorosulfonic acid (PFSA) was an electrically neutral ionic monomer with partially ionized cations discovered by Willard T. Grubb in 1959.[85] The sole purpose of developing PFSA membranes was to provide proton conductivity while isolating gases for PEMFCs. Currently, the most widely used membrane is still PFSA, a commonly applied solid electrolyte in PEMFCs. Nafion is a perfluoropolymer that was discovered by Walter Grote in 1960 and represents the first commercially produced perfluorinated ion-exchange membrane. It was introduced by DuPont under the trade name Nafion®.[86] Due to its unique hydrophilic/hydrophobic nanophase structure and excellent ionic properties, Nafion has been used as the PEM since its inception. Although Nafion has made progress in achieving ideal chemical/mechanical stability and considerable proton conductivity, it still has some drawbacks as PEM. In addition to its high cost (commercial Nafion® series > \$500 m^{-2}), it also suffers from low proton conductivity, high fuel crossover, excessive swelling, and water loss at high temperatures, which reduces proton conductivity and degrades performance.[87] Moreover, it is challenging to simultaneously maintain mechanical and chemical stability while increasing proton conductivity, as any alteration in the assembly or chain structure may disrupt the nanophase structure of Nafion. Improving the overall performance of Nafion membranes has become a challenging issue for the further development of high-performance PEM.

It is worth mentioning that advanced PEMs should meet the following requirements: (1) excellent chemical and thermal stability to ensure the continuous and stable operation of PEMFCs; (2) high proton conductivity to provide an efficient transport pathway for protons; (3) superior mechanical properties for easy manufacturing of the MEA; (4) low fuel permeability to prevent direct contact between reactants; (5) low cost and excellent durability; and (6) good surface adhesion to ensure better contact with the CL and reduce interfacial resistance.[86]

7.4.2 Advanced Manufacturing Technology

7.4.2.1 Bipolar Plates

Bipolar plates are a crucial component of PEMFCs, forming the backbone of the hydrogen fuel cell stack. They conduct current among fuel cells, facilitate water and thermal management through the fuel cells, and provide pathways for the transport of reactant gases (i.e., hydrogen and oxygen). These bipolar plates significantly influence the overall performance of the fuel cell and affect the manufacturing cost (accounting for 30%–45% of the stack total cost). In the design of PEMFCs, bipolar plates are manufactured for large-scale production using materials with excellent manufacturability and that are suitable for cost-effective, high-volume automated production systems.[88] Currently, graphite composites are considered the standard material for PEMFCs bipolar plates due to their low surface contact resistance and high corrosion resistance. However, compared to metals such as aluminum, stainless steel, nickel, and titanium, graphite and graphite composites are classified as brittle and gas-permeable, with higher costs in large-scale manufacturing.[89] Compared to carbon-based materials, metals offer higher mechanical strength, better resistance to impact and vibration, impermeability, and superior manufacturability and cost-effectiveness. However, the main issue with metallic bipolar plates is the lack of corrosion resistance in the highly acidic and humid environment within PEMFCs, which can lead to the formation of oxidants, passivation layers, and metal ions, thereby causing significant degradation of PEMFCs performance. To enhance the corrosion resistance of metals without compromising surface contact resistance and cost-effectiveness, extensive research is being conducted, including the use of noble metals, stainless steels, and various nitride- and carbide-based alloy coatings.[90]

The ideal characteristics of bipolar plate materials include high corrosion resistance and low surface contact resistance, like graphite, as well as high mechanical strength, impermeability to reactant gases, and non-brittleness, like metals, stainless steel, aluminum, and titanium. However, the main challenge is that corrosion-resistant metallic bipolar plates tend to form passive oxide layers on their surfaces, which can protect the bulk metal from corrosion but also lead to the undesirable effect of high surface contact resistance. This results in the dissipation of some electrical energy as heat, reducing the overall efficiency of the fuel cell stack. Various uncoated and coated metals, metal foams, and nonmetallic graphite composites are currently being investigated to produce cost-effective and durable PEMFCs bipolar plates.[91]

7.4.2.2 Flow Fields

The flow field within the PEMFCs is formed by channels in the bipolar plates that transport reactants to the reaction zone and simultaneously remove water generated within the fuel cells. Common flow field designs in PEMFCs include straight parallel, serpentine, pin, and interdigitated flow fields. The flow field in PEMFCs serves multiple functions, including isolating reactant gases, diffusing them to the CL, and conducting current to the external circuit. Manso et al. conducted a comprehensive evaluation of studies on the most common flow field configurations and geometric features, such as channel width and rib width.[92] Liu et al. conducted experiments with different types of flow field designs, and the results indicated that a single cell with a single serpentine flow field outperformed all other configurations. In addition to water and thermal management, flow field design also affects the effective distribution of reactants and the voltage stability of the fuel cells.[93]

The flow field has a significant impact on PEMFCs performance, as the channel structure of the flow field determines water migration and reactant gas distribution. A rational channel structure ensures the uniform distribution of reactant gases in the reaction zone of the bipolar plate and effectively removes water. Therefore, improvements in channel structure are necessary to enhance power density and reduce the manufacturing cost of PEMFCs. Research has primarily focused on traditional flow fields, such as parallel and serpentine flow fields.[94] In recent years, new types of flow fields have emerged, including phase-interleaved, spiral, dendritic, and radial flow fields. Additionally, the Toyota Mirai developed a three-dimensional complex flow field that offers excellent water management and oxygen delivery. However, the intricate channel structure leads to higher costs and increased manufacturing challenges. Zhu et al. designed a circular radial flow field that combines the flow capacity of serpentine channels with the flow distribution characteristics of parallel channels. This design indicates that radial flow fields have broad application prospects in PEMFCs.[95]

7.5 SUMMARY AND OUTLOOK

In conclusion, fuel cells offer a viable solution for large-scale electricity generation, with their high efficiency, low emissions, and grid compatibility making them a key technology for the transition to a sustainable energy future. By understanding the general aspects of fuel cells applied in grid applications, stakeholders can leverage the benefits of this technology to enhance grid reliability, reduce the carbon footprint, and promote energy independence. Continued research and collaboration in the field of electrochemical energy storage will further accelerate the adoption of fuel cells for grid applications, ensuring a reliable and sustainable energy supply for generations to come. The future of fuel cells in large-scale electricity generation holds promising developments and challenges. Emerging trends such as electrolyzer–fuel cell systems, hybrid energy storage solutions, and grid-scale fuel cell deployments are shaping the landscape of electrochemical energy storage for the grid. Advancements in materials science and manufacturing processes are driving innovation in fuel cell research and development, paving the way for a sustainable and resilient grid infrastructure.

REFERENCES

1. Rathoure, A. K.; Kapoor, A.; Devnani, G. L.; Pal, D. B. Transforming energy storage with unitized regenerative fuel cells: Challenges and future potential. *Journal of Energy Storage* **2025**, *114*, 115773.
2. Liao, M.; Marocco, P.; Gandiglio, M.; Liu, C.; Santarelli, M. Flexible dispatch strategy for electric grid integrating PEM electrolyzer for hydrogen generation. *Frontiers in Energy* **2025**, *19*, 1–14.
3. Okundamiya, M. S. Size optimization of a hybrid photovoltaic/fuel cell grid connected power system including hydrogen storage. *International Journal of Hydrogen Energy* **2021**, *46* (59), 30539–30546.
4. Zhu, Z.; Jiang, T.; Ali, M.; Meng, Y.; Jin, Y.; Cui, Y.; Chen, W. Rechargeable batteries for grid scale energy storage. *Chemical Reviews* **2022**, *122* (22), 16610–16751.
5. Ding, D.; Wu, X.-Y. Evaluating the economic and carbon emission reduction potential of fuel cell electric vehicle-to-grid. *DeCarbon* **2025**, *7*, 100096.
6. Jiang, Y.; Xie, H.; Han, L.; Zhang, Y.; Ding, Y.; Shen, S.; Chen, B.; Ni, M. Advances in TiS_2 for energy storage, electronic devices, and catalysis: A review. *Progress in Natural Science: Materials International* **2023**, *33* (2), 133–150.
7. Kebede, A. A.; Kalogiannis, T.; Van Mierlo, J.; Berecibar, M. A comprehensive review of stationary energy storage devices for large scale renewable energy sources grid integration. *Renewable and Sustainable Energy Reviews* **2022**, *159*, 112213.
8. Pei, H.; Liu, Z.; Chen, J.; Sun, L.; Wang, B.; Xing, L.; Cai, S.; Tu, Z. Dynamic response study of air-cooled proton exchange membrane fuel cell stack. *Progress in Natural Science: Materials International* **2024**, *34* (6), 1318–1326.
9. Gandiglio, M.; Marocco, P.; Nieminen, A.; Santarelli, M.; Kiviaho, J. Energy and environmental performance from field operation of commercial-scale SOFC systems. *International Journal of Hydrogen Energy* **2024**, *85*, 997–1009.
10. Wehrle, L.; Ashar, A.; Deutschmann, O.; Braun, R. J. Evaluating high power density, direct-ammonia SOFC stacks for decarbonizing heavy-duty transportation applications. *Applied Energy* **2024**, *372*, 123646.
11. Blanco, E. C.; Sánchez, A.; Martín, M.; Vega, P. Methanol and ammonia as emerging green fuels: Evaluation of a new power generation paradigm. *Renewable and Sustainable Energy Reviews* **2023**, *175*, 113195.
12. Valera-Medina, A.; Amer-Hatem, F.; Azad, A. K.; Dedoussi, I. C.; de Joannon, M.; Fernandes, R. X.; Glarborg, P.; Hashemi, H.; He, X.; Mashruk, S.; et al. Review on ammonia as a potential fuel: From synthesis to economics. *Energy & Fuels* **2021**, *35* (9), 6964–7029.
13. Di Florio, G.; Macchi, E. G.; Mongibello, L.; Baratto, M. C.; Basosi, R.; Busi, E.; Caliano, M.; Cigolotti, V.; Testi, M.; Trini, M. Comparative life cycle assessment of two different SOFC-based cogeneration systems with thermal energy storage integrated into a single-family house nanogrid. *Applied Energy* **2021**, *285*, 116378.
14. McQueen, S.; Stanford, J.; Satyapal, S.; Miller, E.; Stetson, N.; Papageorgopoulos, D.; Rustagi, N.; Arjona, V.; Adams, J.; Randolph, K.; et al. *Department of Energy*

Hydrogen Program Plan; DOE/EE-2128; 8387; US Department of Energy (USDOE), Washington, DC (United States), United States, 2020.

15. Miao, Z.; Li, S.; Priest, C.; Wang, T.; Wu, G.; Li, Q. Effective approaches for designing stable M–N/C oxygen-reduction catalysts for proton-exchange-membrane fuel cells. *Advanced Materials* **2022**, *34* (52), 2200595.
16. Hossain, M. B.; Islam, M. R.; Muttaqi, K. M.; Sutanto, D.; Agalgaonkar, A. P. Advancement of fuel cells and electrolyzers technologies and their applications to renewable-rich power grids. *Journal of Energy Storage* **2023**, *62*, 106842.
17. Abdin, Z. Shaping the stationary energy storage landscape with reversible fuel cells. *Journal of Energy Storage* **2024**, *86*, 111354.
18. Exner, J.; Nazarenus, T.; Kita, J.; Moos, R. Dense Y-doped ion conducting perovskite films of $BaZrO_3$, $BaSnO_3$, and $BaCeO_3$ for SOFC applications produced by powder aerosol deposition at room temperature. *International Journal of Hydrogen Energy* **2020**, *45* (16), 10000–10016.
19. Murtza Qamar, H. G.; Guo, X.; Seif Ghith, E.; Tlija, M.; Siddique, A. Assessment of energy management and power quality improvement of hydrogen based microgrid system through novel PSO-MWWO technique. *Scientific Reports* **2025**, *15*(1), 863.
20. Wang, Y.; Pang, Y.; Xu, H.; Martinez, A.; Chen, K. S. PEM fuel cell and electrolysis cell technologies and hydrogen infrastructure development-a review. *Energy & Environmental Science* **2022**, *15* (6), 2288–2328.
21. Perry, M. L.; Fuller, T. F. A historical perspective of fuel cell technology in the 20th century. *Journal of The Electrochemical Society* **2002**, *149* (7), S59–S67.
22. Kanuri, S. V. UTC Power and the PureCell Model 400 fuel cell power plant finding wide application. *Fuel Cells Bulletin* **2012**, *2012* (2), 12–15.
23. Qasem, N. A. A.; Abdulrahman, G. A. Q. A recent comprehensive review of fuel cells: History, types, and applications. *International Journal of Energy Research* **2024**, *2024* (1), 7271748.
24. Kinnon, M. M.; Razeghi, G.; Samuelsen, S. The role of fuel cells in port microgrids to support sustainable goods movement. *Renewable and Sustainable Energy Reviews* **2021**, *147*, 111226.
25. Thomas, J. M.; Edwards, P. P.; Dobson, P. J.; Owen, G. P. Decarbonising energy: The developing international activity in hydrogen technologies and fuel cells. *Journal of Energy Chemistry* **2020**, *51*, 405–415.
26. Laughlin, R. B.; Freund, S. Chapter 8 – Economics of hydrogen fuel. In *Machinery and Energy Systems for the Hydrogen Economy*, Brun, K., Allison, T. Eds.; Elsevier, Amsterdam, Netherlands, 2022; pp. 305–330.
27. Ito, H. Economic and environmental assessment of residential micro combined heat and power system application in Japan. *International Journal of Hydrogen Energy* **2016**, *41* (34), 15111–15123.
28. Duerr, M.; Gair, S.; Cruden, A.; McDonald, J. Hydrogen and electrical energy from organic waste treatment. *International Journal of Hydrogen Energy* **2007**, *32* (6), 705–709.
29. Hong, J.; Yang, J.; Weng, Z.; Ma, F.; Liang, F.; Zhang, C. Review on proton exchange membrane fuel cells: Safety analysis and fault diagnosis. *Journal of Power Sources* **2024**, *617*, 235118.
30. Manzo, D.; Thai, R.; Le, H. T.; Venayagamoorthy, G. K. Fuel cell technology review: Types, economy, applications, and vehicle-to-grid scheme. *Sustainable Energy Technologies and Assessments* **2025**, *75*, 104229.
31. Long, Q.; Sha, R.; Wang, R.; Xu, B.; Men, H.; Wang, Q.; Hou, J. Research progress of composite cathode materials for solid oxide fuel cells. *Progress in Natural Science: Materials International* **2023**, *33* (3), 267–278.

32. Jawad, N. H.; Yahya, A. A.; Al-Shathr, A. R.; Salih, H. G.; Rashid, K. T.; Al-Saadi, S.; AbdulRazak, A. A.; Salih, I. K.; Zrelli, A.; Alsalhy, Q. F. Fuel cell types, properties of membrane, and operating conditions: A review. *Sustainability* **2022**, *14* (21), 14653.
33. Coduri, M.; Karlsson, M.; Malavasi, L. Structure–property correlation in oxide-ion and proton conductors for clean energy applications: Recent experimental and computational advancements. *Journal of Materials Chemistry A* **2022**, *10*(10), 5082–5110.
34. Bronicki, L.; Nett, C. N.; Nordquist, J. Electricity generation from fuel cell waste heat using an organic rankine cycle. In *ASME 2014 12th International Conference on Fuel Cell Science, Engineering and Technology Collocated with the ASME 2014 8th International Conference on Energy Sustainability*, Boston, MA, USA, **2014**.
35. Sinha, A. A.; Srivastava, K.; Rajpoot, A. S.; Choudhary, T.; Pandey, S. P.; Sanjay. A thermodynamic approach to analyze energy, exergy, emission, and sustainability (3E-S) performance by utilizing low temperature waste heat in SOFC-CHP-TEG system. *International Journal of Hydrogen Energy* **2024**, *63*, 1088–1104.
36. Yuan, Q.; Li, X.; Han, S.; Wang, S.; Wang, M.; Chen, R.; Kudashev, S.; Wei, T.; Chen, D. Performance analysis and optimization of SOFC/GT hybrid systems: A review. *Energies* **2024**, *17* (5), 1265.
37. Wu, Z.; Zhu, P.; Huang, Y.; Yao, J.; Yang, F.; Zhang, Z.; Ni, M. A comprehensive review of modeling of solid oxide fuel cells: From large systems to fine electrodes. *Chemical Reviews* **2025**, *125* (4), 2184–2268.
38. Peng, S. Molten carbonate fuel cells. In *Hydrogen Fuel Cells*, Peng, S. Ed.; Springer Nature, Singapore, 2024; pp. 167–187.
39. Wang, Y.; Ed. *Nanomaterials for Direct Alcohol Fuel Cell.* CRC Press, Boca Raton, FL, **2016.**
40. de Figueiredo Luiz, D.; Boon, J.; Otero Rodriguez, G.; van Sint Annaland, M. Review of the molten salt technology and assessment of its potential to achieve an energy efficient heat management in a decarbonized chemical industry. *Chemical Engineering Journal* **2024**, *498*, 155819.
41. Yakubu, E.; Ikeuba, A. I.; Njoku, C. N.; Adeniyi, O. D. Electrochemical performance of locust bean waste and coal in a molten carbonate direct carbon fuel cell. *Sustainable Chemistry One World* **2025**, *5*, 100041.
42. Gao, L.; Selman, J. R.; Nash, P. Materials and wetting issues in molten carbonate fuel cell technology: A review. *Journal of Materials Science* **2023**, *58* (41), 15936–15972.
43. Giorgi, L.; Leccese, F. Fuel cells: Technologies and applications. *The Open Fuel Cells Journal* **2013** *6*(1), 1–20.
44. Verma, V.; Choudhury, S. R.; Rathour, V.; Choudhury, S. R.; Ganesan, V. Hollow core mesoporous carbon spheres as catalyst support for improved platinum utilization in phosphoric acid fuel cells. *Microporous and Mesoporous Materials* **2024**, *367*, 113005.
45. Biebl, M.; Roes, J.; Hoster, H. Investigation of the technical potential of a hydrogen powered phosphoric acid fuel cell (PAFC) for CHP. In *Journal of Physics: Conference Series* 2024, 2689, 012014.
46. Agyekum, E. B.; Ampah, J. D.; Wilberforce, T.; Afrane, S.; Nutakor, C. Research progress, trends, and current state of development on PEMFC-new insights from a bibliometric analysis and characteristics of two decades of research output. *Membranes* **2022**, *12*, 1103.
47. Parekh, A. Recent developments of proton exchange membranes for PEMFC: A review. *Frontiers in Energy Research* **2022**, *10*: 956132.
48. Salam, A.; Zholobko, O.; Wu, X.-F. Roles of functionalized nanoparticles in the performance improvement of proton-exchange membranes used in low- and intermediate-temperature hydrogen fuel cells: A review. *Progress in Natural Science: Materials International* **2024**, *34* (3), 437–453.

49. Kwon, L.; Kang, J.-G.; Baik, K. D.; Kim, K.; Ahn, C. Advancement and applications of PEMFC energy systems for large-class unmanned underwater vehicles: A review. *International Journal of Hydrogen Energy* **2024**, *79*, 277–294.
50. Erdinc, O.; Uzunoglu, M. Recent trends in PEM fuel cell-powered hybrid systems: Investigation of application areas, design architectures and energy management approaches. *Renewable and Sustainable Energy Reviews* **2010**, *14* (9), 2874–2884.
51. Jiao, K.; Xuan, J.; Du, Q.; Bao, Z.; Xie, B.; Wang, B.; Zhao, Y.; Fan, L.; Wang, H.; Hou, Z.; et al. Designing the next generation of proton-exchange membrane fuel cells. *Nature* **2021**, *595* (7867), 361–369.
52. Das, G.; Choi, J.-H.; Nguyen, P. K. T.; Kim, D.-J.; Yoon, Y. S. Anion exchange membranes for fuel cell application: A review. *Polymers* **2022**, *14* (6), 1197.
53. Ferriday, T. B.; Middleton, P. H. Alkaline fuel cell technology-A review. *International Journal of Hydrogen Energy* **2021**, *46* (35), 18489–18510.
54. Hamada, A. T.; Orhan, M. F.; Kannan, A. M. Alkaline fuel cells: Status and prospects. *Energy Reports* **2023**, *9*, 6396–6418.
55. Hossain, M. J.; Gaikwad, P.; Shin, Y. K.; Schulze, J. A.; Penrod, K. A.; Li, M.; Lin, Y.; Pawar, G.; van Duin, A. C. T. eReaxFF force field development for $BaZr_{0.8}Y_{0.2}O_{3-\delta}$ solid oxide electrolysis cells applications. *Computational Materials* **2024**, *10* (1), 136.
56. Jemai, R.; Wederni, M. A.; Amorri, O.; Hzez, W.; Martín-Palma, R. J.; Khirouni, K.; Alaya, S. Effects of doping by copper on electrical properties of $LaCrO_3$ based perovskite. *Ceramics International* **2022**, *48* (10), 14050–14059.
57. Wang, L.; Tseng, X. Y.; Hsu, Y. C.; Li, T. Y.; Mokhlis, H.; Chua, K. H.; Tripathy, M. Grid-resilience enhancement of a multimachine power system with a DC microgrid using a vanadium redox flow battery-based energy storage system. In *2024 IEEE Power & Energy Society General Meeting (PESGM)*, Seattle, WA, USA, **2024**; pp. 1–5.
58. Zhao, Y.; Xue, H.; Jin, X.; Xiong, B.; Liu, R.; Peng, Y.; Jiang, L.; Tian, G. System level heat integration and efficiency analysis of hydrogen production process based on solid oxide electrolysis cells. *International Journal of Hydrogen Energy* **2021**, *46* (77), 38163–38174.
59. Sadeghian, O.; Shotorbani, A. M.; Ghassemzadeh, S.; Mohammadi-Ivatloo, B. Energy management of hybrid fuel cell and renewable energy based systems - A review. *International Journal of Hydrogen Energy* **2024**, *107*, 135–163.
60. Evro, S.; Oni, B. A.; Tomomewo, O. S. Carbon neutrality and hydrogen energy systems. *International Journal of Hydrogen Energy* **2024**, *78*, 1449–1467.
61. Dang, Y.; Wang, W. Low-carbon economic scheduling of hydrogen-integrated energy systems with enhanced bilateral supply–demand response considering vehicle to grid under power-to-gas–carbon capture system coupling. *Processes* **2025**, *13* (3), 636.
62. Rao, P.; Liu, Y.; Shi, X.; Yu, Y.; Zhou, Y.; Li, R.; Liang, Y.; Wu, D.; Li, J.; Tian, X.; et al. Protection of Fe single-atoms by Fe clusters for chlorine-resistant oxygen reduction reaction. *Advanced Functional Materials* **2024**, *34* (46), 2407121.
63. Wang, X.; Li, Z.; Qu, Y.; Yuan, T.; Wang, W.; Wu, Y.; Li, Y. Review of metal catalysts for oxygen reduction reaction: From nanoscale engineering to atomic design. *Chem* **2019**, *5* (6), 1486–1511.
64. Zhang, Q.; Yu, X.; Ling, Y.; Cai, W.; Yang, Z. Ultrathin nitrogen doped carbon layer stabilized Pt electrocatalyst supported on N-doped carbon nanotubes. *International Journal of Hydrogen Energy* **2017**, *42* (15), 10354–10362.
65. Zhang, Q.; Zhang, Y.; Cai, W.; Yu, X.; Ling, Y.; Yang, Z. Nitrogen doped carbon layer coated platinum electrocatalyst supported on carbon nanotubes with enhanced stability. *International Journal of Hydrogen Energy* **2017**, *42* (26), 16773–16781.
66. Zhou, Z.; Zhang, H.-J.; Feng, X.; Ma, Z.; Ma, Z.-F.; Xue, Y. Progress of Pt and iron-group transition metal alloy catalysts with high ORR activity for PEMFCs. *Journal of Electroanalytical Chemistry* **2024**, *959*, 118165.

67. Li, X.; Fu, L.; Cao, L.; Lei, L.; Zhan, X.; Hu, C.; Tong, X.; Tian, J. High electrocatalytic activity of crystallization-modulated ordered $PtCu_3/C$ nanoalloy catalysts for oxygen reduction reaction. *Chemical Engineering Journal* **2025**, *507*, 160445.
68. Zhu, J.; Yang, Y.; Chen, L.; Xiao, W.; Liu, H.; Abruña, H. D.; Wang, D. Copper-induced formation of structurally ordered Pt–Fe–Cu ternary intermetallic electrocatalysts with tunable phase structure and improved stability. *Chemistry of Materials* **2018**, *30* (17), 5987–5995.
69. Chen, Z. W.; Li, J.; Ou, P.; Huang, J. E.; Wen, Z.; Chen, L.; Yao, X.; Cai, G.; Yang, C. C.; Singh, C. V.; et al. Unusual Sabatier principle on high entropy alloy catalysts for hydrogen evolution reactions. *Nature Communications* **2024**, *15* (1), 359.
70. Miao, Z.; Xia, Y.; Liang, J.; Xie, L.; Chen, S.; Li, S.; Wang, H.-L.; Hu, S.; Han, J.; Li, Q. Constructing Co–N–C catalyst via a double crosslinking hydrogel strategy for enhanced oxygen reduction catalysis in fuel cells. *Small* **2021**, *17* (29), 2100735.
71. Miao, Z.; Wang, X.; Tsai, M.-C.; Jin, Q.; Liang, J.; Ma, F.; Wang, T.; Zheng, S.; Hwang, B.-J.; Huang, Y.; et al. Atomically dispersed Fe-N/C electrocatalyst boosts oxygen catalysis via a new metal-organic polymer supramolecule strategy. *Advanced Energy Materials* **2018**, *8* (24), 1801226.
72. Zang, J.; Wang, F.; Cheng, Q.; Wang, G.; Ma, L.; Chen, C.; Yang, L.; Zou, Z.; Xie, D.; Yang, H. Cobalt/zinc dual-sites coordinated with nitrogen in nanofibers enabling efficient and durable oxygen reduction reaction in acidic fuel cells. *Journal of Materials Chemistry A* **2020**, *8* (7), 3686–3691.
73. Miao, Z.; Wang, X.; Zhao, Z.; Zuo, W.; Chen, S.; Li, Z.; He, Y.; Liang, J.; Ma, F.; Wang, H.-L.; et al. Improving the stability of non-noble-metal M–N–C catalysts for proton-exchange-membrane fuel cells through M–N bond length and coordination regulation. *Advanced Materials* **2021**, *33* (39), 2006613.
74. Sun, W.; Liu, G.; Zou, H.; Wang, S.; Duan, X. Site-designed dual-active-center catalysts for co-catalysis in advanced oxidation processes. *Materials Sustainability* **2025**, *3* (1), 2.
75. Matter, P. H.; Ozkan, U. S. Non-metal catalysts for dioxygen reduction in an acidic electrolyte. *Catalysis Letters* **2006**, *109* (3), 115–123.
76. Gong, K.; Du, F.; Xia, Z.; Durstock, M.; Dai, L. Nitrogen-doped carbon nanotube arrays with high electrocatalytic activity for oxygen reduction. *Science* **2009**, *323* (5915), 760–764.
77. Chung, H. T.; Cullen, D. A.; Higgins, D.; Sneed, B. T.; Holby, E. F.; More, K. L.; Zelenay, P. Direct atomic-level insight into the active sites of a high-performance PGM-free ORR catalyst. *Science* **2017**, *357* (6350), 479–484.
78. Tateyama, S.; Suzuki, T.; Nasu, M.; Hirayama, N.; Watanabe, M.; Uchida, M.; Iiyama, A.; Tsushima, S. J. E. T. Effect of GDL structure and operating conditions on PEMFC performance and liquid water removal. *ECS Transactions* **2024**, *114* (5), 353.
79. Shi, Q.; Feng, C.; Li, B.; Ming, P. Investigation of the structural characteristics of the gas diffusion layer using micro-X-ray computed tomography. *Energies* **2025**, *18* (2), 381.
80. Ma, T.; Guo, H.; Gu, Z.; Lin, W.; Qi, J.; Yu, C.; Li, J. Performance analysis and experimental study of titanium GDL in proton exchange membrane fuel cell. *International Journal of Hydrogen Energy* **2024**, *83*, 604–613.
81. Zhao, J.; Liu, H.; Li, X. J. E. E. R. Structure, property, and performance of catalyst layers in proton exchange membrane fuel cells. *Electrochemical Energy Reviews* **2023**, *6* (1), 13.
82. Gao, J.; Wang, H.; Zhang, Z.; Chen, Y.; Li, D.; Zhou, M.; Chai, M.; Li, Y. Semi-ordered catalyst layer with ultra-low Pt loading for proton exchange membrane fuel cells. *Journal of Power Sources* **2024**, *606*, 234516.
83. Middelman, E. Improved PEM fuel cell electrodes by controlled self-assembly. *Fuel Cells Bulletin* **2002**, *2002* (11), 9–12.
84. Qasem, N. A. A. A recent overview of proton exchange membrane fuel cells: Fundamentals, applications, and advances. *Applied Thermal Engineering* **2024**, *252*, 123746.

85. Kraytsberg, A.; Ein-Eli, Y. Review of advanced materials for proton exchange membrane fuel cells. *Energy & Fuels* **2014**, *28* (12), 7303–7330.
86. Deng, X.; Ma, L.; Wang, C.; Ye, H.; Cao, L.; Zhan, X.; Tian, J.; Tong, X. Recent progress in materials design and fabrication techniques for membrane electrode assembly in proton exchange membrane fuel cells. *Catalysts* **2025**, *15* (1), 74.
87. Motz, A. R.; Kuo, M.-C.; Horan, J. L.; Yadav, R.; Seifert, S.; Pandey, T. P.; Galioto, S.; Yang, Y.; Dale, N. V.; Hamrock, S. J.; et al. Heteropoly acid functionalized fluoroelastomer with outstanding chemical durability and performance for vehicular fuel cells. *Energy & Environmental Science* **2018**, *11* (6), 1499–1509.
88. Zhang, Q.; Lin, R.; Técher, L.; Cui, X. Experimental study of variable operating parameters effects on overall PEMFC performance and spatial performance distribution. *Energy* **2016**, *115*, 550–560.
89. Kim, J.; Song, S. G.; You, N.-H.; Kim, S. G.; Ku, B.-C. Enhanced electrical and thermal properties of expanded graphite-polyphenylene sulfide (PPS) composites through in-situ polymerization for fuel cell bipolar plates. *Composites Communications* **2025**, *53*, 102252.
90. Mishra, S.; Kumar, R.; Verma, V. A review of thermoplastic-based low-cost composite bipolar plates for proton exchange membrane fuel cells. *Polymer Bulletin* **2025**, *82* (4), 1057–1084.
91. Chen, X.; Chen, Y.; Liu, Q.; Xu, J.; Liu, Q.; Li, W.; Zhang, Y.; Wan, Z.; Wang, X. Performance study on a stepped flow field design for bipolar plate in PEMFC. *Energy Reports* **2021**, *7*, 336–347.
92. Lim, B. H.; Majlan, E. H.; Daud, W. R. W.; Husaini, T.; Rosli, M. I. Effects of flow field design on water management and reactant distribution in PEMFC: A review. *Ionics* **2016**, *22* (3), 301–316.
93. Liu, H.; Li, P.; Juarez-Robles, D.; Wang, K.; Hernandez-Guerrero, A. Experimental study and comparison of various designs of gas flow fields to PEM fuel cells and cell stack performance. *Frontiers in Energy Research* **2014**, *2*, 2.
94. Liu, Z.; Li, Q.; Yang, S.; Zhang, H.; Chen, X.; Xie, N.; Deng, C.; Du, W. Numerical investigation of PEMFC performance based on different multistage serpentine flow field designs. *Chemical Engineering Journal* **2024**, *500*, 156951.
95. Qi, W.; Chen, X.; Zhang, Z. G.; Ge, S.; Wang, H.; Deng, R.; Liu, Z.; Tuo, J.; Guo, S.; Cheng, J. Performance of the multi-U-style structure based flow field for polymer electrolyte membrane fuel cell. *Scientific Reports* **2024**, *14* (1), 23318.

8 Solid Oxide Fuel Cells

Chuangang Yao, Haixia Zhang, and Kedi Cai

8.1 INTRODUCTION

Energy is the cornerstone of modern society. A review of the evolution of human civilization reveals that each transformation in energy technology has significantly accelerated the progress of modern civilization. From the advent of the traditional steam engine to the development of steam turbines, internal combustion engines, and gas turbines, these milestones in energy technology have been pivotal in driving humanity's advancement [1,2]. Entering the 21st century, rapid industrial and societal development worldwide has led to an ever-increasing demand for energy. Despite this, traditional fossil fuels, such as coal, oil, and natural gas, are expected to remain the dominant sources of energy consumption globally for the foreseeable future [3,4].

In recent decades, the supply of traditional energy has been stable and consistent, with well-established technologies for extraction, transportation, and utilization. The relatively low cost of these energy sources has made them a crucial foundation for economic development, contributing significantly to global growth and prosperity [5]. However, the extensive use of traditional energy sources has also led to a series of serious issues. First, the reserves of these fossil fuels are finite, and with the rapid increase in global energy demand and continuous extraction, the depletion of these resources is accelerating, thereby leading to an impending energy shortage crisis. Second, the use of traditional energy generates a significant amount of exhaust gases and pollutants, including carbon dioxide (CO_2), sulfur dioxide (SO_2), nitrogen oxides (NO_x), and particulate matter. These pollutants contribute to air pollution, which poses serious risks to human health. Furthermore, the extraction of traditional energy is often accompanied by large-scale land reclamation and mining activities, resulting in environmental degradation and the loss of biodiversity [6–8]. Therefore, while the use of traditional energy sources has provided essential support for human society, it has also led to severe environmental, social, and health issues [9]. To address the series of problems associated with the use of conventional fossil fuels and achieve sustainable development, it is imperative for the global community to take measures to reduce dependence on traditional energy sources and transition toward the use of clean, renewable energy [10]. This shift is necessary to mitigate the adverse effects of conventional energy consumption. Currently, countries around the world are actively developing and promoting the use of renewable energy, with solar, wind, and hydropower being recognized as key directions for future energy development. These clean energy sources offer advantages such as renewability and environmental benefits, helping to reduce dependence on finite fossil fuels and, in turn, minimizing the risks associated with energy shortages and supply instability [11]. At the same

DOI: 10.1201/9781003492511-8

time, researchers worldwide are actively exploring new, clean, and efficient energy utilization and conversion technologies. Through technological innovation, efforts are being made to improve energy efficiency and reduce pollutant emissions, paving the way for more sustainable energy systems [12–16]. Solid oxide fuel cells (SOFCs) represent the fourth generation of power generation technology following hydroelectric, thermal, and nuclear power, offering an efficient and environmentally friendly means of energy conversion [17]. As a high-performance, all-solid-state device, SOFCs directly convert the chemical energy stored in fuel gases and oxidants into electrical energy without the need for combustion [18]. This technology is widely recognized as one of the most promising energy solutions of the 21st century, playing a crucial role in meeting electricity demands, alleviating the energy crisis, protecting the ecological environment, and ensuring national security [19–21].

8.2 WORKING PRINCIPLE OF SOFC

SOFC is a type of all-solid-state electrochemical device, structured similarly to a sandwich, consisting of three main components. The central layer is the electrolyte, with the anode and cathode layers on either side. The electrolyte layer is dense, while the anode and cathode layers are porous [22]. SOFCs can be categorized by the primary ionic species responsible for charge transport within the electrolyte, leading to classifications such as oxygen-ion conducting SOFCs (O-SOFCs) and proton-conducting SOFCs (H-SOFCs) [23]. Taking the O-SOFCs as an example, their working principle is illustrated in Figure 8.1.

In the anode side of the SOFC, hydrogen gas (H_2) is introduced. The hydrogen molecules are adsorbed onto the catalytic surface of the anode and gradually diffuse to the interface between the anode and the electrolyte. At this interface, hydrogen molecules lose electrons and are oxidized to form hydrogen ions (H^+). Meanwhile, on the cathode side of the SOFC, oxygen gas (O_2) or air is supplied. The oxygen molecules are adsorbed onto the porous and loosely structured surface of the cathode and dissociate into oxygen atoms (O). Subsequently, these oxygen atoms combine with external electrons to form oxygen ions (O^{2-}). Driven by the electrochemical

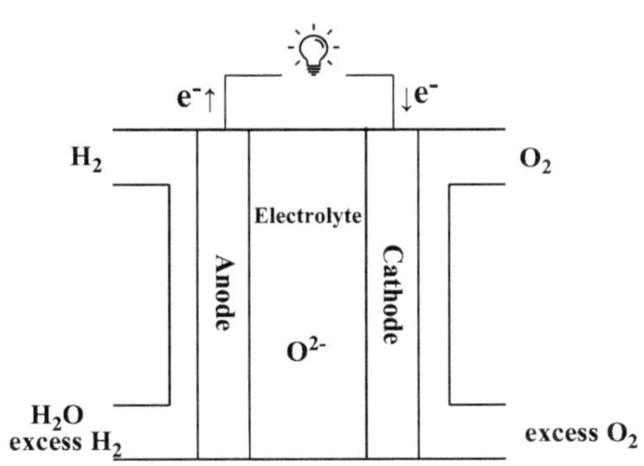

FIGURE 8.1 Schematic diagram of the operating principle of solid oxide fuel cell.

potential difference and concentration gradient across the electrolyte, the oxygen ions diffuse through oxygen vacancies in the solid electrolyte toward the anode side. On the anode side, the oxygen ions react with hydrogen ions, producing water (H_2O) and releasing electrons. These electrons flow through the external circuit, performing work on the load and thereby enabling the efficient conversion of the chemical energy stored in the fuel into electrical energy [24].

The electrochemical reactions occurring at the anode, cathode, and the overall reaction in a SOFC are expressed as follows:

Anode reaction:

$$H_2 + O^{2-} \rightarrow H_2O + 2e^- \tag{8.1}$$

Cathode reaction:

$$\frac{1}{2}O_2 + 2e^- \rightarrow O^{2-} \tag{8.2}$$

Overall reaction:

$$H_2 + \frac{1}{2}O_2 \rightarrow H_2O \tag{8.3}$$

8.3 ADVANTAGES OF SOFC

As a third-generation fuel cell, SOFCs hold significant potential for applications in various fields, including distributed power generation, portable power sources, and aerospace. The notable advantages of SOFCs are primarily reflected in the following aspects:

1. All-solid-state structure: SOFCs feature solid-state anodes, cathodes, and electrolytes, which eliminates the corrosion issues associated with acidic or molten salt electrolytes. This all-solid-state configuration contributes to enhanced durability and stability over time [25,26].
2. High energy conversion efficiency: The energy conversion efficiency of SOFCs can reach up to 65%. When utilized in combined heat and power (CHP) systems, the efficiency can exceed 80%, offering a distinct advantage over traditional power generation technologies [27].
3. Environmental friendliness: SOFCs operate with low noise and produce zero or minimal emissions of nitrogen oxides (NO_x), sulfur oxides (SO_x), and other pollutants, making them highly environmentally friendly [28].
4. Fuel versatility: SOFCs can operate on a wide range of fuels, including hydrogen, natural gas, and methanol. This versatility enhances their applicability across different energy resource environments, making them suitable for diverse applications and operational scenarios [29].
5. No need for precious metal catalysts: As high-temperature fuel cells, SOFCs exhibit rapid fuel oxidation and electrode reactions at elevated operating

temperatures, which eliminates the need for costly precious metals as catalysts. This feature significantly reduces manufacturing costs [30].

6. Ease of installation and maintenance: The simple structure of SOFCs enables modular installation, making them suitable for flexible deployment across various scales and locations. Moreover, their straightforward design facilitates easier maintenance and management [31].

8.4 ANODE MATERIALS FOR SOFCS

The anode plays a critical role in SOFCs, as its primary function is to provide a site for electrochemical oxidation reactions and facilitate the transfer of electrons generated during fuel oxidation to the external circuit, thus enabling the crucial step of energy conversion. Therefore, anode materials must possess several key characteristics to ensure efficient operation of the fuel cell.

Firstly, the anode material must exhibit high electronic conductivity to effectively transfer electrons produced during the fuel oxidation process to the external circuit. Additionally, the anode needs to demonstrate high stability under reducing atmospheres to ensure long-term, stable operation. This stability not only affects the lifespan of the cell but also directly influences its reliability and sustained performance [32].

Secondly, the anode material should have an appropriate level of porosity to ensure sufficient triple-phase boundaries (TPBs) and fuel gas diffusion. This contributes to maintaining a high reaction rate within the cell and facilitates efficient fuel gas transport between the anode and the electrolyte. A well-designed porous structure enhances the surface area of the TPB, thereby improving the power density of the fuel cell [33].

8.4.1 Ni/YSZ Composite Anode Materials

Nickel (Ni), owing to its high catalytic activity and low cost, is widely used as an anode material in SOFCs. However, the thermal expansion coefficient (TEC) mismatch between metallic Ni and commonly used electrolyte materials such as yttria-stabilized zirconia (YSZ) can lead to issues such as cracking and delamination.

To address this mismatch in TEC, Ni is typically combined with YSZ to form Ni/YSZ composite anodes. This approach not only effectively mitigates the thermal expansion mismatch between the anode and electrolyte during sintering and operation but also increases the length of the TPB within the material. This, in turn, enhances the active sites for the electrochemical oxidation of fuel gases at the anode [34].

Despite its widespread use in SOFCs operating with hydrogen as fuel, Ni/YSZ composite anodes still exhibit certain drawbacks. For instance, prolonged high-temperature operation leads to the coarsening of Ni particles, which results in a decline in electrochemical performance. Studies have shown that after 250 hours of operation, significant morphological changes occur in the Ni/YSZ anode, with a 17.9% decrease in specific surface area and an increase in polarization resistance from 0.54 to 1.82 $\Omega \cdot cm^2$ [35]. Additionally, when Ni is used for hydrocarbon cracking, the main product, carbon, tends to deposit on the Ni surface, leading to a rapid degradation in performance.

Research has shown that modifying the Ni/YSZ composite anode with BaO significantly enhances its stability. Han et al. [36] employed a phase transformation-impregnation method to modify the surface of the composite anode with BaO. Density functional theory calculations indicated that BaO effectively dissociates water molecules to generate hydroxyl ions, thereby serving as an effective medium for carbon elimination. In experiments with wet hydrogen and methane fuels at 800°C, the corresponding peak power densities were 0.30 W·cm^2, respectively.

8.4.2 Cu/CeO$_2$ Composite Anode Materials

Cu/CeO$_2$ composite materials represent another class of anode materials for SOFCs. Studies have shown that when hydrogen is used as the fuel at 800°C, the power density of a Cu/CeO$_2$-based SOFC reaches 0.608 W·cm^{-2} [37]. Furthermore, extensive research by Gorte et al. [38,39] has been conducted on Cu-CeO$_2$/YSZ composite anodes. When hydrogen is used as the fuel, the maximum power density of a Cu-CeO$_2$/YSZ-based SOFC reaches 300 mW·cm^{-2}. In contrast, when hydrocarbon fuels are used, the maximum power density of the cell ranges between 100 and 120 mW·cm^{-2}. In Cu-CeO$_2$/YSZ anode materials, Cu functions primarily as an electronic conductor but exhibits limited catalytic activity for hydrocarbons. On the other hand, CeO$_2$ not only provides excellent catalytic properties for hydrocarbons but also enhances both the electronic and ionic conductivity of the anode [40,41]. Additionally, research indicates that doping CeO$_2$ with transition metals such as Fe and Mn can increase the concentration of oxygen vacancies and improve the mobility of lattice oxygen. Doping CeO$_2$ with elements such as Gd and Sm not only enhances the ionic conductivity of the material but also effectively mitigates the carbon deposition problem on the anode. Surface modification techniques such as impregnation can further improve the Cu/CeO$_2$ composite materials by increasing the length of the TPB, thereby reducing the polarization resistance [42].

8.4.3 Perovskite-Type Anode Materials

Perovskite-type anode materials, which are essential for SOFCs, must exhibit high catalytic activity for fuels such as hydrogen (H$_2$), alongside adequate electronic and ionic conductivity. Additionally, they should possess resistance to sulfur poisoning and carbon deposition. In recent years, several perovskite-structured oxides have shown promising electronic and ionic conductivity in reducing atmospheres, as well as catalytic activity for hydrocarbons, making them potential candidates for SOFC anode applications. Among these, SrTiO$_3$-based, LaCrO$_3$-based, and Sr$_2$MgMoO$_6$-based materials have been the focus of extensive research [43].

In SrTiO$_3$-based anode materials, the transition of Ti^{4+} to Ti^{3+} contributes to their inherent electronic conductivity. Studies have shown that the performance of such anodes can be significantly enhanced by substituting La and Mn at the A-site and B-site, respectively [44]. It has been reported that the SOFC with La$_{0.4}$Sr$_{0.6}$Ti$_{0.4}$Mn$_{0.6}$O$_3$ as the anode exhibited a polarization resistance of only 0.82 Ω·cm^2 at 810°C in a CH$_4$ atmosphere [44]. Furthermore, La$_{0.75}$Sr$_{0.25}$Cr$_{0.5}$Mn$_{0.5}$O$_3$, at 900°C, showed polarization resistances of 0.85 and 0.26 Ω·cm$_2$ in CH$_4$ and H$_2$ atmospheres, respectively, which are comparable to the electrochemical performance of the Ni/YSZ anode [45]. Studies also indicate that Mn-doped $Sr_2Mg_{1-x}Mn_xMoO_6$ demonstrates excellent stability and sulfur resistance as an SOFC anode material [46].

8.5 ELECTROLYTE MATERIALS FOR SOFCS

In SOFCs, the key properties of the electrolyte layer have a critical impact on the overall system performance and stability. The electrolyte layer primarily functions to conduct oxygen ions while simultaneously acting as a barrier to separate the fuel gas and the oxidant gas [47]. One of the most important factors influencing SOFC performance is the ionic conductivity of the electrolyte material. Under both oxidizing and reducing conditions, the electrolyte must exhibit a high ionic conductivity to ensure the efficient transport of oxygen ions within the electrolyte layer. At the same time, the electronic conductivity should be as low as possible to ensure that electrons are only transported through the external circuit. This design ensures the efficient transfer of oxygen ions or hydrogen ions within the SOFC system while preventing electron leakage, thereby maintaining the separation function of the electrolyte layer [48].

In addition, physical compatibility plays a crucial role in the selection of electrolyte materials. Physical compatibility requires that the electrolyte material matches the TEC of other cell components, such as the electrodes, to ensure the structural integrity of the entire SOFC system during high-temperature operation. Only when the TECs of the components are well-matched can potential damage or failure due to temperature-induced stresses be effectively avoided [49].

Chemical compatibility is another vital consideration in the selection of electrolyte materials. The electrolyte must exhibit good chemical stability in both oxidizing and reducing atmospheres to ensure that material degradation does not occur during prolonged operation of the SOFC system. Additionally, the electrolyte must not undergo undesirable chemical reactions with other components in the fuel cell, such as the anode and cathode, to maintain the overall chemical stability of the system [50]. Furthermore, density is an essential factor in the design of electrolyte materials. Typically, the electrolyte material undergoes high-temperature sintering to form a dense ceramic layer. A highly dense electrolyte not only enhances the transport efficiency of oxygen or hydrogen ions within its structure but also effectively isolates the fuel gas and oxidant gas, preventing gas leakage and thereby improving the efficiency and safety of the SOFC system [51].

8.5.1 ZrO_2-BASED ELECTROLYTE MATERIALS

ZrO_2 is a typical fluorite-structured (Figure 8.2) compound and is one of the earliest studied and most widely used solid electrolyte materials. Pure ZrO_2 undergoes a phase transformation from monoclinic to tetragonal at around 1,100°C, which leads to significant volumetric changes. However, doping ZrO_2 with ions such as Ca^{2+} and Y^{3+} can not only stabilize the structure at lower temperatures but also significantly increase the concentration of oxygen vacancies, thereby enhancing its ionic conductivity [52].

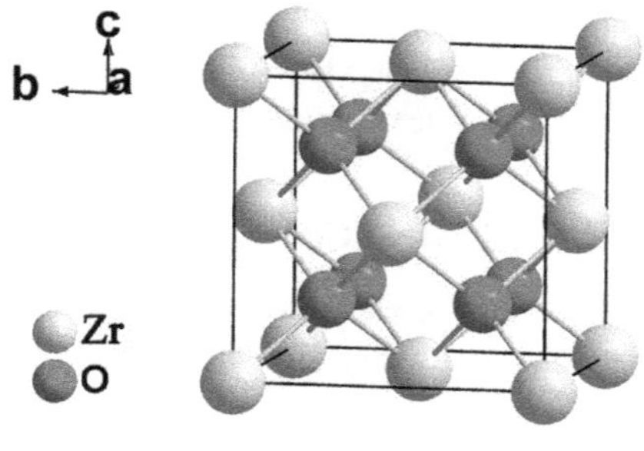

FIGURE 8.2 Schematic diagram of the fluorite structure.

YSZ, where Y_2O_3 is used as a stabilizer, is one of the most commonly applied solid electrolyte materials. Studies have shown that the optimal doping level of Y_2O_3 is 8 mol%, at which point the ionic conductivity of YSZ reaches its maximum. When the doping concentration of Y_2O_3 exceeds 8 mol%, the ionic conductivity of YSZ begins to decline. This decrease is attributed to the excessive doping causing defect association in the material, which reduces the mobility of oxygen vacancies. It is worth noting that although YSZ demonstrates high oxygen ion conductivity, stability, and mechanical strength at high temperatures, its ionic conductivity in the intermediate temperature range is only 0.001–0.003 $S{\cdot}cm^{-1}$, which does not meet the requirements for intermediate-temperature SOFCs [53].

8.5.2 CeO_2-Based Electrolyte Materials

CeO_2-based oxides represent another class of commonly used electrolyte materials for SOFCs. Pure CeO_2 adopts a fluorite structure, exhibiting low ionic conductivity, with a conductivity of approximately 10^{-5} $S{\cdot}cm^{-1}$ at 600°C. Extensive studies have demonstrated that doping with low-valence cations can significantly enhance its ionic conductivity [54].

Samarium-doped ceria (SDC) and gadolinium-doped ceria (GDC) are two widely studied materials in the field of SOFC electrolytes. Both materials have been extensively researched theoretically and experimentally, with the goal of improving the performance of SOFC electrolytes. Despite the excellent characteristics exhibited by solid electrolyte materials such as SDC and GDC, their application in reducing atmospheres presents several challenges. When SDC and GDC are exposed to a reducing atmosphere, some of the Ce^{4+} ions in the electrolyte undergo reduction to Ce^{3+}, leading to a degree of electronic conductivity in the electrolyte. This reduction not only reduces the open-circuit voltage of the cell but also decreases its power density. Furthermore, the reduction of Ce^{4+} to Ce^{3+} results in lattice expansion, which may cause poor contact between the electrolyte and the electrode layers, potentially leading to cracking. This lattice expansion has a significant negative impact on the output performance of the cell [55].

To address this issue, researchers have proposed doping strategies to suppress the reduction of Ce^{4+} to Ce^{3+} in CeO_2-based electrolytes. For example, Kang et al. [56] employed a mixed valence state of Pr^{4+}/Pr^{3+} to inhibit the reduction of Ce ions, and cation co-doping was found to improve ionic conductivity by suppressing the ordering of oxygen vacancies. Consequently, a Sm, Pr, and Y co-doped CeO_2-based electrolyte was developed. The results demonstrated that the conductivity of the co-doped electrolyte remained almost unchanged over a 60-hour period, indicating that doping can effectively maintain the stability of CeO_2-based electrolytes in air. At 800°C, the open-circuit voltage of the single cell was 0.74 V, and the maximum power density reached 674.6 mW·cm^2.

8.5.3 $LaGaO_3$-Based Electrolyte Materials

At room temperature, pure $LaGaO_3$ adopts an orthorhombic crystal structure. Ion doping has been shown to significantly enhance its ionic conductivity. Studies indicate that the conductivity increases in the order of Sr > Ba > Ca when various alkaline earth cations are doped at the La site. Therefore, doping the La site with Sr is the most effective approach for improving the ionic conductivity of $LaGaO_3$-based electrolyte materials. Although theoretically, increasing the Sr doping content could enhance the oxygen vacancy concentration and thereby improve the ionic conductivity, the solubility limit of Sr in $LaGaO_3$ is relatively low. When the doping level exceeds 10 mol%, the formation of a secondary phase is observed. Consequently, simply increasing the Sr doping content has a limited effect on enhancing the material's ionic conductivity [57].

Since $LaGaO_3$ belongs to the perovskite structure, doping can also be performed at the Ga site. Research shows that doping with Mg at the Ga site can significantly improve the ionic conductivity of $LaGaO_3$. The ionic conductivity reaches its maximum at a doping level of 20 mol%. This improvement is attributed to the relatively large ionic radius of Mg^{2+}, which, compared to Ga^{3+}, causes an increase in the lattice constant. This, in turn, enhances the solubility of Sr in $LaGaO_3$, raising the solubility limit from 10% to 20% [58].

The introduction of Sr^{2+}- and Mg^{2+}- co-doped $LaGaO_3$ (LSGM) has garnered significant attention since its initial report. Studies show that the ionic conductivity of $(La_{0.9}Ln_{0.1})_{0.8}Sr_{0.2}Ga_{0.8}Mg_{0.2}O_{3-\delta}$ (Ln = Y, Nd, Sm, Gd, and Yb) increases in the following order: Y < Yb < Gd < Sm < Nd. This is because as the radius of the dopant ions decreases, the tolerance factor gradually deviates from the ideal cubic perovskite structure, which leads to a reduction in oxygen ion mobility [59]. It has been found that $La_{0.8}Sr_{0.2}Ga_{0.8}Mg_{0.2}O_{3-\delta}$ exhibits an ionic conductivity of 0.137 S·cm^{-1} at 800°C, which is comparable to that of 8YSZ at 1,000°C [60]. Furthermore, LSGM electrolytes show negligible electronic conductivity under both oxidizing and reducing atmospheres and exhibit excellent stability in high-temperature redox environments [61]. The introduction of small amounts of transition metals such as Ba, Gd, and Ca into the La site of LSGM reduces the activation energy by altering the octahedral tilt angles, resulting in higher conductivity at lower temperatures. Additionally, studies have reported that doping small amounts of transition metals like Fe, Co, and Ni at the Mg site in LSGM further enhances its ionic conductivity [62–64].

In addition to the LSGM series, several novel oxide systems have also attracted widespread attention, such as $Ba_2In_2O_5$ [65], $La_2Zr_2O_7$ [66], $La_2Mo_2O_9$ [67], $M_{10}(XO_4)_6O_{2+y}$ (Ln is a rare earth or alkaline earth element, and X is Si or Ge) [68], and $Ca_{12}Al_{14}O_{33}$ [69]. While these materials show potential as SOFC electrolyte materials, they face challenges that limit their practical applications. Further research is required to address these issues and improve their performance.

8.5.4 Proton-Conducting Electrolyte Materials

In 1981, Iwahara et al. [70] first discovered that $SrCeO_3$-based perovskite ABO_3-type oxides exhibited significant proton conductivity under intermediate temperature conditions, which attracted widespread attention. Compared to oxygen-ion conducting electrolytes, proton-conducting electrolytes demonstrate higher proton conductivity at lower temperatures, making them more suitable for applications in the intermediate to low-temperature range. In recent years, proton-conducting electrolytes represented by $BaCeO_3$ and $BaZrO_3$ have been extensively studied.

It has been established that cerium-based perovskite oxides possess the highest proton conductivity among all ceramic oxides. However, the metal elements present in $BaCeO_3$-based proton-conducting electrolytes tend to be highly basic, which makes them susceptible to element segregation under acidic conditions, such as in environments containing CO_2 and SO_2. This phenomenon can lead to the formation of impurity phases, such as $BaCO_3$ and $Ba(OH)_2$, at the interfaces, which are detrimental to proton conduction and, consequently, degrade the performance of the materials [71,72].

Rare earth elements, such as Yb, Sc, Y, and La, are commonly doped into proton-conducting electrolytes. When Ce^{4+} is substituted by trivalent metal cations, additional oxygen vacancies are generated, which enhances the performance of the material [73]. Matsumoto et al. [74] conducted a study to investigate the effects of partially substituting Ce with various trivalent metal ions (including Y, Tm, Yb, Lu, In, and Sc) on the chemical stability and conductivity of $BaCeO_3$. They found that as the ionic radius of the dopant ions decreased, conductivity decreased, but chemical stability improved. Gu et al. explored doping Gd, Y, and Yb into $BaCeO_3$ to enhance its conductivity. The results showed that Y-doped $BaCeO_3$ exhibited the highest conductivity [75].

Despite the significant improvement in conductivity after doping $BaCeO_3$, its stability remains relatively poor. To enhance the stability of such electrolyte materials, doping with highly electronegative particles, such as Nb, Ta, and Zr, is commonly employed [76]. Among them, $BaZrO_3$, as one of the most widely studied high-temperature proton-conducting materials, offers advantages such as low activation energy and good stability. However, its conductivity is relatively low, limiting its practical application. To combine the advantages of both types of electrolytes, Zr is typically introduced at the Ce site to improve the material's stability. Research has shown that BZCY ($BaZr_{0.1}Ce_{0.7}Y_{0.2}O_{3-\delta}$) exhibits excellent conductivity performance in intermediate- to low-temperature environments [77].

By doping Yb into BZCY electrolyte, $BaCe_{0.7}Zr_{0.1}Y_{0.1}Yb_{0.1}O_{3-\delta}$ (BCZYYb) was synthesized, showing good conductive properties [78]. Duan et al. [79] studied the application of BCZYYb in proton-conducting SOFCs and found that at 500°C under a hydrogen atmosphere, the maximum power density reached 455 mW cm^2. Chen et al. [80] employed 2 mol% NiO as a sintering aid in the performance study of BZCY electrolyte. The results showed that with the assistance of NiO, BZCY could achieve densification, with Ni entering the B-site of the BZCY lattice. After sintering, the BZCY grains were larger, and the grain boundaries were fewer, which improved the material's conductivity. The peak power density of proton-conducting SOFCs using BZCY reached 855 mW cm^2.

8.6 CATHODE MATERIALS FOR SOFCS

The cathode is the primary site for the oxygen reduction reaction (ORR). Depending on the properties of the cathode materials, they can be classified into three main categories: electronic conductor cathodes, mixed electronic and ionic conductor (MIEC) cathodes, and composite cathodes. The oxygen reduction process on the cathode varies according to the nature of the cathode material.

1. Electronic conductor cathode materials: As shown in Figure 8.3a, the oxygen reduction process on an electronic conductor cathode is illustrated. These cathode materials are purely electronic conductors. Oxygen molecules (O_2) first diffuse to the surface of the cathode material, where they undergo adsorption and dissociation to form oxygen atoms (O). These oxygen atoms then diffuse across the cathode surface to the TPB formed by the cathode, electrolyte, and air. At the TPB, the oxygen atoms gain electrons and are reduced to oxygen ions (O^{2-}), which then diffuse into the electrolyte. In electronic conductor cathode materials, the electrochemical reaction primarily occurs at the TPB.
2. MIEC cathode materials: For MIEC cathode materials, which can conduct both electrons and ions, the ORR is not confined to the TPB but extends across the entire cathode. The oxygen reduction process on MIEC cathodes involves two diffusion pathways: surface diffusion and bulk diffusion, as illustrated in Figure 8.3b.

   Surface diffusion pathway: Oxygen first diffuses to the cathode surface, where it adsorbs and dissociates into oxygen atoms. The oxygen atoms then diffuse across the electrode surface to the TPB, where they gain electrons and are reduced to oxygen ions. The oxygen ions subsequently diffuse into the electrolyte.

   Bulk diffusion pathway: Oxygen first adsorbs and dissociates into oxygen atoms on the cathode surface. The oxygen atoms then combine with electrons at active sites, where they are reduced to oxygen ions. These oxygen ions enter the bulk of the MIEC cathode material, where they diffuse from the cathode surface to the interface with the electrolyte and finally diffuse into the electrolyte.
3. Composite cathode materials: To enhance the ionic conductivity of SOFC cathode materials, single-phase electronic conductor or MIEC cathode materials are often combined with electrolyte materials to form composite cathodes. Figure 8.3c illustrates the oxygen reduction process in a composite cathode formed by a single-phase electronic conductor and electrolyte materials. The incorporation of the electrolyte in the composite cathode significantly increases the length of the TPB and the number of active sites

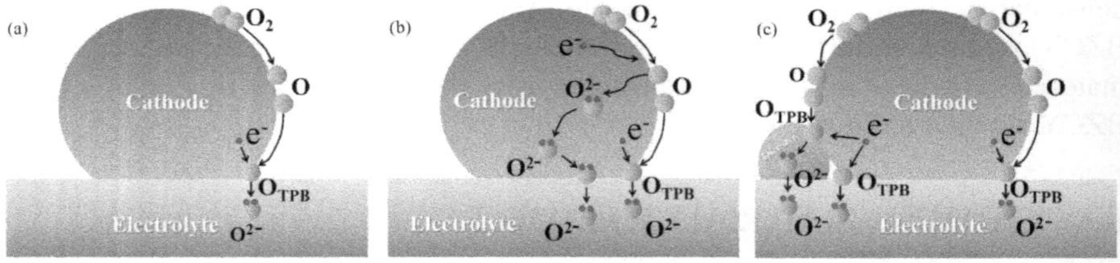

FIGURE 8.3 Schematic diagram of the oxygen reduction reaction process in different types of cathode materials. (a) The reaction scenario of electronic conductor cathode materials. (b) The reaction scenario of MIEC cathode materials. (c) The reaction scenario of composite cathode materials.

for the ORR, thereby improving the cathode performance. Additionally, the presence of the electrolyte in composite cathodes helps reduce the polarization resistance and enhances the thermal expansion matching between the cathode and the electrolyte layers.

The cathode is one of the key components in SOFCs. As the electrolyte layer becomes thinner, the cathode has become the primary source of polarization loss in SOFCs. Therefore, enhancing the performance of cathode materials and identifying high-performance cathode materials have remained focal points in SOFC research. Currently, research on SOFC cathode materials is primarily focused on perovskite-type oxides, double perovskite-type oxides, and composite cathode materials.

8.6.1 Perovskite-Type Cathode Materials

The general formula for simple perovskite-type oxides is ABO_3, with an ideal crystal structure of a face-centered cubic lattice, as shown in Figure 8.4. The A-site cations occupy the corner positions of the cubic unit cell, with a coordination number of 12. These cations are typically rare earth elements or alkaline earth metals. The B-site cations, with a coordination number of 6, are located at the body center of the cubic unit cell and are usually transition metal cations with relatively smaller ionic radii, or some metal cations from the main groups. Oxygen anions occupy the face-centered positions of the cubic unit cell and form BO_6 octahedra with the B-site cations at the body center.

Although the crystal structure of ideal perovskite oxides is very simple, in practice, the crystal lattice often deviates from the ideal structure due to the differing electronic configurations of constituent ions and their interactions, resulting in distortions that reduce the symmetry of the structure. To quantitatively describe the structural stability of ABO_3-type perovskite compounds, Goldschmidt introduced the concept of the tolerance factor, which establishes a relationship between the structural stability of perovskite compounds and the ionic radii of their constituent ions. The tolerance factor is defined as follows [81]:

$$t = \frac{r_A + r_O}{\sqrt{2}\left(r_B + r_O\right)} \tag{8.4}$$

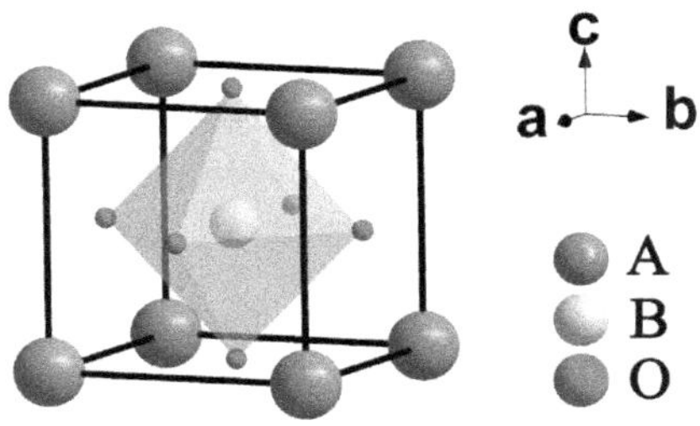

FIGURE 8.4 Schematic diagram of the unit cell structure of the ideal ABO_3 perovskite oxide.

where r_A, r_B, and r_O represent the radii of the A-site and B-site metal cations and the oxygen anion in the perovskite compound, respectively. Typically, when $0.71 < t < 0.9$, the structure is orthorhombic or rhombohedral. When $0.9 < t < 1.0$, the structure is cubic. When $t > 1.0$, the structure is hexagonal, and when $t < 0.71$, the structure is either ilmenite or corundum.

By doping cations at the A or B sites in ABO_3-type perovskite compounds, the crystal structure and properties of the material can be tuned. This approach enhances structural stability, improves electrical and electrochemical catalytic performance, and increases the potential application of the material as a cathode in SOFCs [82].

The Sr-doped $LaMnO_3$ series compounds $La_{1-x}Sr_xMnO_3$ (LSM) are among the most extensively studied cathode materials. At high temperatures, LSM cathode materials exhibit high electronic conductivity, excellent oxygen reduction catalytic activity, and good thermal and chemical compatibility with electrolyte materials, making them widely used in SOFCs with YSZ, GDC, and LSGM as electrolytes [83]. Zheng et al. [84] systematically investigated the effect of Sr^{2+} content on the crystal structure of LSM materials. The study found that when $0 < x < 0.2$, the material adopts an orthorhombic perovskite structure. When $0.2 \leq x \leq 0.3$, it adopts a monoclinic perovskite structure, and when $x > 0.3$, the structure reverts to orthorhombic. Additionally, the study showed that the conductivity of LSM cathode materials can reach up to 200 S cm^{-1}, with an ionic conductivity of approximately 10^{-6} to 10^{-7} S cm^{-1} [85].

Although LSM exhibits good electrochemical catalytic performance as an SOFC cathode material at high temperatures, its electrochemical performance rapidly deteriorates as the temperature decreases. For instance, Jiang et al. reported that when the temperature drops from 900°C to 700°C, the interface polarization resistance of the LSM cathode material increases sharply from 0.39 to 55.7 Ω cm^2, leading to a rapid decline in its electrochemical performance [86].

Research has shown that $La_{1-x}Sr_xCoO_3$ (LSC) cathode materials exhibit superior ionic conductivity, electronic conductivity, and electrocatalytic performance compared to $La_{0.8}Sr_{0.2}MnO_3$ (LSM) cathode materials at lower temperatures [87]. At 800°C, the electronic and ionic conductivities of LSC can reach up to 1,000 S cm^{-1} and 0.22 S cm^{-1}, respectively, far exceeding the conductivity requirements for SOFC cathode materials. However, a significant drawback of LSC cathode materials is their relatively high TEC. For instance, when $x=0.7$, the TEC can reach as high as 26×10^{-6} K^{-1}, approximately twice that of traditional SOFC electrolytes such as YSZ and SDC [88]. The large disparity in TECs between LSC cathode materials and electrolyte materials can lead to cracking between the electrode and electrolyte layers during the heating and cooling processes of cell fabrication or operation, severely affecting the cell's output performance.

To address the issue of excessive TECs in LSC cathode materials, Fe^{3+} ions have been used to partially substitute Co^{3+} in LSC, leading to a significant reduction in the material's TEC. This substitution effectively alleviates the thermal expansion mismatch between LSC and common electrolyte materials that existed prior to the modification [89].

Furthermore, studies have revealed that the electronic and ionic conductivities of $La_{1-x}Sr_xCo_{1-y}Fe_yO_3$ cathode materials are strongly dependent on the doping levels of Sr and Fe. For instance, when the Fe content is $y=0.8$ and the Sr content is approximately $x=0.4$, the material exhibits the highest electronic and ionic conductivities. When the Sr content reaches $x=0.7$, the electrochemical catalytic performance of the material improves as the Co content increases [90]. In addition, LSCF (LaSrCoFe material) cathode materials demonstrate enhanced resistance to Cr poisoning. However, the mechanical properties of LSCF cathode materials remain relatively poor and require further improvement.

$Ba_{1-x}Sr_xCo_{1-y}Fe_yO_3$ (BSCF) was initially widely studied as an oxygen permeable membrane material, and later, Shao et al. [91] investigated its use as a cathode material for SOFCs. Studies have shown that BSCF possesses excellent electrochemical catalytic performance and oxygen ion conductivity. For instance, at 500°C and 600°C, the interface polarization resistance of the BSCF cathode was found to be 0.135 and 0.021 Ω cm^2, respectively. At 775°C and 900°C, its oxygen vacancy diffusion rates were as high as 7.3×10^{-5} and $1.31\times10^{-4} cm^2 s^{-1}$, respectively. When used as a cathode with Ni/SDC as the anode in SOFCs fueled by H_2, the cell output power reached 402 and 1,010 mW cm^{-2} at 500°C and 600°C, respectively. Additionally, BSCF also demonstrates excellent catalytic performance for propane, with a single SOFC cell powered by propane reaching an output power of 440 mW cm^{-2} at 500°C.

Despite its excellent electrochemical properties, BSCF cathode materials exhibit some drawbacks. Firstly, the TEC of BSCF is relatively high, with an average TEC of 19.7×10^{-6} K^{-1} between 50°C and 900°C, which is significantly higher than that of conventional SOFC electrolyte materials such as SDC (TEC $=12.8\times10^{-6}$ K^{-1}). Secondly, the conductivity of BSCF is relatively low, ranging from 20 to 60 S cm^{-1}, which limits its practical application [92]. To address these issues, Li et al. [93] doped the A-site of BSCF with Sm^{3+} and investigated the electrical and electrochemical properties of the doped $(Ba_{0.5}Sr_{0.5})_{1-x}Sm_xCo_{0.8}Fe_{0.2}O_{3-\delta}$ (BSSCF) series cathode materials. The results indicated that the conductivity and electrochemical performance of the doped BSSCF cathodes were superior to those of BSCF. For example, at 500°C, the conductivity and interface polarization resistance of the BSSCF cathode were 85.6 S cm^{-1} and 2.98 Ω cm^2, respectively, significantly higher than the conductivity (27.4 S cm^{-1}) and interface polarization resistance (6.04 Ω cm^2) of BSCF at the same temperature.

8.6.2 Double Perovskite Cathode Materials

Double perovskite compounds of the general formula $AA'B_2O_6$ or $AA'BB'O_6$ are derived from the simple perovskite structure ABO_3. In recent years, the $LnBaCo_2O_{5+\delta}$ (Ln = La, Pr, Nd, Sm, and Gd) series of double perovskite compounds has attracted significant attention. In these $LnBaCo_2O_{5+\delta}$ compounds, Ln and Ba occupy the A-sites of the unit cell, while Co occupies the B-site. The structure alternates along the *c*-axis direction with layers of CoO/BaO/CoO/LnO/CoO stacked in succession, as illustrated in Figure 8.5. In these A-site-ordered layered double perovskite compounds, oxygen vacancies are confined to the LnO layers. This unique distribution of oxygen vacancies provides a pathway for rapid oxygen ion diffusion within the bulk material, resulting in a higher bulk diffusion coefficient for these materials.

Kim et al. [94] conducted relevant studies on the transport dynamics of PBCO materials, and their findings indicate that this material exhibits a high surface exchange coefficient for oxygen and excellent oxygen transport performance. Seymour et al. [95] performed theoretical simulations to calculate the oxygen ion transport properties in PBCO materials, and the results revealed that oxygen ion transport in this material exhibits a pronounced two-dimensional transport characteristic. Joung et al. [96] systematically investigated the electrical conductivity of $LnBaCo_2O_{5+\delta}$ (Ln = Pr, Nd, Sm, and Gd) series of double perovskite compounds. The results show that all these compounds exhibit very high electronic conductivity. PBCO demonstrates the highest electronic conductivity among this series of compounds, with a maximum conductivity of 1,323 S cm^{-1} at 100°C and a minimum conductivity of 310 S cm^{-1} at 900°C. GBCO exhibits a maximum conductivity of 655 S cm^{-1} and a minimum conductivity of 163 S cm^{-1}. At 900°C, NBCO shows the lowest electronic conductivity among this series, with a value of 132 S cm^{-1}, which still meets the conductivity requirement for traditional intermediate-temperature SOFC cathode materials ($\sigma >$ 100 S cm^{-1}).

Tarancon et al. [97] conducted extensive research on the performance of GBCO as a cathode material for SOFCs, demonstrating its excellent electrochemical properties. For instance, when GDC was used as the electrolyte, the polarization resistance of GBCO at 625°C was only 0.25 Ω cm^{-2}. Zhu et al. [98] investigated the

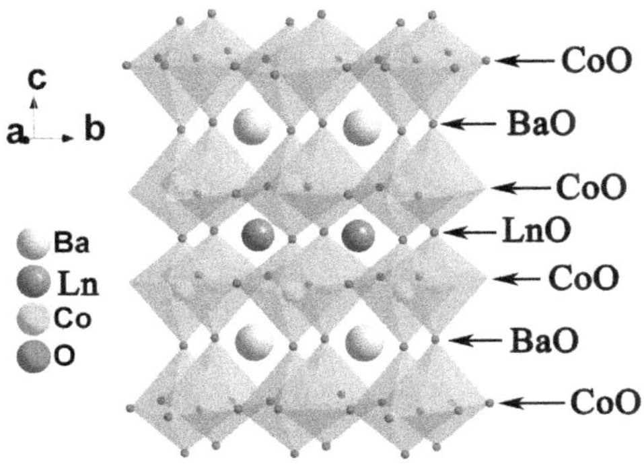

FIGURE 8.5 Schematic diagram of the crystal structure of $LnBaCo_2O_{5+\delta}$ double perovskite.

performance of PBCO as the cathode material, with SDC and Ni/SDC as the electrolyte and anode, respectively. Their study showed that the output power densities of the SOFC single cells at 600°C and 650°C were 583 and 866 mW cm^{-2}, respectively. Zhou et al. [99] found that SBCO also exhibited good electrochemical performance as a cathode material for SOFCs. Specifically, when SDC and LSGM were used as the electrolytes, the polarization resistances at 750°C were 0.098 and 0.054 Ω cm^{-2}, respectively, and the maximum output power densities of the cells at 800°C reached 641 and 777 mW cm^{-2}, respectively.

Although $LnBaCo_2O_{5+\delta}$ (Ln = La, Pr, Nd, Sm, and Gd) series double perovskite oxides demonstrate good electrochemical catalytic properties as cathode materials, they share the common drawback of having a high coefficient of thermal expansion, similar to other cobalt-based materials. For example, Kim et al. [100] reported that the average CTEs of LBSC, NBSC, SBSC, and GBSC between 80°C and 900°C were 24.3×10^{-6} K^{-1}, 19.1×10^{-6} K^{-1}, 17.1×10^{-6} K^{-1}, and 16.1×10^{-6} K^{-1}, respectively, all of which are higher than those of conventional electrolytes such as YSZ (10.8×10^{-6} K^{-1}) and SDC (12.0×10^{-6} K^{-1}).

Chen et al. [101] reported the Cu-doped $NdBa_{0.5}Sr_{0.5}Co_2O_{5+\delta}$ (NBSC) material. By introducing Cu ions into the Co sites of the NBSC lattice, the concentration of oxygen vacancies and the Co^{3+}/Co^{4+} ratio can be effectively controlled. As a result, the charge transfer, oxygen dissociation, and the average bond energy during the ORR process are significantly tuned, leading to a simultaneous enhancement in both ORR catalytic activity and CO_2 tolerance. Specifically, $NdBa_{0.5}Sr_{0.5}Co_{1.8}Cu_{0.2}O_{5+\delta}$ exhibits a polarization resistance of 0.04 Ω cm^2 and a maximum power density (PPD peak power density) of 1.03 W cm^{-2} at 750°C.

8.6.3 Composite Cathode Materials

To improve the performance of SOFC cathode materials, electrolyte materials are often incorporated into the cathode to form composite cathodes. The addition of electrolytes in composite cathode materials can significantly enhance the ionic conductivity, adjust the TEC, improve the compatibility between the cathode and the electrolyte, optimize the microstructure of the cathode, and enhance the electrochemical catalytic performance of the cathode material. Leng et al. [102] performed performance tests on an SOFC single cell with LSCF/GDC composite material as the cathode, GDC as the electrolyte, and Ni/GDC as the anode, achieving output power densities of 167 and 578 mW cm^{-2} at 500°C and 600°C, respectively. According to Murray et al. [103], at 600°C and 750°C, the interfacial polarization resistance of the LSCF/GDC composite cathode was 0.33 and 0.01 Ω cm^2, respectively, which is significantly lower than that of the single-phase LSCF cathode material.

In addition to developing new electrode materials, optimizing the microstructure of existing cathode materials is an effective approach to enhance their performance. Materials such as nanopowders, nanowires, and nanotubes have been used as SOFC cathode materials and have shown excellent performance. Cathode materials at the nanoscale, with their high specific surface area, provide more sites for the adsorption and dissociation of oxygen, significantly increasing the length of the internal TPBs. This can effectively improve the catalytic performance of the cathode material for

ORRs. However, traditional sintering methods are not conducive to controlling the microstructure of cathode materials. Therefore, to achieve precise microstructural control of the cathode material, special preparation methods are required.

Recent studies have shown that nanocomposite cathode materials prepared by impregnation methods exhibit excellent cathode performance. Jiang et al. [104] found that impregnating GDC particles into LSM cathode materials not only significantly enhances the electrochemical catalytic activity of the cathode but also improves the electrode's resistance to Cr poisoning. Simner et al. [105] reported that SOFC cells with Pd-impregnated LSF/SDC composite cathodes demonstrated a 50% increase in output power density at 700°C. Zhang et al. [106] synthesized $NdBa_{0.5}Sr_{0.5}Co_2O_{5+\delta}$ (NBSC)-$x$$Gd_{0.1}Ce_{0.9}O_{2-\delta}$ (GDC) (x=0–30 wt%) composite materials through in situ self-assembly using a one-pot method. The incorporation of GDC promotes the migration of oxygen ions within the NBSC particles, which are oriented in different directions. This effectively alters the anisotropic ionic conductivity of NBSC, forming a three-dimensional conduction pathway. Compared to previous studies on $LnBaCo_2O_{5+\delta}$ single-phase cathode materials and mechanically mixed composite cathodes, our in situ self-assembled NBSC-xGDC composite materials exhibit superior electrochemical performance. Notably, the NBSC-10GDC composite cathode shows remarkable performance at 750°C, with a low area-specific resistance of 0.029 Ω cm^2 and an impressive maximum power density of 1.36 W cm^{-2}.

8.7 CURRENT STATUS OF SOFC DEVELOPMENT

Currently, developed countries and regions, including the United States, Japan, and Europe, lead the global research and development (R&D) efforts in the field of SOFCs. These regions have not only made significant progress in fundamental research but have also achieved breakthroughs in technological innovations, material development, battery design, and system integration [107].

In the United States, government agencies, research institutions, and leading energy companies have invested substantial resources into SOFC research. Westinghouse, an American company, is one of the earliest enterprises globally to focus on tubular SOFC research. As early as 1986, the company successfully developed the world's first kilowatt-class tubular SOFC stack, which operated stably for thousands of hours, earning significant recognition for its excellent performance. In 1997, the company scaled up its SOFC stack to 150 kW and successfully commercialized it, demonstrating its outstanding capabilities in technological innovation and industrialization [108].

Bloom Energy, a representative enterprise in the SOFC sector, offers mature and reliable technology. The company's flagship SOFC product, a 50-kW module, can be combined into multi-module systems capable of reaching several megawatts. This technology has been successfully applied in various renowned companies, such as Apple and Google. Bloom Energy has also partnered with South Korea's SK Group to increase its installation capacity by at least 500 MW between 2022 and 2025 [109]. Another prominent enterprise in the global fuel cell market is FuelCell Energy, an American company. Its 250 kW SOFC system offers exceptional versatility, functioning not only as a hydrogen electrolyzer for efficient hydrogen production but also as a power generation fuel cell. This multifunctionality positions the system for broad applications within the energy sector [110].

In contrast to the United States, where the SOFC market primarily focuses on medium to large-scale commercial power systems, the European market places more emphasis on the promotion of micro combined heat and power (Micro-CHP) systems. Notable companies in this sector include Sunfire, Ceres Power, Solid Power, Hexis, Elcogen, Convion, Topose, and Bosch. Germany's Sunfire, a global leader in industrial electrolyzer production based on alkaline and solid oxide technologies, successfully installed the world's first 2.6 MW high-temperature electrolyzer system at Neste's refinery in Rotterdam, the Netherlands, in 2022 [111]. Ceres Power, based in the UK, has developed the Steel Cell technology, which features fast start-up times and high-power density, making it highly applicable in residential, commercial power generation, and transportation sectors. Solid Power, an Italian company, focuses on the development and application of Micro-CHP systems. Its 2.5 kW ENGEN2500 system demonstrates outstanding performance, with a total efficiency of approximately 90% [112]. Swiss company Hexis specializes in designing and manufacturing fuel cell-based micro-CHP systems for single-family homes, multi-apartment buildings, and small commercial applications [113]. Elcogen, an Estonian company, produces SOFC single cells and stacks that excel in performance and offer wide-ranging application potential [114]. Bosch, based in Germany, is actively participating in and strategically positioning itself within the SOFC industry chain. In 2018 and 2019, Bosch made two significant investments in the UK-based stack manufacturer Ceres Power, laying a solid foundation for the commercialization of SOFC technology [115].

In Japan, with strong governmental support, the SOFC program has achieved remarkable success. Japan's small-scale residential SOFC cogeneration technology has reached a mature and reliable level, with the country holding the largest market share globally. This success has set a model for the widespread adoption and application of clean energy technologies [116]. Japan is also home to numerous well-known companies, including Kyocera, Osaka Gas, Mitsubishi Hitachi Power Systems, and Aisin Seiki. These companies have made significant strides in SOFC R&D and applications. Kyocera began fuel cell technology development in 1985 and has been tackling technological challenges in the small-scale SOFC field. The company now offers third-generation, more compact products that generate up to 700 W of power, with a service life of up to 90,000 hours and a design life of 12 years. Mitsubishi Heavy Industries, which began SOFC R&D in the 1980s, made remarkable breakthroughs and, by 2018, had commercialized 250 kW and 1 MW CHP products, marking a significant step in the commercialization of SOFC technology [117].

8.8 APPLICATION PROSPECTS

SOFC technology demonstrates multifaceted application potential across energy conversion systems, principally attributable to its superior thermodynamic efficiency, exceptional fuel adaptability, and near-zero criteria pollutant emissions. These intrinsic merits position SOFCs as a transformative platform technology capable of addressing critical challenges in sustainable energy infrastructure.

The integration of high-temperature SOFCs with integrated gasification combined cycle (IGCC) power generation systems presents a promising pathway towards clean and highly efficient electricity generation. In an IGCC system, coal is first subjected to a gasification process, yielding syngas (H_2/CO). This syngas is then fed into the SOFC for electricity generation. This integrated approach not only significantly enhances power generation efficiency (<60%) but also effectively mitigates pollutant emissions, including nitrogen oxides (NO_x), sulfur oxides (SO_x), and particulate matter. Furthermore, SOFC-IGCC systems possess the potential for carbon capture, contributing to the goal of low-carbon emission energy production. The coupling of SOFC-IGCC systems with carbon capture and sequestration technologies positions them as a potentially vital option for future clean energy generation [118].

The high-temperature operation of SOFCs allows for the direct utilization of the syngas produced by coal gasification, circumventing the combustion stage inherent in conventional power generation processes. This reduces thermodynamic losses, thereby increasing power generation efficiency. Simultaneously, the electrochemical reaction within SOFCs avoids the production of nitrogen oxides that are associated with combustion, further minimizing emissions. Future development of large-scale SOFC power plants will focus on improving system efficiency, reducing costs, and enhancing reliability. This includes the development of higher-performance SOFC materials, optimization of system integration designs, and the exploration of more effective carbon capture technologies [119].

SOFCs demonstrate significant advantages in distributed generation applications. Their high electrical efficiency maximizes fuel utilization and minimizes energy consumption. Concurrently, ultra-low emissions align with increasingly stringent environmental regulations. The modular design of SOFCs provides flexibility in adjusting power output to meet diverse user requirements. With power ranges spanning from several kilowatts to several megawatts, SOFCs can provide a reliable electricity supply. Application scenarios include commercial buildings, industrial parks, and remote areas lacking grid access [120]. SOFCs can also enable CHP systems, recovering and utilizing waste heat generated during electricity production for heating, cooling, or industrial steam production. This comprehensive energy utilization approach substantially enhances energy efficiency, thereby decreasing energy costs. Additionally, SOFCs can be integrated with renewable energy sources, such as solar and wind power, to construct hybrid energy systems. SOFCs can serve as a complement to renewable energy sources, providing electricity during periods of insufficient renewable energy generation, thereby increasing the stability and reliability of the energy supply.

Micro-CHP systems utilizing SOFCs can supply both electricity and heat to residential dwellings, meeting the daily electricity, heating, and hot water demands. SOFC-based micro-CHP systems offer the benefits of high efficiency and flexible power adjustment, enabling energy optimization based on individual household needs [121]. By providing self-sufficient energy supplies, SOFC micro-CHP systems can significantly reduce household energy costs. Future development in the area of household SOFC micro-CHP systems will focus on lowering costs, improving reliability, and implementing intelligent control systems. This includes the development of more compact and easily maintainable SOFC systems, as well as the integration of such systems with smart home technologies [122,123].

SOFCs exhibit potential in the transportation sector, particularly in heavy-duty vehicles, ships, and trains. Currently, the application of SOFCs in transportation faces several challenges, including slow start-up times, volume and weight constraints, and the need for improved shock resistance and durability. If these challenges can be addressed, SOFCs are poised for widespread adoption within the transportation sector [124]. For example, SOFC auxiliary power units (APUs) can be developed to supply electricity to vehicles, thereby reducing engine idling and improving fuel efficiency and emissions [125,126].

The reverse operation of SOFCs produces a solid oxide electrolytic cell (SOEC), which is utilized for hydrogen production through water electrolysis. SOEC water electrolysis achieves a higher efficiency compared to other electrolysis technologies, such as alkaline electrolysis and proton exchange membrane electrolysis. This is because SOECs operate at high temperatures, reducing the energy barrier for water electrolysis, thereby enhancing efficiency [127]. SOECs are not only efficient but can also utilize various energy sources as drivers, including electricity, thermal energy, and nuclear energy. Furthermore, SOECs can leverage industrial waste heat, further enhancing energy utilization efficiency [128]. SOECs represent an important technological direction for large-scale hydrogen production in the future. By integrating with renewable energy sources, SOECs can realize clean and efficient hydrogen production, thus supporting the development of a hydrogen-based economy [129].

In summary, SOFC technology, with its distinctive advantages, holds vast application prospects in electricity generation, CHP, transportation, and hydrogen production. Continued technological advancements and cost reductions will enable SOFCs to play a more prominent role within future energy systems. To fully realize the potential of SOFCs, sustained investment in R&D is required to address technological challenges, along with the implementation of supportive policies to facilitate their commercial deployment.

REFERENCES

1. Goudsblom, J. (2012) Energy and civilization, *International Review of Sociology*, 22 (3), 405–411.
2. Issawi, C. (1991) Technology, energy, and civilization: some historical observations, *International Journal of Middle East Studies*, 23 (3), 281–289.
3. Dechamps, P. (2023) The IEA world energy outlook 2022-a brief analysis and implications, *The European Energy and Climate Journal*, 11 (3), 100–103.
4. Yolcan, O. O. (2023) World energy outlook and state of renewable energy: 10-Year evaluation, *Innovation and Green Development*, 2 (4), 100070.
5. Abas, N., Kalair, A., Khan, N. (2015) Review of fossil fuels and future energy technologies, *Futures*, 69, 31–49.
6. Capellán-Pérez, I., Mediavilla, M., de Castro, C., Carpintero, Ó., Miguel, L. J. (2014) Fossil fuel depletion and socio-economic scenarios: An integrated approach, *Energy*, 77, 641–666.
7. Höök, M., Tang, X. (2013) Depletion of fossil fuels and anthropogenic climate change-A review, *Energy Policy*, 52, 797–809.
8. Kreps, B. H. (2020) The rising costs of fossil-fuel extraction: An energy crisis that will not go away, *The American Journal of Economics and Sociology*, 79 (3), 695–717.

9. Hassan, A., Ilyas, S. Z. A., Jalil, Z. U. (2021) Monetization of the environmental damage caused by fossil fuels, *Environmental Science and Pollution Research*, 28 (17), 21204–21211.
10. Abbasi, K. R., Zhang, Q., Ozturk, I., Alvarado, R., Musa, M. (2024) Energy transition, fossil fuels, and green innovations: Paving the way to achieving sustainable development goals in the United States, *Gondwana Research*, 130, 326–341.
11. Solomon, B. D., Krishna, K. (2011) The coming sustainable energy transition: History, strategies, and outlook, *Energy Policy*, 39 (11), 7422–7431.
12. Islam, M. M., Hasanuzzaman, M., Pandey, A. K., Rahim, N. A. (2020) Chapter 2 – Modern energy conversion technologies, In: M. D. Hasanuzzaman, N. A. Rahim (Eds.) *Energy for Sustainable Development*, Academic Press, New York, 19–39.
13. Nižetić, S., Arıcı, M., Hoang, A. T. (2023) Smart and sustainable technologies in energy transition, *Journal of Cleaner Production*, 389, 135944.
14. Nowotny, J., Dodson, J., Fiechterand, S. (2018) Towards global sustainability: Education on environmentally clean energy technologies, *Renewable and Sustainable Energy Reviews*, 81, 2541–2551.
15. Vujanović, M., Wang, Q., Mohsen, M., Duić, N., Yan, J. (2021) Recent progress in sustainable energy-efficient technologies and environmental impacts on energy systems, *Applied Energy*, 283, 116280.
16. Yu, Z., Kamran, H. W., Amin, A., Ahmed, B., Peng, S. (2023) Sustainable synergy via clean energy technologies and efficiency dynamics, *Renewable and Sustainable Energy Reviews*, 187, 113744.
17. Qasem, N. A. A., Abdulrahman, G. A. Q. (2024) A recent comprehensive review of fuel cells: History, types, and applications, *International Journal of Energy Research*, 2024 (1), 7271748.
18. Chen, R., Gao, Y., Gaoand, J. (2024) From concept to commercialization: A review of tubular solid oxide fuel cell technology, *Journal of Energy Chemistry*, 97, 79–109.
19. Chun, O., Jamshaid, F., Khanand, M. Z. (2024) Advances in low-temperature solid oxide fuel cells: An explanatory review, *Journal of Power Sources*, 610, 234719.
20. Corigliano, O., Pagnotta, L., Fragiacomo, P. (2022) On the technology of solid oxide fuel cell (SOFC) energy systems for stationary power generation: A review, *Sustainability*, 14 (22), 15276.
21. Li, J., Cheng, J., Zhangand, Y. (2024) Advancements in solid oxide fuel cell technology: Bridging performance gaps for enhanced environmental sustainability, *Advanced Energy and Sustainability Research*, 5 (11), 2400132.
22. Zakaria, Z., Awang Mat, Z., Abu Hassan, S. H., Boon Kar, Y. (2020) A review of solid oxide fuel cell component fabrication methods toward lowering temperature, *International Journal of Energy Research*, 44 (2), 594–611.
23. Wu, Z., Zhu, P., Huangand, Y. (2025) A comprehensive review of modeling of solid oxide fuel cells: From large systems to fine electrodes, *Chemical Reviews*, 125 (4), 2184–2268.
24. Yousaf, M., Lu, Y., Akbarand, M. (2024) Advances in solid oxide fuel cell technologies: lowering the operating temperatures through material innovations, *Materials Today Energy*, 44, 101633.
25. Badwal, S. P. S. (2001) Stability of solid oxide fuel cell components, *Solid State Ionics*, 143 (1), 39–46.
26. Mehran, M. T., Khan, M. Z., Songand, R.-H. (2023) A comprehensive review on durability improvement of solid oxide fuel cells for commercial stationary power generation systems, *Applied Energy*, 352, 121864.
27. Braun, R. J., Klein, S. A., Reindl, D. T. (2006) Evaluation of system configurations for solid oxide fuel cell-based micro-combined heat and power generators in residential applications, *Journal of Power Sources*, 158 (2), 1290–1305.

28. Hart, N., Brandon, N., Shemilt, J. (1999) The environmental impact of solid oxide fuel cell manufacturing, *Fuel Cells Bulletin*, 2 (15), 4–7.
29. Weber, A. (2021) Fuel flexibility of solid oxide fuel cells, *Fuel Cells*, 21 (5), 440–452.
30. Song, C. (2002) Fuel processing for low-temperature and high-temperature fuel cells: Challenges, and opportunities for sustainable development in the 21st century, *Catalysis Today*, 77 (1), 17–49.
31. Tan, L., Dong, X., Chen, C., Gong, Z., Wang, M. (2021) Diverse system layouts promising fine performance demonstration: A comprehensive review on present designs of SOFC-based energy systems for building applications, *Energy Conversion and Management*, 245, 114539.
32. Ge, X.-M., Chan, S.-H., Liu, Q.-L., Sun, Q. (2012) Solid oxide fuel cell anode materials for direct hydrocarbon utilization, *Advanced Energy Materials*, 2 (10), 1156–1181.
33. Lee, J. H., Heo, J. W., Leeand, D. S. (2003) The impact of anode microstructure on the power generating characteristics of SOFC, *Solid State Ionics*, 158 (3), 225–232.
34. Shri Prakash, B., Senthil Kumar, S., Aruna, S. T. (2014) Properties and development of Ni/YSZ as an anode material in solid oxide fuel cell: A review, *Renewable and Sustainable Energy Reviews*, 36, 149–179.
35. Jiao, Z., Takagi, N., Shikazono, N., Kasagi, N. (2011) Study on local morphological changes of nickel in solid oxide fuel cell anode using porous Ni pellet electrode, *Journal of Power Sources*, 196 (3), 1019–1029.
36. Han, Z., Dong, H., Yang, Y., Yu, H., Yang, Z. (2024) Novel BaO-decorated carbon-tolerant Ni-YSZ anode fabricated by an efficient phase inversion-impregnation approach, *Journal of Power Sources*, 591, 233869.
37. Ye, X.-F., Wang, S. R., Huand, Q. (2009) Improvement of Cu-CeO_2 anodes for SOFCs running on ethanol fuels, *Solid State Ionics*, 180 (2), 276–281.
38. Gorte, R. J., Park, S., Vohs, J. M., Wang, C. (2000) Anodes for direct oxidation of dry hydrocarbons in a solid-oxide fuel cell, *Advanced Materials*, 12 (19), 1465–1469.
39. Gorte, R. J., Vohs, J. M. (2003) Novel SOFC anodes for the direct electrochemical oxidation of hydrocarbons, *Journal of Catalysis*, 216 (1), 477–486.
40. Atkinson, A., Barnett, S., Gorteand, R. J. (2004) Advanced anodes for high-temperature fuel cells, *Nature Materials*, 3 (1), 17–27.
41. Gorte, R. (2004) Recent developments on anodes for direct fuel utilization in SOFC, *Solid State Ionics*, 175 (1–4), 1–6.
42. Ramírez-Cabrera, E., Atkinson, A., Chadwick, D. (2000) The influence of point defects on the resistance of ceria to carbon deposition in hydrocarbon catalysis, *Solid State Ionics*, 136–137, 825–831.
43. Shaikh, S. P. S., Muchtar, A., Somalu, M. R. (2015) A review on the selection of anode materials for solid-oxide fuel cells, *Renewable and Sustainable Energy Reviews*, 51, 1–8.
44. Li, X., Zhao, H., Gao, F., Chen, N., Xu, N. (2008) La and Sc co-doped $SrTiO_3$ as novel anode materials for solid oxide fuel cells, *Electrochemistry Communications*, 10 (10), 1567–1570.
45. Tao, S., Irvine, J. T. S. (2003) A redox-stable efficient anode for solid-oxide fuel cells, *Nature Materials*, 2 (5), 320–323.
46. Huang, Y.-H., Dass, R. I., Denyszyn, J. C., Goodenough, J. B. (2006) Synthesis and Characterization of $Sr_2MgMoO_{6-\delta}$: An anode material for the solid oxide fuel cell, *Journal of The Electrochemical Society*, 153 (7), A1266.
47. Vinchhi, P., Khandla, M., Chaudhary, K., Pati, R. (2023) Recent advances on electrolyte materials for SOFC: A review, *Inorganic Chemistry Communications*, 152, 110724.
48. Dwivedi, S. (2020) Solid oxide fuel cell: Materials for anode, cathode and electrolyte, *International Journal of Hydrogen Energy*, 45 (44), 23988–24013.

49. Shah, N., Xu, X., Loveand, J. (2024) Mitigating thermal expansion effects in solid oxide fuel cell cathodes: A critical review, *Journal of Power Sources*, 599, 234211.
50. Kostogloudis, G. C., Ftikos, C., Ahmad-Khanlou, A., Naoumidis, A., Stöver, D. (2000) Chemical compatibility of alternative perovskite oxide SOFC cathodes with doped lanthanum gallate solid electrolyte, *Solid State Ionics*, 134 (1), 127–138.
51. Hussain, S., Li, Y. P. (2020) Review of solid oxide fuel cell materials: cathode, anode, and electrolyte, *Energy Transitions*, 4 (2), 113–126.
52. Fergus, J. W. (2006) Electrolytes for solid oxide fuel cells, *Journal of Power Sources*, 162 (1), 30–40.
53. Figueiredo, F. M. L., Marques, F. M. B. (2013) Electrolytes for solid oxide fuel cells, *WIREs Energy and Environment*, 2 (1), 52–72.
54. Kashyap, D., Patro, P. K., Lenka, R. K., Mahata, T., Sinha, P. K. (2014) Effects of Gd and Sr co-doping in CeO_2 for electrolyte application in solid oxide fuel cell (SOFC), *Ceramics International*, 40 (8, Part A), 11869–11875.
55. Zha, S., Xia, C., Meng, G. (2003) Effect of Gd (Sm) doping on properties of ceria electrolyte for solid oxide fuel cells, *Journal of Power Sources*, 115 (1), 44–48.
56. Kang, J., Wen, C., Wangand, B. (2021) Structure and performance of Pr, Sm, Y co-doped cerium-based electrolyte for intermediate temperature solid oxide fuel cells, *Materials Letters*, 305, 130855.
57. Morales, M., Roa, J. J., Tartaj, J., Segarra, M. (2016) A review of doped lanthanum gallates as electrolytes for intermediate temperature solid oxides fuel cells: From materials processing to electrical and thermo-mechanical properties, *Journal of the European Ceramic Society*, 36 (1), 1–16.
58. Harrison, C. J., Hatton, P. V., Gentile, P., Miller, C. A. (2021) Nanoscale strontium-substituted hydroxyapatite pastes and gels for bone tissue regeneration, *Nanomaterials*, 11 (6), 1611.
59. Ishihara, T., Matsuda, H., Takita, Y. (1995) Effects of rare earth cations doped for La site on the oxide ionic conductivity of $LaGaO_3$-based perovskite type oxide, *Solid State Ionics*, 79, 147–151.
60. Yang, L., Zuo, C., Wang, S., Cheng, Z., Liu, M. (2008) A novel composite cathode for low-temperature SOFCs based on oxide proton conductors, *Advanced Materials*, 20 (17), 3280–3283.
61. Van Gestel, T., Sebold, D., Buchkremer, H. P. (2015) Processing of 8YSZ and CGO thin film electrolyte layers for intermediate- and low-temperature SOFCs, *Journal of the European Ceramic Society*, 35 (5), 1505–1515.
62. Ishihara, T., Ishikawa, S., Hosoi, K., Nishiguchi, H., Takita, Y. (2004) Oxide ionic and electronic conduction in Ni-doped $LaGaO_3$-based oxide, *Solid State Ionics*, 175 (1), 319–322.
63. Khorkounov, B. A., Näfe, H., Aldinger, F. (2006) Relationship between the ionic and electronic partial conductivities of co-doped LSGM ceramics from oxygen partial pressure dependence of the total conductivity, *Journal of Solid State Electrochemistry*, 10 (7), 479–487.
64. Stevenson, J. W., Hasinska, K., Canfield, N. L., Armstrong, T. R. (2000) Influence of cobalt and iron additions on the electrical and thermal properties of (La,Sr)(Ga,Mg) $O_{3-\delta}$, *Journal of The Electrochemical Society*, 147 (9), 3213.
65. Goodenough, J. B., Ruiz-Diaz, J. E., Zhen, Y. S. (1990) Oxide-ion conduction in $Ba_2In_2O_5$ and $Ba_3In_2MO_8$ (M=Ce, Hf, or Zr), *Solid State Ionics*, 44 (1), 21–31.
66. Wang, J.-D., Xie, Y.-H., Zhang, Z.-F., Liu, R.-Q., Li, Z.-J. (2005) Protonic conduction in Ca^{2+}-doped $La_2M_2O_7$ (M=Ce, Zr) with its application to ammonia synthesis electrochemically, *Materials Research Bulletin*, 40 (8), 1294–1302.
67. Lacorre, P., Goutenoire, F., Bohnke, O., Retoux, R., Laligant, Y. (2000) Designing fast oxide-ion conductors based on $La_2Mo_2O_9$, *Nature*, 404 (6780), 856–858.

68. Kendrick, E., Islam, M. S., Slater, P. R. (2007) Developing apatites for solid oxide fuel cells: Insight into structural, transport and doping properties, *Journal of Materials Chemistry*, 17 (30), 3104–3111.
69. Hosono, H., Hayashi, K., Kajihara, K., Sushko, P. V., Shluger, A. L. (2009) Oxygen ion conduction in $12CaO_7Al_2O_3$:O^{2-} conduction mechanism and possibility of O^- fast conduction, *Solid State Ionics*, 180 (6–8), 550–555.
70. Iwahara, H., Esaka, T., Uchida, H., Maeda, N. (1981) Proton conduction in sintered oxides and its application to steam electrolysis for hydrogen production, *Solid State Ionics*, 3, 359–363.
71. Batool, M., Sattar, M., Khan, Z. S. (2018) Influence of temperature and synthesis dependent changes upon structure, morphology and electrical properties of $BaCeO_3$-based proton conducting electrolyte, *Materials Chemistry and Physics*, 213, 213–219.
72. Xie, K., Yan, R., Chenand, X. (2009) A new stable $BaCeO_3$-based proton conductor for intermediate-temperature solid oxide fuel cells, *Journal of Alloys and Compounds*, 472 (1), 551–555.
73. Gu, Y., Luo, G., Chen, Z., Huo, Y., Wu, F. (2022) Enhanced chemical stability and electrochemical performance of $BaCe_{0.8}Y_{0.1}Ni_{0.04}Sm_{0.06}O_{3-\delta}$ perovskite electrolytes as proton conductors, *Ceramics International*, 48 (8), 10650–10658.
74. Matsumoto, H., Kawasaki, Y., Ito, N., Enoki, M., Ishihara, T. (2007) Relation between electrical conductivity and chemical stability of $BaCeO_3$-based proton conductors with different trivalent dopants, *Electrochemical and Solid-State Letters*, 10 (4), B77.
75. Gu, Y.-J., Liu, Z.-G., Ouyang, J.-H., Yan, F.-Y., Zhou, Y. (2013) Structure and electrical conductivity of $BaCe_{0.85}Ln_{0.15}O_{3-\delta}$ (Ln=Gd, Y, Yb) ceramics, *Electrochimica Acta*, 105, 547–553.
76. Bhella, S. S., Fürstenhaupt, T., Paul, R., Thangadurai, V. (2011) Synthesis, structure, chemical stability, and electrical properties of Nb–, Zr–, and Nb-codoped $BaCeO_3$ perovskites, *Inorganic Chemistry*, 50 (14), 6493–6499.
77. Konwar, D., Nguyen, N. T. Q., Yoon, H. H. (2015) Evaluation of $BaZr_{0.1}Ce_{0.7}Y_{0.2}O_{3-\delta}$ electrolyte prepared by carbonate precipitation for a mixed ion-conducting SOFC, *International Journal of Hydrogen Energy*, 40 (35), 11651–11658.
78. Zhou, C., Sunarso, J., Songand, Y. (2019) New reduced-temperature ceramic fuel cells with dual-ion conducting electrolyte and triple-conducting double perovskite cathode, *Journal of Materials Chemistry A*, 7 (21), 13265–13274.
79. Duan, C., Tong, J., Shangand, M. (2015) Readily processed protonic ceramic fuel cells with high performance at low temperatures, *Science*, 349 (6254), 1321–1326.
80. Chen, M., Zhou, M., Liu, Z., Liu, J. (2022) A comparative investigation on protonic ceramic fuel cell electrolytes $BaZr_{0.8}Y_{0.2}O_{3-\delta}$ and $BaZr_{0.1}Ce_{0.7}Y_{0.2}O_{3-\delta}$ with NiO as sintering aid, *Ceramics International*, 48 (12), 17208–17216.
81. Richter, J., Holtappels, P., Graule, T., Nakamura, T., Gauckler, L. J. (2009) Materials design for perovskite SOFC cathodes, *Monatshefte für Chemie-Chemical Monthly*, 140 (9), 985–999.
82. Bilal Hanif, M., Motola, M., Qayyumand, S. (2022) Recent advancements, doping strategies and the future perspective of perovskite-based solid oxide fuel cells for energy conversion, *Chemical Engineering Journal*, 428, 132603.
83. Bidrawn, F., Kim, G., Aramrueang, N., Vohs, J. M., Gorte, R. J. (2010) Dopants to enhance SOFC cathodes based on Sr-doped $LaFeO_3$ and $LaMnO_3$, *Journal of Power Sources*, 195 (3), 720–728.
84. Zheng, F., Pederson, L. R. (1999) Phase behavior of lanthanum strontium manganites, *Journal of The Electrochemical Society*, 146 (8), 2810.
85. Yasuda, I., Ogasawara, K., Hishinuma, M., Kawada, T., Dokiya, M. (1996) Oxygen tracer diffusion coefficient of $(La,Sr)MnO_{3\pm\delta}$, *Solid State Ionics*, 86–88, 1197–1201.

86. Jiang, S. P., Zhang, J. P., Foger, K. (2001) Deposition of chromium species at Sr-doped $LaMnO_3$ electrodes in solid oxide fuel cells: III. Effect of air flow, *Journal of The Electrochemical Society*, 148 (7), C447.
87. Yang, Y. L., Chen, C. L., Chen, S. Y., Chu, C. W., Jacobson, A. J. (2000) Impedance studies of oxygen exchange on dense thin film electrodes of $La_{0.5}Sr_{0.5}CoO_{3-\delta}$, *Journal of The Electrochemical Society*, 147 (11), 4001.
88. Huang, Y., Ahn, K., Vohs, J. M., Gorte, R. J. (2004) Characterization of Sr-doped $LaCoO_3$_YSZ composites prepared by impregnation methods, *Journal of The Electrochemical Society*, 151 (10), A1592.
89. Esquirol, A., Brandon, N. P., Kilner, J. A., Mogensen, M. (2004) Electrochemical characterization of $La_{0.6}Sr_{0.4}Co_{0.2}Fe_{0.8}O_3$ cathodes for intermediate-temperature SOFCs, *Journal of The Electrochemical Society*, 151 (11), A1847.
90. Ullmann, H., Trofimenko, N., Tietz, F., Stöver, D., Ahmad-Khanlou, A. (2000) Correlation between thermal expansion and oxide ion transport in mixed conducting perovskite-type oxides for SOFC cathodes, *Solid State Ionics*, 138 (1), 79–90.
91. Shao, Z., Haile, S. M. (2004) A high-performance cathode for the next generation of solid-oxide fuel cells, *Nature*, 431 (7005), 170–173.
92. Wang, H., Wang, R., Liang, D. T., Yang, W. (2004) Experimental and modeling studies on $Ba_{0.5}Sr_{0.5}Co_{0.8}Fe_{0.2}O_{3-\delta}$ (BSCF) tubular membranes for air separation, *Journal of Membrane Science*, 243 (1), 405–415.
93. Li, S., Lü, Z., Weiand, B. (2006) A study of $(Ba_{0.5}Sr_{0.5})_{1-x}Sm_xCo_{0.8}Fe_{0.2}O_{3-\delta}$ as a cathode material for IT-SOFCs, *Journal of Alloys and Compounds*, 426 (1), 408–414.
94. Kim, G., Wang, S., Jacobsonand, A. J. (2007) Rapid oxygen ion diffusion and surface exchange kinetics in $PrBaCo_2O_{5+\delta}$ with a perovskite related structure and ordered A cations, *Journal of Materials Chemistry*, 17 (24), 2500–2505.
95. Seymour, I. D., Tarancón, A., Chroneosand, A. (2012) Anisotropic oxygen diffusion in $PrBaCo_2O_{5.5}$ double perovskites, *Solid State Ionics*, 216, 41–43.
96. Joung, Y.-H., Kang, H. I., Choi, W. S., Kim, J. H. (2013) Investigation of X-ray photoelectron spectroscopy and electrical conductivity properties of the layered perovskite $LnBaCo_2O_{5+\delta}$ (Ln=Pr, Nd, Sm, and Gd) for IT-SOFC, *Electronic Materials Letters*, 9 (4), 463–465.
97. Tarancón, A., Peña-Martínez, J., Marrero-Lópezand, D. (2008) Stability, chemical compatibility and electrochemical performance of $GdBaCo_2O_{5+\delta}$ layered perovskite as a cathode for intermediate temperature solid oxide fuel cells, *Solid State Ionics*, 179 (40), 2372–2378.
98. Zhu, C., Liu, X., Yi, C., Yan, D., Su, W. (2008) Electrochemical performance of $PrBaCo_2O_{5+\delta}$ layered perovskite as an intermediate-temperature solid oxide fuel cell cathode, *Journal of Power Sources*, 185 (1), 193–196.
99. Zhou, Q., He, T., Ji, Y. (2008) $SmBaCo_2O_{5+\delta}$ double-perovskite structure cathode material for intermediate-temperature solid-oxide fuel cells, *Journal of Power Sources*, 185 (2), 754–758.
100. Kim, J. H., Manthiram, A. (2008) $LnBaCo_2O_{5+\delta}$ oxides as cathodes for intermediate-temperature solid oxide fuel cells, *Journal of The Electrochemical Society*, 155 (4), B385.
101. Chen, S., Zhang, H., Yao, C. G. (2023) Tailored double perovskite with boosted oxygen reduction kinetics and CO_2 durability for solid oxide fuel cells, *ACS Sustainable Chemistry & Engineering*, 11 (35), 13198–13208.
102. Leng, Y. J., Chan, S. H., Jiang, S. P., Khor, K. A. (2004) Low-temperature SOFC with thin film GDC electrolyte prepared in situ by solid-state reaction, *Solid State Ionics*, 170 (1), 9–15.
103. Perry Murray, E., Sever, M. J., Barnett, S. A. (2002) Electrochemical performance of (La,Sr)(Co,Fe)O_3-(Ce,Gd)O_3 composite cathodes, *Solid State Ionics*, 148 (1), 27–34.

104. Jiang, S. P., Wang, W. (2005) Fabrication and performance of GDC-impregnated (La,Sr) MnO_3 cathodes for intermediate temperature solid oxide fuel cells, *Journal of The Electrochemical Society*, 152 (7), A1398.
105. Simner, S. P., Bonnett, J. F., Canfieldand, N. L. (2003) Development of lanthanum ferrite SOFC cathodes, *Journal of Power Sources*, 113 (1), 1–10.
106. Zhang, Z., Chen, S., Zhang, H. X. (2024) In situ self-assembled $NdBa_{0.5}Sr_{0.5}Co_2O_{5+\delta}$/$Gd_{0.1}Ce_{0.9}O_{2-\delta}$ hetero-interfaces enable enhanced electrochemical activity and CO_2 durability for solid oxide fuel cells, *Journal of Colloid and Interface Science*, 655, 157–166.
107. Oluleye, G., Gandiglio, M., Santarelli, M., Hawkes, A. (2021) Pathways to commercialisation of biogas fuelled solid oxide fuel cells in European wastewater treatment plants, *Applied Energy*, 282, 116127.
108. Williams, M. C., Vora, S. D., Jesionowski, G. (2020) Worldwide status of solid oxide fuel cell technology, *ECS Transactions*, 96 (1), 1.
109. Choudhury, A., Chandra, H., Arora, A. (2013) Application of solid oxide fuel cell technology for power generation-A review, *Renewable and Sustainable Energy Reviews*, 20, 430–442.
110. Ghezel-Ayagh, H., Borglum, B. P. (2017) Review of progress in solid oxide fuel cells at fuelcell energy, *ECS Transactions*, 78 (1), 77.
111. Leah, R., Bone, A., Selcukand, A. (2021) Commercialization of the ceres power SteelCell® technology: Latest update, *ECS Transactions*, 103 (1), 679.
112. Bertoldi, M., Bucheli, O. F., Ravagni, A. (2017) Development, manufacturing and deployment of SOFC-based products at SOLIDpower, *ECS Transactions*, 78 (1), 117.
113. Mai, A., Grolig, J. G., Doldand, M. (2019) Progress in HEXIS' SOFC development, *ECS Transactions*, 91 (1), 63.
114. Noponen, M., Torri, P., Göösand, J. (2015) Status of solid oxide fuel cell development at elcogen, *ECS Transactions*, 68 (1), 151.
115. Gupta, P., Toksha, B., Rahaman, M. (2024) A critical review on hydrogen based fuel cell technology and applications, *The Chemical Record*, 24 (1), e202300295.
116. Nirasawa, H. (2017) Current status of national SOFC projects in Japan, *ECS Transactions*, 78 (1), 33.
117. Fernandes, M. D., Bistritzki, V., Dominguesand, R. Z. (2020) Solid oxide fuel cell technology paths: National innovation system contributions from Japan and the United States, *Renewable and Sustainable Energy Reviews*, 127, 109879.
118. Duan, L., Yang, Y., Xu, E. (2007) Study on SOFC-IGCC hybrid power system with high efficiency, In *ASME turbo expo 2007: Power for land, Sea, and Air*, Montreal, Canada, pp. 223–230.
119. Hassmann, K. (2001) SOFC power plants, the Siemens-Westinghouse approach, *Fuel Cells*, 1 (1), 78–84.
120. Ramadhani, F., Hussain, M. A., Mokhlis, H., Hajimolana, S. (2017) Optimization strategies for Solid Oxide Fuel Cell (SOFC) application: A literature survey, *Renewable and Sustainable Energy Reviews*, 76, 460–484.
121. Farhad, S., Hamdullahpur, F., Yoo, Y. (2010) Performance evaluation of different configurations of biogas-fuelled SOFC micro-CHP systems for residential applications, *International Journal of Hydrogen Energy*, 35 (8), 3758–3768.
122. Longo, S., Cellura, M., Guarino, F., Brunaccini, G., Ferraro, M. (2019) Life cycle energy and environmental impacts of a solid oxide fuel cell micro-CHP system for residential application, *Science of The Total Environment*, 685, 59–73.
123. Pfeifer, T., Nousch, L., Lieftink, D., Modena, S. (2013) System design and process layout for a SOFC micro-CHP unit with reduced operating temperatures, *International Journal of Hydrogen Energy*, 38 (1), 431–439.
124. Boldrin, P., Brandon, N. P. (2019) Progress and outlook for solid oxide fuel cells for transportation applications, *Nature Catalysis*, 2 (7), 571–577.

125. Gu, X., Wang, Y., Shi, Y., Cai, N. (2023) Analysis of a gas turbine auxiliary power unit system based on a fuel cell combustor, *International Journal of Hydrogen Energy*, 48 (4), 1540–1551.
126. Khanafer, K., Al-Masri, A., Vafai, K., Preethichandra, P. (2022) Heat up impact on thermal stresses in SOFC for mobile APU applications: Thermo-structural analysis, *Sustainable Energy Technologies and Assessments*, 52, 102159.
127. Nechache, A., Hody, S. (2021) Alternative and innovative solid oxide electrolysis cell materials: A short review, *Renewable and Sustainable Energy Reviews*, 149, 111322.
128. Jolaoso, L. A., Bello, I. T., Ojelade, O. A. (2023) Operational and scaling-up barriers of SOEC and mitigation strategies to boost H_2 production- a comprehensive review, *International Journal of Hydrogen Energy*, 48 (85), 33017–33041.
129. Zong, S., Zhao, X., Jewell, L. L., Zhang, Y., Liu, X. (2024) Advances and challenges with SOEC high temperature co-electrolysis of CO_2/H_2O: Materials development and technological design, *Carbon Capture Science & Technology*, 12, 100234.

Index

For Product Safety Concerns and Information please contact our EU representative GPSR@taylorandfrancis.com
Taylor & Francis Verlag GmbH, Kaufingerstraße 24, 80331 München, Germany

www.ingramcontent.com/pod-product-compliance
Ingram Content Group UK Ltd.
Pitfield, Milton Keynes, MK11 3LW, UK
UKHW022220190926
475702UK00003B/469

* 9 7 8 1 0 3 2 7 9 5 2 5 6 *

www.ingramcontent.com/pod-product-compliance
Ingram Content Group UK Ltd.
Pitfield, Milton Keynes, MK11 3LW, UK
UKHW020131290926
15471UKWH00003B/417

9 789403 518725